Darney's Circuit Modelling for Electromagnetic Compatibility, 2nd Edition

Other related titles:

You may also like

- SBEW571 | Darney's Circuit Theory and Modelling: Updated and extended for EMC/EMI | Ian B. Darney and Chris M. Hewitt | 2023
- SBEW552 | Uncertainty Quantification of Electromagnetic Devices, Circuits, and Systems | Sourajeet Roy | 2021
- SBEW544 | Electromagnetic Reverberation Chambers: Recent advances and innovative applications | Guillaume Andrieu | 2020
- SBEW510 | EMI Troubleshooting Cookbook for Product Designers | Patrick G. André and Kenneth Wyatt | 2014

We also publish a wide range of books on the following topics:
Computing and Networks
Control, Robotics and Sensors
Electrical Regulations
Electromagnetics and Radar
Energy Engineering
Healthcare Technologies
History and Management of Technology
IET Codes and Guidance
Materials, Circuits and Devices
Model Forms
Nanomaterials and Nanotechnologies
Optics, Photonics and Lasers
Production, Design and Manufacturing
Security
Telecommunications
Transportation

All books are available in print via https://shop.theiet.org or as eBooks via our Digital Library https://digital-library.theiet.org.

SCITECH ELECTROMAGNIETIC WAVES SERIES 576

Darney's Circuit Modelling for Electromagnetic Compatibility, 2nd Edition

Ian B. Darney and Chris M. Hewitt

The Institution of Engineering and Technology

About the IET

This book is published by the Institution of Engineering and Technology (The IET).

We inspire, inform and influence the global engineering community to engineer a better world. As a diverse home across engineering and technology, we share knowledge that helps make better sense of the world, to accelerate innovation and solve the global challenges that matter.

The IET is a not-for-profit organisation. The surplus we make from our books is used to support activities and products for the engineering community and promote the positive role of science, engineering and technology in the world. This includes education resources and outreach, scholarships and awards, events and courses, publications, professional development and mentoring, and advocacy to governments.

To discover more about the IET please visit https://www.theiet.org/.

About IET books

The IET publishes books across many engineering and technology disciplines. Our authors and editors offer fresh perspectives from universities and industry. Within our subject areas, we have several book series steered by editorial boards made up of leading subject experts.

We peer review each book at the proposal stage to ensure the quality and relevance of our publications.

Get involved

If you are interested in becoming an author, editor, series advisor, or peer reviewer please visit https://www.theiet.org/publishing/publishing-with-iet-books/ or contact author_support@theiet.org.

Discovering our electronic content

All of our books are available online via the IET's Digital Library. Our Digital Library is the home of technical documents, eBooks, conference publications, real-life case studies and journal articles. To find out more, please visit https://digital-library.theiet.org.

In collaboration with the United Nations and the International Publishers Association, the IET is a Signatory member of the SDG Publishers Compact. The Compact aims to accelerate progress to achieve the Sustainable Development Goals (SDGs) by 2030. Signatories aspire to develop sustainable practices and act as champions of the SDGs during the Decade of Action (2020–30), publishing books and journals that will help inform, develop, and inspire action in that direction.

In line with our sustainable goals, our UK printing partner has FSC accreditation, which is reducing our environmental impact to the planet. We use a print-on-demand model to further reduce our carbon footprint.

Published by SciTech Publishing an imprint of The Institution of Engineering and Technology, London, United Kingdom

The Institution of Engineering and Technology (the "**Publisher**") is registered as a Charity in England & Wales (no. 211014) and Scotland (no. SC038698).

First published 2013

Second edition published 2025

The Institution of Engineering and Technology
Futures Place,
Kings Way,
Stevenage,
Herts, SG1 2UA,
United Kingdom

www.theiet.org

British Library Cataloguing in Publication Data
A catalogue record for this product is available from the British Library

ISBN 978-1-83724-206-1 (hardback)
ISBN 978-1-83724-207-8 (PDF)

Typeset in India by MPS Limited

Cover image: Original by Alberto Menendez Cervero modified and enhanced by Chris M. Hewitt

Contents

Foreword by Chris M. Hewitt xiii
Figure border colour key, Internet link and Bibliographic references xv
Acknowledgements and In Memory of xvii
About the authors xix

1 Introduction 1
1.1 Background 1
1.1.1 The need for EMC 1
1.1.2 Pragmatic approach 1
1.1.3 Academic approach 2
1.1.4 Managerial approach 2
1.1.5 Misleading concepts 3
1.1.6 Circuit modelling 3
1.1.7 Computations 4
1.1.8 Testing 4
1.1.9 Essence of the approach 4
1.2 Developing the model 5
1.2.1 Basic model 5
1.2.2 Parameter types 6
1.2.3 Derivation process 7
1.2.4 Composite conductors 8
1.2.5 Proximity effect 9
1.2.6 Electrical length 9
1.2.7 Distributed parameters 10
1.3 Intra-system interference 12
1.3.1 The signal link 12
1.3.2 Simulating the structure 12
1.3.3 Equivalent circuits 13
1.3.4 Conducted emission 14
1.3.5 Conducted susceptibility 14
1.3.6 Voltage transformer 15
1.3.7 Current transformer 15
1.3.8 Representative circuit model 16
1.4 Inter-system interference 17
1.4.1 Dipole model 17
1.4.2 The virtual conductor 18

1.4.3 The threat voltage 19
1.4.4 Worst-case analysis 20
1.5 Transients 20
1.6 The importance of testing 22
1.7 Practical design techniques 23
1.8 System design 24
1.8.1 Guidelines 24
1.8.2 Top-down approach 25
1.8.3 Formal EMC requirements 25

2 Lumped parameter models 27
2.1 Primitive capacitance 29
2.2 Primitive inductance 32
2.3 Duality of L and C 36
2.4 Loop parameters 37
2.5 Circuit parameters 40
2.5.1 Inductance 40
2.5.2 Capacitance 41
2.5.3 Maintaining duality 42
2.5.4 Resistance 43
2.5.5 Basic assumption 44
2.6 Twin-conductor model 44
2.7 Three-conductor model 47
2.8 Optimum coupling 51
2.9 Transfer admittance 54
2.10 Co-axial coupling 57
2.11 The ground plane 59

3 Other cross-sections 63
3.1 Single composite conductor 65
3.2 The composite pair 68
3.3 The screened pair 76

4 Transmission line models 83
4.1 Single-T model 84
4.2 Triple-T model 88
4.3 Cross-coupling 91
4.4 Bench test models 99

5 Antenna models 105
5.1 The half-wave dipole 106
5.1.1 Radiated power 106
5.1.2 Power density 109
5.1.3 Field strength 109
5.1.4 Power received 110

5.2 The virtual conductor 111
5.3 The threat voltage 118
5.4 The threat current 122
5.5 Coupling via the structure 128
5.6 Radiation susceptibility 136
5.7 Radiated emission 138

6 Transient analysis **141**
6.1 Time-step analysis 143
6.1.1 Basic concept 143
6.1.2 Basic equations 143
6.1.3 Series LCR circuit 144
6.1.4 Parallel LCR circuit 146
6.2 Delay line model 149
6.3 Line characteristics 156
6.4 Antenna-mode current 163
6.5 Radiated emission 170
6.5.1 Current linking the transformer 170
6.5.2 Line voltage 172
6.5.3 Source current and voltage 173
6.5.4 Radiated current 174
6.5.5 Cable losses 175
6.5.6 Line parameter measurements 176
6.6 Transient emission model 176

7 Bench testing **185**
7.1 Voltage transformer 186
7.2 Current transformer 193
7.3 Tri-axial cable 199
7.4 The isolated conductor 201
7.5 Cable characterisation 208
7.6 Cable transients 221
7.7 Capacitor characterisation 227
7.8 When theory meets reality: analysis 230

8 Practical design **233**
8.1 Grounding 234
8.2 Conductor pairing 238
8.3 Ground loops 241
8.4 Common-mode rejection 243
8.4.1 Differential amplifier 243
8.4.2 Differential logic driver 244
8.4.3 Differential analogue driver 245
8.4.4 Common-mode choke 246
8.4.5 Transformer coupling 247

8.4.6 Centre-tapped transformer 248
8.4.7 Opto-isolator 248
8.5 Differential-mode damping 248
8.5.1 Transient damping 249
8.5.2 Mains filtering 250
8.5.3 Solenoid switching 251
8.5.4 Commercial filters 252
8.5.5 Use of carbon 253
8.6 Common-mode damping 253
8.6.1 Common-mode resistor 253
8.6.2 Minimising pickup 254
8.6.3 Tri-axial cable 255
8.6.4 Transformer screen-winding 256
8.6.5 Common-mode filter 257
8.6.6 Transformer-coupled resistors 258
8.7 Shielding 259
8.7.1 Equipment shielding 260
8.7.2 Shielding of buildings 261
8.7.3 Use of carbon 263
8.8 Filling in a few blanks and clarification of Practical Design using PCB example 263

9 System design 267
9.1 Design guidelines 268
9.1.1 Structure as a shield 268
9.1.2 Return conductors 269
9.1.3 Ground loops 269
9.1.4 Current balance 270
9.1.5 Differential-mode damping 270
9.1.6 Common-mode damping 271
9.1.7 System assessment 271
9.1.8 Bench tests 271
9.2 Relating the diagrams 272
9.2.1 Circuit diagrams 272
9.2.2 Wiring diagrams 273
9.2.3 Block diagrams 275
9.2.4 Interface diagrams 275
9.2.5 Circuit models 276
9.2.6 Deriving component values 277
9.2.7 Analysing the signal link 278
9.2.8 Testing the link 279
9.3 Printed circuit boards 279
9.4 Susceptibility requirements 280
9.5 Emission requirements 284
9.6 Planning 286

9.6.1 Performance requirements 286
9.6.2 Bench test equipment 287
9.6.3 Software 287
9.6.4 Critical signal links 288
9.6.5 Critical frequencies 288
9.6.6 Characterisation 289
9.6.7 General approach 291

Appendix A: Mathcad worksheets **293**
Appendix B: MATLAB® **297**
Appendix C: The hybrid equations **299**
Appendix D: Definitions **303**
Appendix E: Links to Internet resources and references **307**
References **309**

Index **311**

Foreword by Chris M. Hewitt

Shortly after I was able to fulfil Ian Brook Darney's wish for the culmination of his life's work to be published 'Darney's Circuit Theory and Modelling, Updated and Extended for EMC/EMI'. I was again approached by his son Peter Darney who was interested in the possibility of updating Ian's previous publication 'Circuit Modelling for Electromagnetic Compatibility', to the same standard and format as the aforementioned publication. As can be seen the possibility has become reality in this second edition of the book.

Along with updating the images and the MathCad presentation throughout the books, very little of the main text body has been altered, with the exception of correcting 'typos' and syntax. However, due to an oversight by Ian with respect to Rrad, Sections 7.4–7.6 needed to be rewritten as did the associated MathCad, resulting in the need to amend two paragraphs and the associated MathCad in Section 5.5. Those of you who have access to the first edition of this book will notice that Section 1.1.1 has by request, been extended to emphasise the "obvious" and two sections have been added one to Chapter 7, which provides clarity, whilst detailing the limitations of the models, and one to Chapter 8, which clarifies when SPICE can be used with respect to EMC/EMI and also leaves you, the reader, with a little teaser.

Both personal and professional opinions of the research conducted by Ian can best be described as visionary. Ian's research into EMC began well before EMC became a major issue in an increasingly technological dependent and connected world. The almost exponential increase in the environment of human generated electromagnetic waves has subsequently increased the complexity of using the Maxwell equations for determining EMC and EMI susceptibility of devices, with the increased number of associated variables required to produce results consummate with real-world scenarios, larger and faster computers are required to manipulate the equations in a timely manner to facilitate a device's development and manufacture. An often overlooked, consequence of this is not only that the more complex such equations become and thus more susceptible to induced errors, but much more Energy is required to generate the requisite results. The method employed by Ian throughout this book reduces the complexity of the equations needed to obtain not just satisfactory results but results backed by empirical data, requiring substantially less Energy and thus less impact on the environment. In fact, all the equations demonstrated by Ian have not only been verified by empirical testing but have been continually refined during the testing phases. This is how research and development should be conducted, creating a mathematical model based on the data rather than trying to reinterpret the data to fit an 'established' theoretical model, which is often the case in today's world.

Figure border colour key, Internet link and Bibliographic references

Figure border colour key:

A yellow border indicates the figure is a MathCad worksheet.

A green border indicates the figure is a MathCad or an Excel plct.

A blue border is used to indicate an illustration, a model, a block or circuit diagram.

A purple border indicates a table.

A red border indicates a photographic-derived image.

A red and green border indicates that the figure is a photo/plot combination.

An amber border represents a collection of calculations using MathCad obtained data.

Internet link Identification:
Internet links are indicated in the text body by This Colour Text. The URLs of the links can be found in Appendix E.

Bibliographic References:
Bibliographic references are denoted by **[**x**]**. The reference number "x" bounded by **bold** orange brackets.

Acknowledgements and In Memory of

Acknowledgements:
I would like to thank Ian Brook Darney, along with his former colleagues at British Aerospace and Alistair Duffy (IET), for providing a solid base upon which this second edition was constructed. Respect and appreciation are due to all the IET Publications staff for their professionalism, and especially to Nicki Dennis, for her encouragement and understanding. Thanks to the dedication and support of Ian's wife Frances, his daughter Susan White and last but not least his son Peter Darney, it has been possible to imbue new life into the 2013 publication and reiterate its continuing relevance in today's high-tech world.

In Memory of:
Ian Brook Darney (29 November 1937–26 December 2021) Whose zest, knowledge and commitment to EMC/EMI, was all done in an effort to ensure the safety of passengers and crew of aircraft and to maintain the integrity of aircraft when refuelling or operating in harsh environmental conditions (thunder storms). He enjoyed technical dialog with other EMC experts using social media, keeping up with the latest technologies and innovations. Utterly committed to the care of his wife Frances, who suffers from Multiple Sclerosis and pursuing his love of jazz, mountaineering and open water swimming. See his obituary in IET.

And

Philip Frederick Hewitt (24 December 1953–15 November 2024) He will be greatly missed by his local village pub quiz team, for his extensive knowledge on a great many topics, his quick and incisive wit and love of sport. As a steelworker, he played rugby union but was well known for his support of Sheffield Wednesday, especially after performing a striptease at the 'Ossie Owl Club', for which he won a gallon of beer, though such was his character, he'd probably have done it for a pint! In retirement he pursed many of his diverse interests alongside his wife Nancy with whom he travelled to many parts of the world, which served him admirably on quiz nights, and in a great many aspects of his life. Enjoy continuing to be the 'Freebird', you always were. Goodbye brother, never to be forgotten, your little-un.

About the authors

Ian B. Darney (deceased) was a highly experienced electrical engineer with an extensive background in EMC, particularly in the aircraft industry, who worked for many years with British Aerospace, Airbus and then as an independent EMC consultant. He was a member of the European Organisation for Civil Aviation Equipment (EUROCAE) and of the IET.

Chris M. Hewitt, Independent Technology Consultant. He has nine years' experience installing, repairing and calibrating RADAR and Missile systems along with test and communication equipment whilst serving in HMRAF, where he was awarded the Support Command Certificate of Merit for 'Initiative and Inventiveness towards the improvement of engineering services of the Royal Air Force', with extensive knowledge of EMC/EMI. He is currently working on Concept AE_{42} "a gravity centric and empirical data approach to grand unification".

Chapter 1
Introduction

1.1 Background

1.1.1 The need for EMC

The development of electronic equipment has come a long way since the invention of valves and transistors, to the extent that modern society is highly dependent on the smooth functioning of the myriad of systems now in operation.

Concurrently with that development, Electromagnetic Interference (EMI) has also increased, both in the number of daily incidents and in the severity of the possible consequences. Initially, most of the effects were annoying; for example, crackles on the radio due to a nearby thunderstorm were something one learned to accept. Latterly, some of the effects could be life threatening. The phenomenon described as 'sudden unintended acceleration' could be a case in point.

A succinct definition of Electromagnetic Compatibility (EMC) is: *The ability of a device, unit of equipment, or system to function satisfactorily in its electromagnetic environment without introducing intolerable electromagnetic disturbances to anything in that environment* [1].

Although intrinsic to the preceding definition, it should be noted that 'self-compatibility' of devices, units or systems should operate without introducing 'troublesome' electromagnetic disturbances to anything in its environment, **including itself**!

1.1.2 Pragmatic approach

There is no shortage of regulatory requirements. In fact, the continuous stream of new or revised regulations is enough to keep a team of researchers occupied full time [2].

However, the task of designing equipment to meet the requirements does not seem to be amenable to a simple, systematic approach. Normal practice is to break the phenomenon down into four distinct types: common impedance, electric fields, magnetic fields, and electromagnetic fields [3]. Each type of coupling is treated separately, and examples are provided of the effects which can be expected with different design fixes. The objective is to help readers achieve an overall understanding of the physics involved. Armed with such an understanding, the designer should be able to assess the EMC characteristics of the equipment-under-review.

This pragmatic approach tends to focus on the multitude of methods by which interference can be coupled from one system to another, on layout and grounding, printed circuit layout, cables and connectors, filtering, transient suppression, and shielding. Many useful design techniques are identified in books which adopt this approach [4]. This being so, there is no need to reprise the material they provide.

Whilst such an approach leads to many useful design techniques, it is essentially hit or miss. A technique which works well in one application can cause disastrous effects in another. Since it involves subjective judgment, there is still plenty of scope for disagreement between designers.

This is in marked contrast to requirements such as functional performance, frequency response, power consumption, reliability, mass and size, which are all amenable to rigorous analysis. Care is taken at every stage of the design process to ensure that the requirements will be met. Bench testing is carried out on prototype equipment.

In situ tests are carried out on assembled systems. Test results are compared with predicted performance. Regular design reviews are carried out. If there are any problems, modifications are implemented.

1.1.3 Academic approach

Another approach to the subject is based on the use of computational electromagnetics. There is at least one book which compiles the results of research carried out on this topic [5].

This is essentially a review of the various techniques used in computational electromagnetics, followed by a selection of an appropriate method of analysis. Such techniques are radiation models for wire antennas, diffraction and scattering models for apertures, field coupling using transmission line theory, and shielding models. The problem with this approach is that the mathematical processes require an expertise which is well beyond the capability of the average equipment designer. Moreover, the focus is mostly on the behaviour of electromagnetic fields. This being so, the functional performance of the equipment-under-review is not an aspect that is analysed. It should also be noted that to obtain accurate or even meaningful results from computational modelling software, it is necessary to include **all** variables and constants as inputs to the model. Just as you would when constructing the parameters for a Maxwell equation to obtain accurate results, you must do the same and write out not only the inputs to the model but also the behaviour of the model, so as to permit the mechanism by which EMI conditions are able to manifest, to be identifiable in the results generated by the model.

1.1.4 Managerial approach

In large organisations, the approach to the achievement of EMC is to create a set of Design Rules which are updated periodically by standing committees. The nature of such a management system is that rules which have been formulated decades in the past come to be regarded as sacrosanct. Development of new equipment involves design processes which follow these rules. The equipment is subjected to performance testing, environmental testing, reliability analyses, design reviews and is eventually submitted to the EMC Test House for final testing. If it fails this final test, this fact is discussed at the Critical Review meeting, emergency measures are invoked, design fixes are implemented, and the equipment is resubmitted for EMC testing. Eventually, the equipment muddles through.

1.1.5 Misleading concepts

There is a widespread belief among engineers that the structure can be represented by an equipotential surface. Such an assumption is inherent in every circuit diagram which contains the ubiquitous 'ground' symbol. Attempts to reconcile this concept with the observed interference in the system have led at least one engineer to declare that the subject is a 'black art'.

Another disturbing concept which has been given formal blessing by successive teams of experts is that of the 'single ground point'.

Some point on the structure is designated as the reference point for all signals in the system. Guidance for the design of UK military equipment formalises this concept and provides detailed requirements for its implementation [6].

If this concept is implemented into the design of any electronic equipment, it can be guaranteed that the equipment will suffer from intractable interference problems.

Closely related to the concept of the 'single-point ground' is the stricture to 'avoid earth-loops'.

Although very few of these concepts are to be found in any textbook on electromagnetic theory, they have become firmly entrenched beliefs in the engineering community. There has to be a better way of approaching the task of achieving EMC.

1.1.6 Circuit modelling

The approach described in the following pages enables electronic equipment to be designed to meet the EMC requirements using mathematics which is understandable to every circuit designer. It establishes a clear relationship between the performance of the system-under-review and the requirements placed on the system. The ability to analyse the coupling mechanisms leads to a better understanding of the underlying physics. Given a better understanding, the designer is able to appreciate the value of the detailed advice provided by the existing books and to reject the misleading concepts prevalent in the workplace. The simplifications inherent in the use of circuit theory inevitably mean that some fidelity is lost in the simulation. If high fidelity is required, then the results obtained by the use of circuit modelling can be refined by the techniques of computational electromagnetics.

1.1.7 Computations

Although the simplifications achieved through the use of circuit modelling provide a dramatic reduction in the complexity of the mathematics, it is still necessary to employ the use of personal computers. Since many of the calculations are beyond the capability of Simulation Programs with Integrated Circuit Emphasis (SPICE), another type of General-Purpose software is needed – mathematical software.

A Mathcad worksheet is used to illustrate the details of each computation. Since mathematical notation is used throughout in these programs, they are much easier to follow than ones written in, say, the Java language. A few special features of Mathcad are described in Appendix A. After reading this appendix, it should not be difficult for any reader to understand the contents of the worksheets.

Since every worksheet is fully explained, the reader is not left to formulate his or her own program from a set of mathematical relationships.

In the following pages, the parameter dimensions adhere to those of the SI System.

1.1.8 Testing

The ability to analyse the mechanisms involved in coupling interference into and out of the signal link-under-review leads to the ability to devise tests which measure the actual coupling parameters and correlate them with those of the relevant circuit model. A connection has been established between test and analysis.

Since circuit theory is capable of analysing signals in either the time domain or the frequency domain, it becomes possible to identify the most probable cause of unexpected interference during product development.

It is but a small step to modify the design of the signal link, build a prototype of the new link, check its performance using bench test equipment, create a representative circuit model, and prepare a progress report. More than that, it becomes possible to check functional performance against formal requirements. That is, EMC can be subjected to the same design process that applies to every other performance requirement.

1.1.9 Essence of the approach

The essence of the approach used by the author has always been to carry out tests on a representative assembly, search for a circuit model which allows the results to be simulated, and then to review available literature to relate that model to accepted theory. Having identified a clear relationship between the model and accepted theory, it becomes possible to define that relationship. Since the theory of electromagnetics is based on the use of the Maxwell equations, it can be said that the model is based on those equations.

Since there are always deviations between the model and the test observations, it becomes possible to focus on the deviations and search for ways of refining the model. It is tempting to continue with this process and develop ever more complex models. However, the objective of the exercise is to enable the engineer at the bench to gain an understanding of the mechanisms underlying electromagnetic interference. Once the cause of a particular interference problem is understood, human ingenuity enables a solution to be found. The ability to create a circuit model of a particular signal link allows the user to predict the interference characteristics of that link, enabling tests to be defined to measure those characteristics. Ultimately, the accuracy of the technique is determined by the accuracy of those measurements.

1.2 Developing the model

1.2.1 Basic model

Basic assumptions can be clearly stated. Every conductor (including the structure) possesses the properties of inductance, capacitance and resistance. Any circuit model which simulates interference must represent at least three conductors. Only a three-conductor configuration can be used to simulate the coupling between two independent circuit loops. For example, to simulate:

- Cross-coupling between two signals in a multi-conductor cable or on a printed circuit board (Section 4.3).
- Coupling between the differential-mode and the common-mode loop (Section 4.4).
- Coupling between a transmission line and the environment (Section 5.2).

A circuit model of inductive coupling between three parallel conductors would be as shown in Figure 1.1. Similarly, a circuit model of the capacitive coupling would be as shown in Figure 1.2. Since current must flow along a conductor to flow into the capacitance, and since a cable behaves in the same manner no matter at which end voltages are applied, the most logical way of simulating the combined effect of inductive and capacitive coupling would be to use the circuit of Figure 1.3. This model includes resistors to represent the effect of the series resistance of each conductor.

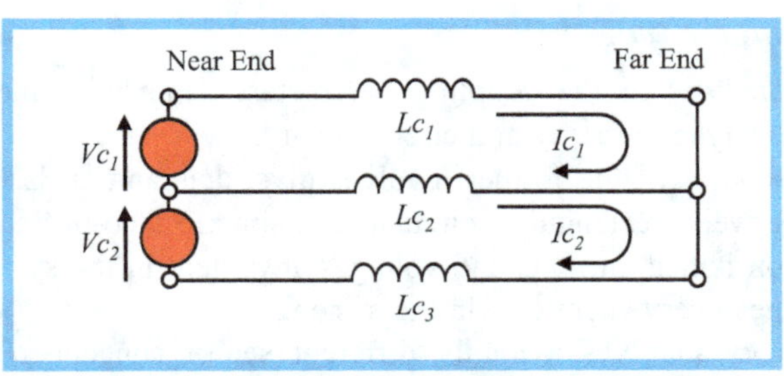

Figure 1.1 Inductive coupling between three conductors

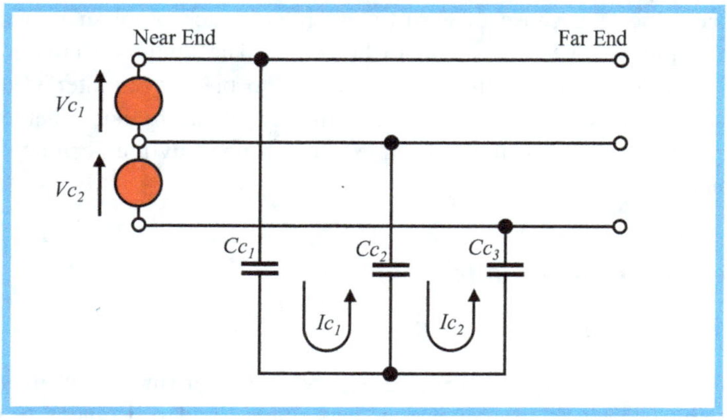

Figure 1.2 Capacitive coupling between three parallel conductors

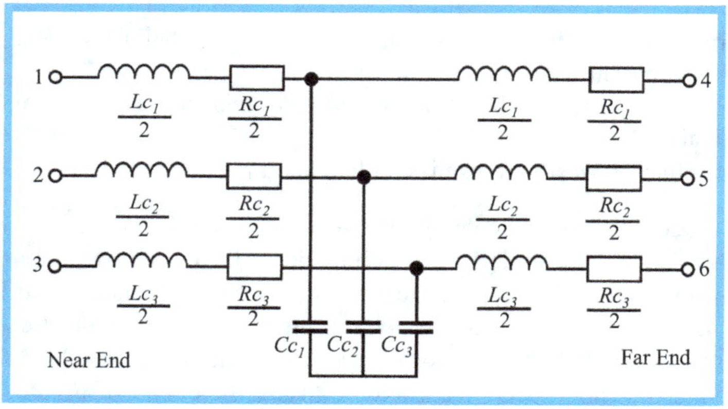

Figure 1.3 Combined effect of inductive and capacitive coupling

1.2.2 Parameter types

Much of the analysis of the coupling parameters deals with the relationships between electromagnetic theory and circuit theory.

The mathematics of the former involves divs, dels and curls, as well as the relationships between rectangular, circular and spherical co-ordinate systems. A basic assumption is that currents and voltages anywhere in the system affect currents and voltages everywhere else in the system.

Circuit theory involves a totally different set of concepts. The system is depicted as a network of nodes and branches and the voltage across each branch is a unique function of the current flowing in that branch. Although circuit theory is much simpler, it still involves a great deal of computation. So, it is essential to ensure at the outset that equations derived using one theory are always distinguishable from equations derived from the other. This can be done by defining different types of parameters and by assigning a unique symbol to each.

Four distinct parameter types can be identified: primitive, partial, loop and circuit:

- A primitive parameter is one which relates the current in a conductor of circular section to the energy level of the electromagnetic field associated with that current.
- A partial parameter is one which relates the current in a conductor of any cross-section to the energy level of the electromagnetic field associated with that current.
- Loop parameters are derived from primitive parameters and are used in equations which relate loop voltages to loop currents. They are the parameters which can be measured directly by electronic test equipment.
- Circuit parameters are those which appear in circuit diagrams.

These parameters are described more fully in Chapters 2 and 3.

Electromagnetic theory invokes the concept of distributed parameters: resistance per metre, capacitance per metre, inductance per metre, and conductance per metre. Chapter 4 identifies a simple transformation formula which obviates the need to use these parameters.

Reflections can occur at transmission line terminations. Incident current flows in one direction along a conductor whilst reflected current flows in the opposite direction. Total current is the sum of these two partial currents. The term 'partial' is also used to identify the associated voltages.

1.2.3 Derivation process

The process used to derive circuit models is based on the method described in textbooks to derive formulae for the equivalent phase inductance [7] and the equivalent phase capacitance [8] of a three-phase power line. It is assumed that three conductors are routed in parallel. It is also assumed that the waveforms of currents and voltages are sinusoidal.

The process can be summarised:

1. Define the length of the assembly, the radii of the conductors and the spacing between the centres.
2. Establish a set of three primitive equations relating the voltage on each conductor to the current in all three conductors.
3. Define the loops in terms of conductor pairs.
4. Derive a set of loop equations.
5. Define the loop inductors and loop capacitors in terms of the primitives.
6. Postulate the existence of a circuit model which creates two mesh equations.
7. Relate the components of the circuit model to the constants in the loop equations.
8. Relate the components of the circuit model to the primitive parameters.

The key feature to note in the above process is that it contains a discontinuity. Step (6) does not follow logically from step (5). Lateral thinking is needed. The purpose of the circuit model is to create a set of mesh equations which correlate precisely with those of the loop equations.

Mesh equations are derived using the rules of circuit theory. Loop equations are derived from the relationships of electromagnetic field theory.

Setting up such a relationship is conceptually the same as defining x as an unknown variable. If circuit theory is treated in this way, it can be utilised to simulate all types of electromagnetic coupling.

It is useful to always bear in mind the fact that Circuit Theory does not define the physical mechanisms. Concepts such as the 'equipotential ground plane' and the 'single point reference' are convenient assumptions which allow circuit models to simulate the behaviour of complex printed circuit boards in signal processing equipment. In so doing, they carry with them the unstated assumption that there is no such thing as electromagnetic interference.

Since mesh analysis caters for the fact that partial currents can flow in both directions along a conductor at the same time, this form of analysis is used in the following pages.

1.2.4 Composite conductors

In practice, there are a great many conductors which are not circular in section. Conduits and cable trays fall into this category. Figure 1.4 illustrates one way of dealing with conductors of any cross-section, which simulate the section as a set of elemental conductors which are short-circuited at each end.

This is a development of a technique pioneered by researchers at Culham Laboratories (as it then was) to analyse the magnetic field distribution in an aircraft wing or fuselage during the occurrence of a direct lightning strike [9].

Since a composite conductor is built up from an array of elemental conductors and since it has inductive and capacitive properties akin to that of $Lp_{i,j}$ and $Cp_{i,j}$, it is necessary to invoke the use of partial parameters; partial inductance $Lq_{m,n}$ and partial capacitance $Cq_{m,n}$, where m and n identify the composite conductors.

This allows a computer program to be compiled which makes a clear distinction between partial parameters and primitive parameters. Chapter 3 describes this technique in greater detail.

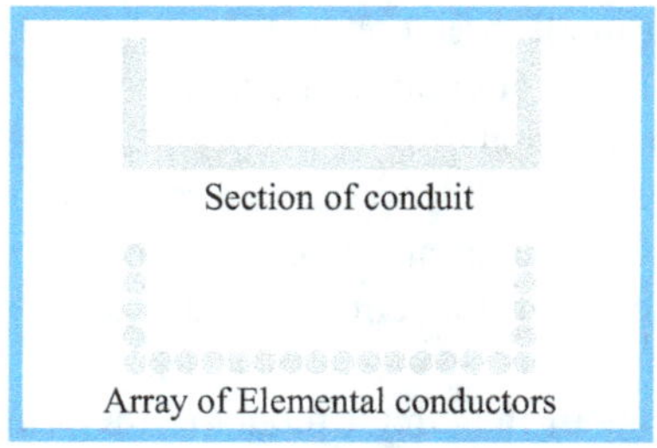

Figure 1.4 Concept of the composite conductor

1.2.5 Proximity effect

The original purpose of elemental conductors was to analyse the indirect effects of a lightning strike on aircraft cables. By determining the magnetic field distribution inside the wing or fuselage when lightning current was flowing in the structure, it was possible to identify regions where the field was less concentrated. This helped in the task of deciding where to route cables.

It is also possible to use the technique to analyse the current distribution in the surface of cables, ducting and conduits, as illustrated by Figures 3.13 and 3.18. These depict the current distribution on the surface of conductors and show that when a loop current is flowing, the current concentrates on adjacent surfaces. These pictures could equally well illustrate the distribution of charge on adjacent charged surfaces.

The technique can be used to simulate the current distribution on the conducting surfaces of printed circuit boards, as illustrated by Figure 8.3. Such a depiction is probably more meaningful to circuit designers than colour images created by full-field simulation techniques. As well as providing results which are easier to interpret, the technique does not require the use of special-purpose software or the ability to understand the underlying mathematics of such software.

Normal analysis of skin effect, as described in Section 2.5.4, assumes that the current is evenly distributed over the surface of the conductor. The proximity effect indicates that, although the current concentrates on the surface, the distribution is not at all uniform. Even so, it does not alter the fact that conductor resistance increases as the frequency increases.

1.2.6 Electrical length

The wavelength λ represents the distance an electromagnetic wave must travel to change phase by 360 degrees. If the frequency of the wave is f and the velocity of propagation is v, then

$$\lambda = \frac{v}{f}$$

The velocity of propagation of a signal along the conductor of a transmission line is the same as that of the associated electromagnetic wave. The electrical length k of a conductor of length l can be defined as

$$k = \frac{l}{\lambda}$$

The usual textbook definition of a short electrical length is that

$$k < 0.1$$

Figure 1.5 illustrates the relationship between short electrical length and wavelength. Experience has shown that the circuit model of Figure 1.3 is reasonably accurate, provided the conductor is electrically short. Equally, it can be said that the model is reasonably accurate up to a frequency at which the wavelength is ten times the length of the assembly-under-review.

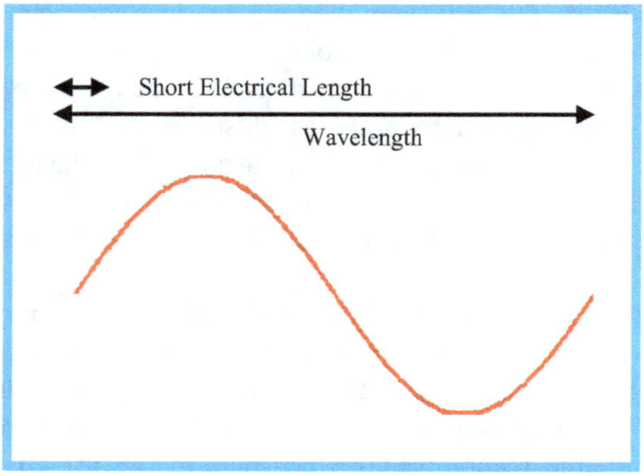

Figure 1.5 Electrical lengths

1.2.7 Distributed parameters

Transmission line theory caters for the fact that currents and voltages vary along the length of a conductor by invoking the concept of distributed parameters. That is, the derivation of formulae is based on the parameters:

$$\frac{Rc}{l}, \frac{Lc}{l}, \frac{Cc}{l} \text{ and } \frac{Gc}{l}$$

where l is the length of the conductor and *Rc, Lc, Cc* and *Gc* are the series resistance, inductance, capacitance and conductance of the line. The analysis results in a pair of hybrid equations which relate the current and voltage at the near end of the line to the current and voltage at the far end. If it is postulated that the complete length of the transmission line can be represented by the circuit model of Figure 1.6, then it is possible to define the impedances *Z1* and *Z2* in terms of the lumped parameters: *Rc, Lc, Cc* and *Gc*.

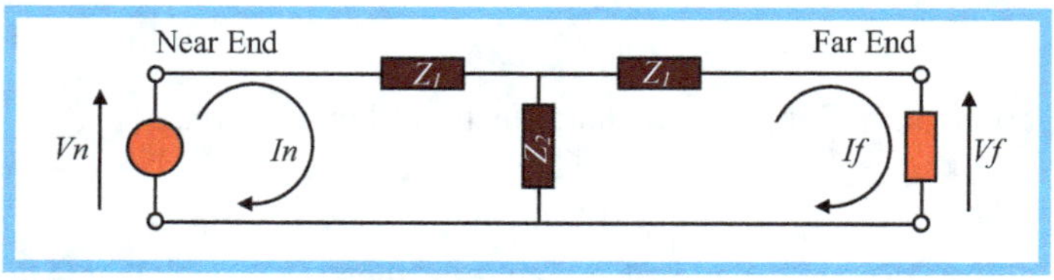

Figure 1.6 Transmission line model

Extending the concept to deal with a three-conductor transmission line gives the circuit model of Figure 1.7. This can be described as a distributed parameter model since all the impedance values are derived via the use of distributed parameters in the derivation process. Although these impedances are derived in an unconventional way, they still evaluate to the form $R + j \cdot X$, where X is the value of the reactance at frequency f.

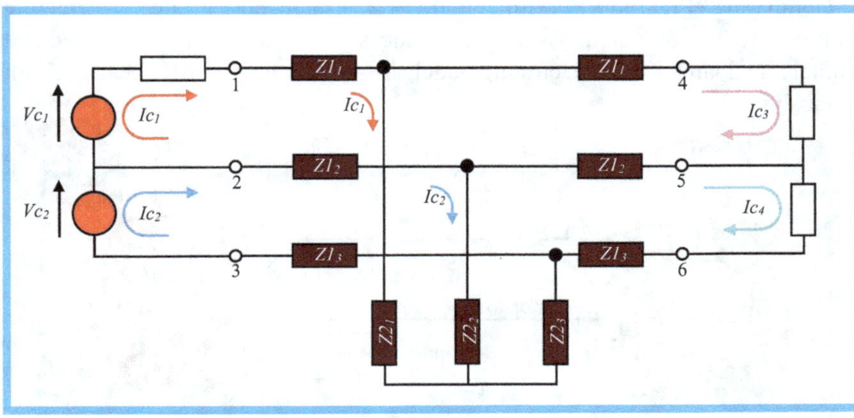

Figure 1.7 Distributed parameter model

Since there is a clear correlation between the models of Figures 1.3 and 1.7, the transformation process is extremely simple. The three-conductor line can be defined in terms of the lumped-parameter model and analysed using the distributed-parameter model. This means that the designer retains visibility of the properties of the model throughout the analytical process.

This method of simulation is much simpler than stringing dozens of lumped-parameter models in series.

Using the distributed parameter model, the maximum frequency of the simulation is no longer limited by the length of the cable. But there is still a limit. It is assumed that action and reaction between adjacent conductors in any cross-section of the cable are instantaneous. By analogy with the limitation defined in Section 1.2.6, the maximum frequency is that at which the wavelength is ten times the maximum spacing between conductors.

Best accuracy is obtained if the cross-section of the cable is exactly uniform along its entire length and the lengths of the three conductors are identical. If these requirements are not met because different types of cable are used or because intermediate connectors are included in the wiring harness, then the accuracy of the simulation will be reduced.

However, accuracy can be restored by carrying out bench testing on a representative assembly and creating a circuit model which replicates the test results.

1.3 Intra-system interference

1.3.1 The signal link

Having derived a method of simulating the coupling between three separate conductors, it becomes possible to relate the model of Figure 1.3 to actual hardware. Figure 1.8 illustrates the general method of transmitting an electrical signal from one unit of equipment to another.

Comparing Figure 1.8 with Figure 1.3 establishes a clear correlation. Terminals 1, 2 and 3 at the near end of the signal link can be correlated with terminals 1, 2 and 3 of the circuit model. Terminals 4, 5 and 6 bear a similar relationship.

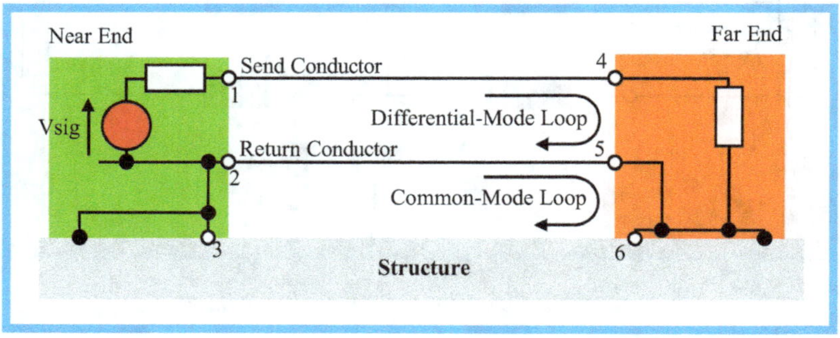

Figure 1.8 Signal link wiring

1.3.2 Simulating the structure

The simplest way of simulating the structure is to treat it as a plane surface and employ the method of images to calculate the relevant parameter values. This technique involves treating the plane surface as a perfectly reflecting mirror, then using the properties of the image conductors to represent the effect of the actual surface as illustrated by Figure 1.9.

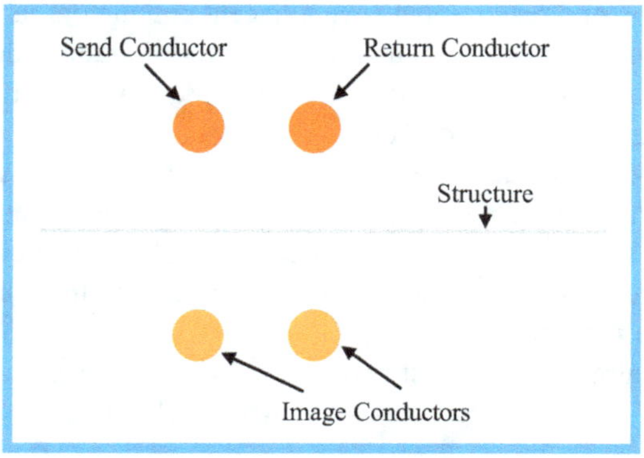

Figure 1.9 Simulating the structure

A set of four primitive equations is created to relate voltages on the four conductors (two real, two images: the structure is transparent) to the currents in those conductors. These can be simplified to two loop equations. Then, the process described in Section 1.2.3 is used to derive formulae for the circuit inductors and capacitors of the triple-T model.

A more accurate method of deriving values for the inductance and capacitance of the structure is to use the method of composite conductors, as shown in Figure 1.10. In this simulation, all other conductors in the vicinity of the send and return conductors are assumed to form a single composite conductor, and the properties of that conductor are assigned to the structure. As far as this method of simulation is concerned, it does not matter that the other signals in the system are interfering with each other. All that matters is how they affect the link-under-review and how the signal in the link-under-review couples with the structure.

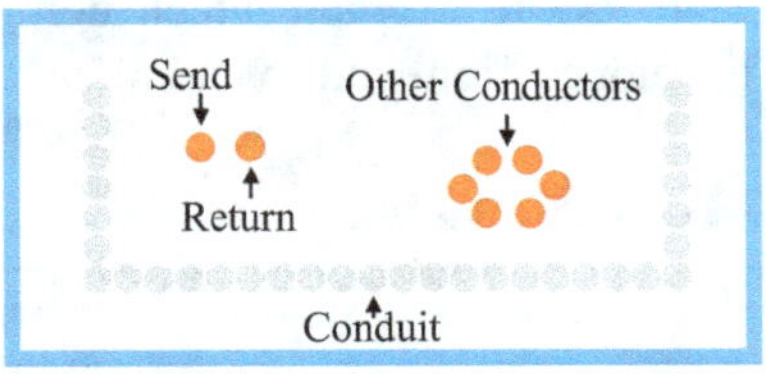

Figure 1.10 Using the method of composite conductors

Chapter 3 derives computer programs for calculating values for the components of Figure 1.3, using data on the geometry of the structure and cables.

1.3.3 Equivalent circuits

One of the concepts inherent in circuit theory is that of the 'equivalent circuit'. Two inductors in series can be represented by a single inductor whose value is the sum of the two inductors it represents. Two capacitors in parallel can be represented by a single capacitor whose value is the sum of the original two.

This means that a series of triple-T networks can be simulated by the single network of Figure 1.3. That is, a triple-T network can be used to simulate any signal link between two units of equipment, no matter how the cable is routed.

With a cable that follows a complex route along different parts of the structure, the task of calculating accurate values for the components of the model can become quite arduous. It is possible to make an initial estimate, but the more complex the route, the less confidence there is in the final result of the calculations.

This is where test equipment comes into the picture. Given a sure knowledge that the link can be represented by a triple-T network, it is not particularly difficult to devise a set of tests to measure the value of every component of the model. An LCR meter could be used. Other ways are described in Chapter 7.

1.3.4 Conducted emission

Adding components to the near and far ends of the triple-T model to represent the interface circuitry within the two units of equipment results in the model of Figure 1.11. The geometry of the signal link provides sufficient data to provide an initial estimate of the value of every component of the triple-T model.

Given this information, values for the four loop currents can be calculated. Current Ic_2 gives a measure of the common-mode current at the near end of the link. Calculating the ratio of current in the common-mode loop to the voltage in the differential-mode loop gives a value for the transfer admittance *YT*.

$$YT = \frac{Ic_2}{Vc_1}$$

Calculating the value of *YT* at a number of spot frequencies allows a graph to be created, relating transfer admittance to frequency. This graph gives a clear indication of the conducted emission characteristics of this particular link.

It is equally possible to create a graph which defines the frequency characteristic of the voltage developed from end to end along the structure.

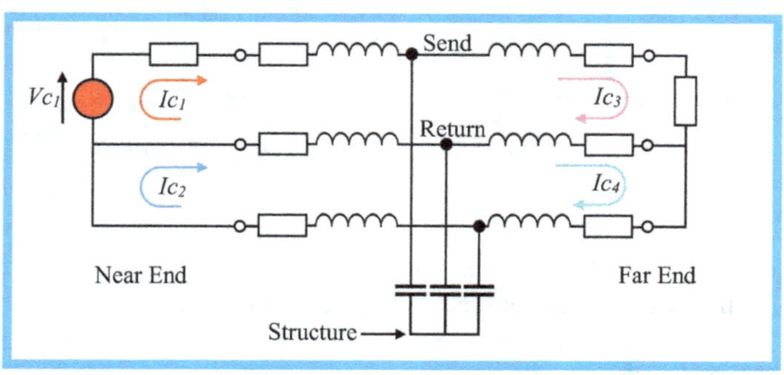

Figure 1.11 Circuit model of signal link

1.3.5 Conducted susceptibility

Given knowledge of the voltages developed along the structure by a number of signal links, the principle of superposition can be used to define the total voltage developed along the structure.

The combined effect of several sources of interference can be represented by a single voltage source, Vc_2, in series with the structure. Figure 1.12 shows how the model can be used to simulate the effect of this interference. Again, the transfer admittance characteristic can be used to define the susceptibility of the link:

$$ZT = \frac{Ic_1}{Vc_2}$$

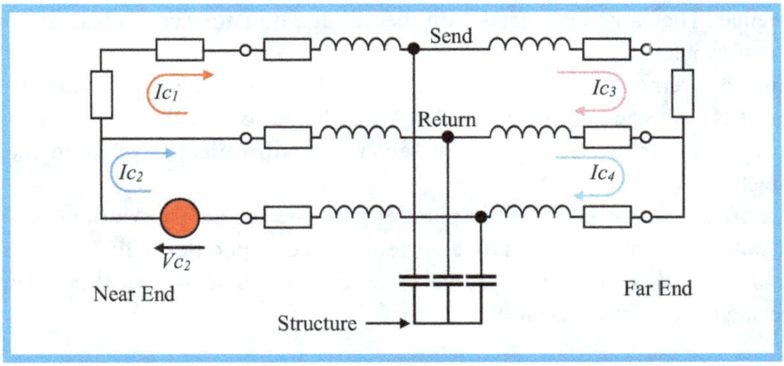

Figure 1.12 Circuit model for analysis of conducted susceptibility

1.3.6 Voltage transformer

Having established a method of simulating the effects of conducted emission and conducted susceptibility, the next step is to carry out bench tests of the signal link.

Comparing the results of such tests with the simulation allows a check to be made on the accuracy of the model. But first, it is necessary to have the right sort of test equipment available.

It is reasonable to assume that anyone responsible for developing electronic equipment will already have available a number of items of general-purpose test equipment and that these items include a signal generator and an oscilloscope.

The first requirement is to be able to induce a fairly constant voltage over a wide range of frequencies into the loop-under-test without the need to physically break into the loop. This can be done using a split-core toroid, similar to those used in EMC test houses. The second requirement is to minimise the impedance reflected into the loop-under-test by the test equipment. This can be done by winding several turns on the primary.

Signal generators usually have an output impedance of 50 Ω. With a turns ratio of 10 to 1, the reflected impedance will be 0.5 Ω. The third requirement is to monitor the output voltage, since the load presented to the transformer can vary dramatically. The addition of a separate monitor turns caters to this requirement.

If the transformer is clamped around both conductors of the signal link, it will induce the same voltage into the send and return conductors. Net voltage induced into the differential-mode loop will be zero. The full voltage is induced into the common-mode loop however. Since current in the common-mode loop flows via the structure, the effect can be simulated by inserting a voltage source in series with the conductor representing the structure.

1.3.7 Current transformer

Current transformers are already in widespread use. However, it is useful to recap on the construction and use of such transformers when used to measure interference. The same core used with the voltage transformer is ideal for the current transformer.

The requirement to reflect minimal impedance into the loop-under-test still applies. It is necessary to use a 50 Ω co-axial cable to link the transformer to the monitoring equipment. So, a ten-turn secondary will reflect 0.5 Ω into the loop-under-test.

If a 50 Ω resistor is placed in parallel with the secondary winding, the transformer output can be simulated by a current source in parallel with 50 Ω. This can also be simulated by a voltage source in series with 50 Ω, the ideal configuration for interfacing with a co-axial cable.

1.3.8 Representative circuit model

Chapter 7 describes several tests which can be carried out on the signal link to determine the frequency response of the transfer admittance. When these are plotted on the same graph as that of the simulated response, it is inevitable that there will be some deviation between the two curves.

However, it will always be possible to modify a few parameters which define the component values of the model to achieve close correlation between the two curves. The number of independent variables is much fewer than the number of components. Experience has shown that it does not take many iterations of the simulation to achieve this objective.

The end result of such an exercise is the creation of a circuit model which defines the coupling characteristics of the signal link. Such a model can be utilised in much the same way as one of the modules in the library of components available in most SPICE software packages.

Advantage can be taken of the fact that every parameter (R, L, C and G) of the model is a function of length. Tests can be carried out on a long test rig at low frequencies and used to create a model of that rig. Knowledge of the length of that rig can be used to modify the component values in the model, to allow the modified model to simulate the behaviour of a much shorter link.

Tests at relatively low frequencies can be used to predict performance at much higher frequencies. Such an approach is analogous to the use of wind-tunnel tests on a small-scale model to predict the behaviour of the actual aircraft.

This means that a circuit model for a 10 m line, tested at frequencies up to 20 MHz, can be used to create a model for a 1m line, valid at frequencies up to 200 MHz. Care would need to be taken that the cross-section of the test rig is constant over the entire length and that the components at the interfaces are suitable for use at the frequency at which the equipment will operate.

The method can be used to simulate the cross-coupling between two 50 mm lengths of printed circuit track, valid up to 4 GHz. Testing the accuracy of this particular model would require the use of sophisticated test equipment.

Most of the tests described in Chapter 7 are limited to the short-wave band. But this is only because the test equipment available was limited to 20 MHz.

The model itself is limited only by the assumptions inherent in the theory of transmission lines. That is, it is assumed that action and reaction between adjacent conductors are instantaneous. In practice, the upper frequency limit is that at which the wavelength is ten times the maximum dimension of any section of the assembly-under-review.

Section 7.7 shows that the technique of circuit modelling can be used to characterise small components such as capacitors, at frequencies over the range 200 kHz to 1 GHz.

1.4 Inter-system interference

1.4.1 Dipole model

Having developed a model which is not restricted to the simulation of conductors of short electrical length, the way is open to extend the approach to the analysis of antenna-mode coupling. A reading of the textbook analysis of the half-wave dipole reveals that a new component can be added to the existing set of primitive parameters: the radiation resistance.

When a sinusoidal voltage is applied between the two dipoles of the antenna and the frequency adjusted to coincide with the quarter-wave frequency of each dipole, the current distribution along each conductor follows a sinusoidal waveform. Maximum current flows at the centre of the antenna, whilst the current at the tips is zero. The current has to go somewhere; it is converted into electromagnetic radiation.

A commonly accepted formula for the average radiated power *Pt* over a spherical surface is:

$$Pt = \frac{1}{2} \cdot Rrad \cdot Ip^2$$

Where *Ip* is the peak amplitude of the current. The parameter *Rrad* is not a resistor in the conventional sense of the word, but a mathematical constant derived from a complex process of integration which happens to have the dimensions of resistance. For a half-wave antenna, the radiation resistance, *Rrad*, is 73 Ω.

Since the current distribution along a monopole of a half-wave dipole is identical to that along an open-circuit transmission line operating at the quarter-wave frequency and since the T-network model of Figure 1.6 simulates the behaviour of the line, it is logical to assume that this model can also simulate the behaviour of a monopole.

This leads to the model of Figure 1.13, where *Lp* and *Cp* are the primitive inductance and primitive capacitance of a monopole.

Tests on a single length of isolated conductor, when it is acting like a dipole, are described in Section 7.4. The fact that there is a close correlation between the responses of the model and the hardware provides a high level of confidence in the soundness of the technique.

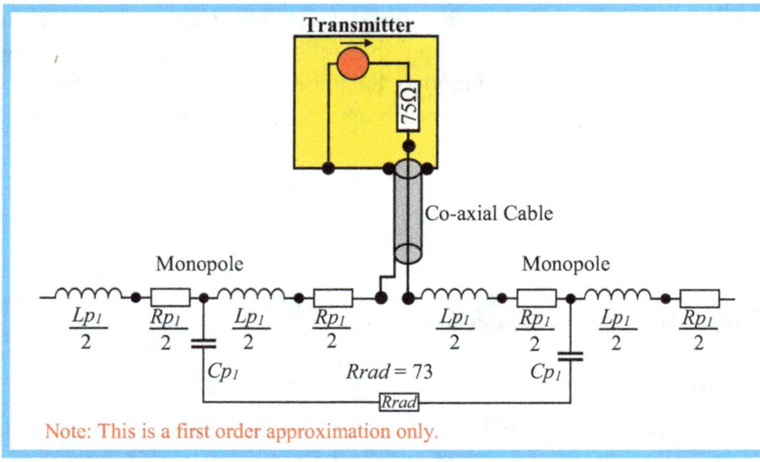

Figure 1.13 Circuit model of a half-wave dipole

1.4.2 The virtual conductor

The common factor in every model developed for analysis of intra-system interference is that every conductor is represented by a T-network of inductors, resistors and capacitors. The concept can be extended further to simulate the coupling between a twin-conductor cable and the environment.

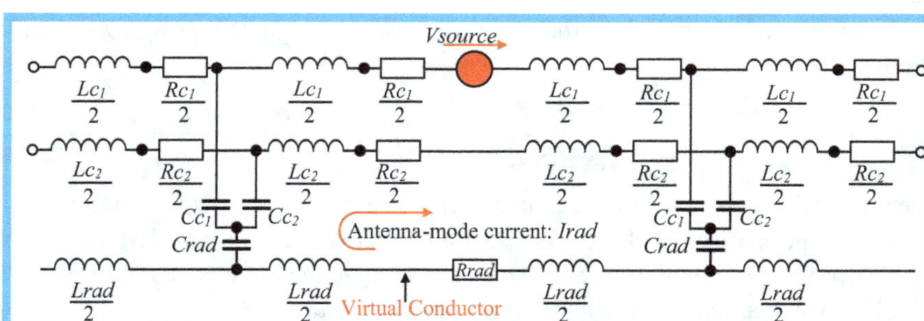

Figure 1.14 The virtual conductor

The model of Figure 1.14 represents a configuration where a voltage source is applied to one conductor of a twin-core cable, at the mid-point along the length of the cable. This conductor acts as a transmitting dipole. The adjacent conductor acts as a receiving dipole.

Although most of the transmitted energy is picked up by the receiving antenna, a significant proportion constitutes radiated interference.

Analysis of the geometry of the twin-conductor cable provides formulae for the inductance *Lrad* and capacitance *Crad*. The other components remain exactly the same as those of the transmission-line model of the cable.

Since the components *Lrad*, *Crad* and *Rrad* are also configured as a T-network, it is logical to give it the name 'virtual conductor', since it does not simulate an actual piece of hardware. It represents the effect of the environment.

The amplitude of the current radiated into the environment *Irad* gives a measure of the radiated field strength. The maximum strength of the magnetic field *H* at a distance *r* is:

$$H = \frac{Irad}{2 \cdot \pi \cdot r}$$

This formula is recognisable as that derived by integrating the Biot–Savart equation along an infinitely long, straight conductor, as (2.2.1) would indicate. However, it can also be derived from the equations of antenna theory. Section 5.7 shows that it is related to the maximum power delivered to the environment by a wire pair routed along the structure. It can be used to predict the results of formal EMC testing of radiated emission of the equipment-under-review, by indicating the maximum strength of the signal which would be picked up by the monitor antenna.

1.4.3 The threat voltage

The model of Figure 1.14 can also be used to simulate the behaviour of a twin-conductor cable when it is exposed to an external field. The only change necessary is to move the location of the voltage source to that shown in Figure 1.15.

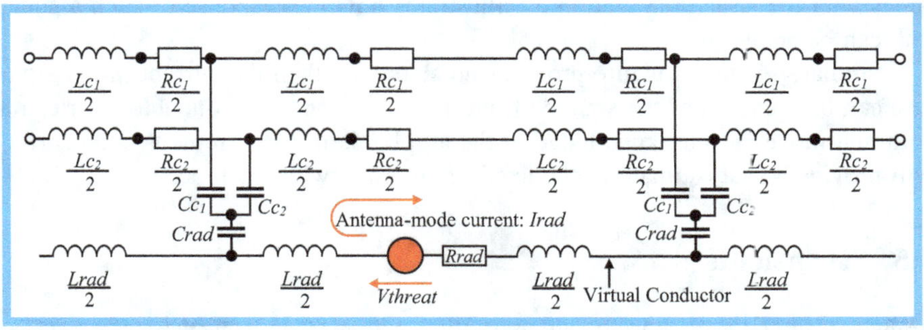

Figure 1.15 The threat voltage

The amplitude of the threat voltage *Vthreat* can be calculated by integrating the value of the electric field strength *E* over the length *l* of the cable. In the case where

$$l = \frac{\lambda}{4} \quad \text{then} \qquad Vthreat = \frac{\lambda}{\pi} \cdot E$$

Vthreat is the maximum voltage injected into the antenna-mode loop by an electromagnetic field of electrical field strength *E*.

Analysis of this model allows the differential-mode current caused by an external field to be calculated.

1.4.4 Worst-case analysis

By assuming that optimum coupling exists between conductors, the end result of any analysis is a calculation of the maximum level of interference. By providing an accurate simulation of the response over the frequency range which includes that of the quarter-wave frequency and the half-wave frequency, the analysis guarantees that the frequency at which interference is a maximum will be predicted.

Skin effect ensures that the amplitude of the interference current reduces with frequency. Also, the amplitude of the threat voltage due to an external field of constant power density decreases with frequency, as illustrated by Figure 5.19. These factors ensure that the first resonant peak is always the highest, at least, in a design that is not intended to act as an antenna.

These factors enable the designer to focus on the performance of each signal link at those frequencies where EMI problems are most likely to arise.

By invoking the simple relationships of (5.32) and (5.40), which define maximum levels, the designer avoids the need to consider the field distribution pattern in the region of the signal link.

By far the majority of signal links in a system are carried by cables which are routed over a conducting structure. If it is assumed that the shielding effectiveness of the structure is zero, then the maximum power which can be delivered to the line can be represented by a voltage source *Vthreat* in series with *Rrad*. One way of simulating the effect of the threat environment is to insert both these components in series with the structure. Section 5.5 illustrates how the response of such a signal link can be analysed.

Similarly, if the structure provides no shielding, then the current in the structure can be regarded as the source of interference. Since this conductor carries the common-mode current, knowledge of the amplitude of the common-mode current will provide a first estimate of the level of emission which can occur.

1.5 Transients

Transients are an ever-present source of glitches in electronic systems. Sources can be relays, switches, motors, and power supplies. They can easily corrupt the data streams handled by microprocessors. Depending on the criticality of the processing circuitry, such events could be inconsequential, annoying, dangerous, or catastrophic.

Since most signal processing is now carried out by digital signals, it is essential that this topic be included in any analysis of interference. The lumped parameter models developed for frequency response analysis can also be used for analysis in the time domain. The textbook approach to such analysis is to invoke the use of Fourier Transforms, Laplace Transforms, and even more complex techniques. The approach adopted here follows the example of SPICE programs and uses time-step analysis.

However, the problems encountered with the analysis of the frequency response of a transmission line also appear in the use of transient analysis. It takes a finite time for a signal to traverse from one end of the line to another. With frequency analysis, it was possible to transform the lumped-parameter model into a distributed-parameter model. With transient analysis, another solution is called for.

In concept, the solution is much simpler; use the computer memory to store the signal applied to the near end of the line for a fixed number of time steps before delivering it to the terminals at the far end.

Books on electromagnetic theory introduce the concept of partial currents and partial voltages to explain the behaviour of transient signals at the interface between cables and equipment terminations. Incident current flows towards the interface; some of it is absorbed by the equipment, and some is reflected back down the line. The total current at any section of the cable is the sum of the incident and reflected currents at that location. Section 6.2 describes the phenomenon in more detail.

The reflected signal is also delayed a fixed number of time steps before appearing back at the near end. A simple program to simulate the propagation of incident and reflected currents is also described in Section 6.2.

An experiment was carried out by applying a square wave to one end of a twin-core cable via a 5Ω resistance, leaving the far end open-circuit and monitoring the current flowing in the line. Since the impedance at the near end was much lower than the characteristic impedance of the line, it was expected that there would be multiple reflections. Attention was focused on the response to the leading edge; that is, on the response to a step input.

Textbook theory predicted that the incident current would be inverted by the open-circuit terminals at the far end and reflected straight back to the near end. At the near end, the inverted current would almost double in amplitude. The response to a step input was expected to be a square waveform of slowly diminishing amplitude. It wasn't.

During the time the leading edge of the pulse took to make the return trip, the current remained steady. This was in accordance with classical theory. However, the trailing edge was not sharp, as one would expect. It took the form of an exponential decay. This process continued, and the waveform underwent a metamorphism, rapidly changing shape into a sine-wave.

A trial-and-error process ensued in an effort to create a circuit model which replicated the observed waveform and which could be explained in terms of electromagnetic theory. It was eventually reasoned that, as the step waveform propagated along the line, it left a residual charge on the conductors and that this charge gradually decayed away via current flow into the environment. This charging current could be simulated.

In addition, it proved to be possible to measure and simulate the proportion of the current which departed into the environment. Results from these transient tests could also be correlated with observations and analysis of tests which had been made using sinusoidal waveforms. Section 6.6 explains the reasoning.

1.6 The importance of testing

Chapters 2 to 6 are essentially concerned with methods of developing circuit models which replicate the relationships defined by electromagnetic theory. Since electromagnetic theory relates the electrical parameters to geometric parameters, the result is the establishment of a theoretical relationship between each circuit model and the hardware it represents.

Electrical tests establish a clear relationship between the response of the model and the response of the signal link it represents.

By correlating the response of the model with that of the test setup, the circuit model can be used to define the characteristics of the link. Electrical tests can be carried out on the configuration-under-review and the results used to define the electromagnetic coupling characteristics of that configuration. The model can be described as a 'representative circuit model'.

Testing and analysis can be carried out concurrently. This eliminates the need for a 'try-it-and-see' approach.

In view of the several chapters of mathematical analysis which precede the introduction of the subject of testing, readers can be led to the belief that theoretical analysis needs to be carried out before testing can begin. This is not strictly true. Testing precedes analysis.

From the point of view of the author, the starting point was the observation of annoying glitches in electronic systems. The sources were easily identifiable, but the coupling mechanisms did not seem to be amenable to analysis. Circuit models were created to simulate the observed phenomena and deviations noted between simulation and observation. These deviations led to speculation as to the probable causes. When a review of electromagnetic effects revealed a plausible explanation, it became possible to establish a firm link between electromagnetic theory and the circuit model. The model was refined and further tests carried out to reveal other deviations.

For example, tests to characterise a twin-conductor cable reveal the fact that antenna-mode current propagates at a higher velocity than differential-mode current. Section 7.5 describes how this can be demonstrated.

In hindsight, the phenomena can be explained by the fact that the antenna-mode waves propagate mainly in the air, whereas the differential-mode waves are mostly in the cable insulation. From the viewpoint of field propagation, this explanation comes quite readily to mind. In this case, however, the evidence comes from observations on the behaviour of currents and voltages rather than the analysis of H-field and E-fields.

To someone who thinks in terms of system function, this is something of a revelation. Although the electromagnetic fields propagate in the insulating medium, the currents which create them flow in the conductors. The antenna-mode current and the differential-mode current are separate entities, just as surely as the incident and reflected currents in transmission lines are independent of each other.

This conclusion is supported by the analysis of the transient tests described in Section 7.6. In the setup of Figure 7.50, a step voltage is injected into the signal conductor of a conductor pair. This creates an antenna-mode current which propagates in the same direction down the pair of conductors, from near end to far end. Current in the signal conductor also causes a current to flow back along the return conductor.

Just behind the leading edge of the antenna-mode current step, current is flowing in both directions along the return conductor. Current flowing in opposite directions in a pair of conductors constitutes differential-mode current. The leading edge of the differential-mode step follows behind the leading edge of the antenna-mode current, at a lower velocity.

From the results of the tests of Section 7.5, the two velocities are:

- antenna-mode current: 230 m/μs
- differential-mode current: 170 m/μs

That is, the use of circuit modelling techniques provides the user with an ever-improving understanding of the mechanisms involved in interference coupling.

The key point here is that progress can only be made by engineers who carry out measurements for themselves.

An equally important aspect of testing is that it provides firm evidence of the validity of the model. Being able to verify theoretical results by comparing them with other theoretical results is encouraging, but it does not provide the confidence achievable when practical measurements correlate with theoretical predictions.

1.7 Practical design techniques

Since the concepts of the 'equipotential ground', the 'single-point reference', and the advice to 'avoid earth loops' have acquired universal acceptance as critically important guidelines, the first three sections of Chapter 8 are devoted to an explanation as to why they are so misleading. The remaining sections identify many of the techniques employed by generations of designers to improve circuit immunity and reduce the level of unwanted emissions.

More than anything else, the process of testing and analysing the different mechanisms involved in interference coupling leads to a much-improved understanding of these mechanisms. This leads to the ability to assess any given wiring assembly and any given interface circuit in light of the underlying physics. Given this ability, it is possible to identify groups of techniques which have the same objective. These are:

- common-mode rejection.
- differential-mode damping.
- common-mode damping.
- shielding.

The design of the interface circuitry and system shielding is critically dependent on an understanding of cable-coupling mechanisms. Every circuit described in Chapter 8 includes a definition of the complete signal link; there are no loose ends where the circuit terminates at a set of plug pins or socket inserts. This allows the design of the interface circuitry to be related to the coupling characteristics of the cable. For example, the performance characteristics of 'grounded' and 'floating' configurations are compared.

Packing density on printed circuit boards (PCBs) precludes the option of carrying out a detailed analysis of every signal link on a board. Even so, anyone who understands the method of modelling will also understand the interference coupling mechanisms and will avoid most of the obvious errors.

It would be good practice to fit a buffer circuit at each PCB interface to decouple the signals on the interconnecting cables from the multiple branches and stubs that must exist on the board. A variety of such buffer circuits are described in Chapter 8.

Any interference experienced would then be due to internal coupling. A well-designed board would not experience such coupling, and this could be confirmed the first time the board is functionally checked.

Since there are books which provide excellent advice on the detailed design of PCBs, there is no point in duplicating their content.

One book which stands out by virtue of the wealth of detailed design information it provides is 'EMC Design Techniques for Electronic Engineers' [10]. Another book which is well worth purchasing is 'Introduction to Electromagnetic Compatibility' [11], since it provides detailed information on the theory underlying electromagnetic field propagation and relates this to practical design problems. The chapters on shielding and electrostatic discharge also provide valuable information.

1.8 System design

1.8.1 Guidelines

The set of guidelines listed below is based on deductions derived from the analysis of electromagnetic coupling. If agreed by members of the design team at the outset of a project, it will avoid the pitfalls suffered by many previous electronic systems.

1. The best use of the structure is as a conducting shield. It should not be used as a convenient return path for any of the signals or power supplies in the system. Any current caused to flow in the structure will be a source of interference.
2. Assign a 'return' conductor to every 'send' conductor and keep the two conductors as close together as possible along the entire route from source to load.
3. Encourage the use of ground loops, since these enhance the shielding properties of the system.
4. Design the interface circuitry to enhance the balance between the current in the 'send' conductor and the current in the return conductor. Ideally, the net current flow through any cable section should be zero.

5. Treat every signal link as a transmission line and use resistors similar in value to the characteristic impedance in the interface circuitry, as far as is practicable.
6. Implement common-mode damping on critical signal links.
7. Test and analyse critical signal links.

1.8.2 Top-down approach

The process described in the following pages starts off by defining the basic building blocks of all circuit models and then develops the models to a point where they can reliably simulate the performance of actual signal links. These signal links can then be regarded as basic building blocks in the design of complete systems.

The modular approach to system design is already well established in the use of block diagrams.

A block diagram of the complete system is defined at the beginning of a project with each block identifying a unit of equipment, or even a set of related equipment units, and with lines between the blocks used to identify the signal and power lines. Each block can then be treated as a separate entity, and its function represented by a separate block diagram.

At the lowest level, the functional block is defined in terms of a circuit diagram such as a logic gate, a J-K flip flop, a band-pass filter or a buffer circuit.

By incorporating the models of the signal links into the process, it becomes possible to invoke the top-down approach to the analysis of the EMC of the system. Section 9.2 describes the relationships between the different types of diagrams.

1.8.3 Formal EMC requirements

Any proper outline of the formal EMC requirements would take many chapters to describe, and is way beyond the scope of this book. However, it is still necessary to establish some sort of relationship between these requirements and the actual performance of the system.

Susceptibility requirements can be analysed by defining the threat environment in terms of a graph relating the maximum amplitude of the external field strength to frequency. Using the circuit model of the link-under-review the response of the differential-mode current to the external field can be compared directly with the characteristics of the actual signal being transmitted. Section 9.4.3 provides an example.

Emission requirements can be analysed by simulating the signal carried by the link-under-review and applying this signal to the differential-mode input of the model. The frequency response of current in the common-mode loop can then be determined. In the worst-case situation, this is the current that delivers interference to the environment. The frequency response of the common-mode current can be compared directly with the maximum acceptable limits defined by the formal requirements.

For both susceptibility and emission, this analysis is based on the assumption that the shielding effectiveness of the structure is zero. Section 8.7 gives some guidance on methods of estimating the shielding effectiveness.

Chapter 2
Lumped parameter models

A review of the method used to calculate circuit component values for three-phase power lines leads to the identification of the first step: the derivation of a formula relating the capacitance of an isolated conductor to its length and radius. This formula can be regarded as a basic building block from which the capacitive parameters of multi-conductor assemblies can be derived.

In computer terminology, a low-level object or operation from which higher-level, more complex objects or operations can be constructed is termed a *primitive*. So, it seems reasonable to use the term *primitive capacitance* to identify the basic relationship.

The first equation quoted in Section 2.1 can be found in textbooks on Electromagnetic Theory [12]. Although the derivation involves some complex integration, the end result is a formula which defines the primitive capacitance, $Cp_{i,j}$. This relates the voltage on conductor i to the charge on conductor j.

Primitive inductance $Lp_{i,j}$ can be regarded as the twin of primitive capacitance. In this case, textbooks can be found that outline the derivation process [13]. However, these earlier derivations do not include the contribution made by internal linkages. Nor do they cater for the fact that in any circuit analysis, the current must flow in a complete loop. The derivation provided by Section 2.2 includes these considerations.

In these earlier textbooks, the name given to the derived parameter is 'partial inductance'. However, in the analysis of Chapter 3, an array of elemental conductors is used to represent a composite conductor. In this treatment, the *partial parameters* associated with each composite conductor are derived from a collection of *primitive parameters*. With any computation process, it is necessary to distinguish between the two types. So, they have been given different names.

Section 2.3 identifies the duality between primitive inductance and primitive capacitance. The square root of the product of these two parameters is the propagation delay; the time it takes for a transient pulse to propagate along the length of a conductor. *Primitive parameters* define the properties of conductors when they are acting as antennae.

Section 2.4 describes how primitive parameters can be combined to derive the capacitance between a pair of parallel conductors and the inductance of a loop formed when the terminals at the far end are connected. *Loop parameters* are those which can be measured with electronic test equipment.

With EMC analysis, an important requirement is to maintain visibility of the voltage developed along each conductor as well as the voltages appearing between conductor terminals.

Section 2.5 derives component values for a circuit model which maintains a one-to-one correlation between each conductor and its inductive and capacitive properties. *Circuit parameters* are those which are used in circuit diagrams.

The resistance of the conductors is also an important factor in determining the performance of any cable. Since this parameter can vary due to skin effect, an equation is derived which relates conductor resistance to frequency.

Each conductor of a two-conductor cable can be represented as a T-network, where each horizontal branch contains an inductor and a resistor, and the vertical branch is formed by a single capacitor. The circuit model of the conductor pair forms a bridge network. Duality between inductors and capacitors means that the bridge is balanced. Voltage across each capacitor is balanced by a voltage across its related inductor, whatever the frequency. Section 2.6 identifies the relationship and indicates that this property can be exploited.

Section 2.7 describes a systematic process which derives relationships between the geometry of a three-conductor cable and the component values of the circuit model which simulates its performance. This process starts with a set of three primitive equations relating currents and voltages of the assembly when it is acting as an antenna. Since signals in an electrical system are carried by loop currents, it is possible to derive two loop equations.

At this point, it is possible to create a circuit model which creates a similar pair of circuit equations. Correlating the parameters of the circuit equations with those of the loop equations allows a relationship to be established with the loop parameters. Since the loop parameters have been derived from primitives, it is possible to define the circuit components in terms of the primitive formulae.

As well as enabling component values to be calculated, the process identifies the essential difference between electromagnetic field theory and circuit theory. The former theory recognises the fact that the voltage of a conductor is affected by the current in every conductor. The latter theory assumes that there is a one-to-one correlation between the current in a circuit branch and the voltage across that branch. This results in a dramatic simplification of the mathematics, and is probably the main reason why circuit models are so useful.

A desirable characteristic of any signal link is that the effect of other signals in the system is minimal. Section 2.8 uses the formulae derived in the previous section to define the most significant feature of the interconnecting cable. The signal and return conductors should be as close together as possible. This ensures maximum magnetic coupling and maximum electric coupling between these two conductors and minimises the coupling between this conductor pair and other conductors in the vicinity. In turn, this minimises any common-mode current created by the signal link and reduces the susceptibility of the link to external interference.

Section 2.9 identifies the most significant parameter associated with EMC: the *Transfer Admittance*. This is the ratio of the current in the victim loop to the voltage in the culprit loop. It is shown that this parameter defines both the conducted susceptibility and the conducted emission.

The general circuit model of the three-conductor assembly can be applied to simulate the coupling between the differential-mode signal carried by a co-axial cable and a common-mode signal in the loop formed by the structure and screen. Section 2.10 shows that for an ideal co-axial cable, the only coupling parameter is the resistance of the screen.

However, the screens of most coaxial cables are braided. At higher frequencies, a number of capacitive and inductive effects due to the screen's construction come into play and dominate the effect that is caused by screen resistance.

The effect can be simulated by representing the screen as a T-network, with very small values assigned to the reactive components. These values can be derived from the manufacturer's data on transfer impedance, or from tests on a cable sample.

Analysing the coupling between two conductors over a ground plane is a problem faced by most designers at one time or another. The method of images can be used to derive formulae for the components of the three-conductor model. The formulation is described in Section 2.11.

A basic assumption inherent in the textbook theory of circuits is that resistors, inductors, capacitors, voltage sources, and current sources are 'lumped parameters'; that is, they are discrete devices which link two nodes of a network. Another assumption is that action and reaction are instantaneous everywhere in the network. Since conventional circuit theory does not cater for the fact that it takes time for current to propagate along a conductor, its use is limited to low-frequency applications, where the length of the cable is less than one-tenth the wavelength of the signal.

This limitation applies to the models derived in this chapter and in Chapter 3. Subsequent chapters describe how this restriction can be overcome.

2.1 Primitive capacitance

Consider a short length, l, of a single conductor of radius $r_{1,1}$ and assume that there is a line charge along that length. If the charge density is ρ, then the intensity of the electric field at a point *P(r,z)* can be calculated. Figure 2.1 illustrates the relationships.

The radial component of the electric field strength at the point P is [12]:

$$E_r = \frac{\rho}{4 \cdot \pi \cdot \varepsilon} \cdot \frac{1}{r} \cdot [\sin(\alpha_1) + \sin(\alpha_2)](\mathrm{V/m}) \tag{2.1}$$

where ε is the permittivity.

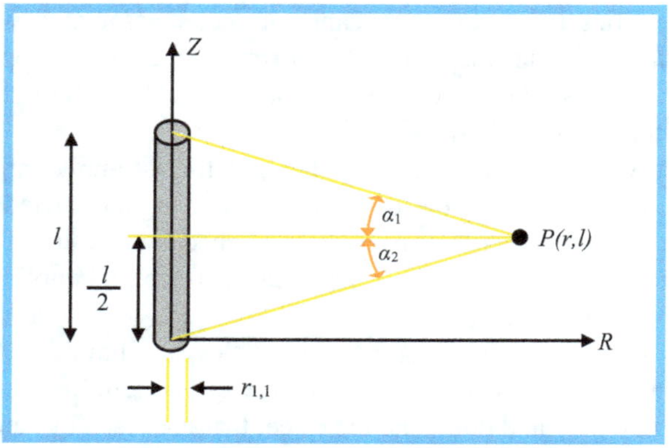

Figure 2.1 Electric field at a point

Since the point P lies on the plane intersecting the mid-point of the conductor, α_1 is equal to α_2. Re-writing (2.1) in terms of radial and axial components gives:

$$E_r = \frac{\rho}{2 \cdot \pi \cdot \varepsilon} \cdot \frac{l/2}{r \cdot \sqrt{(l/2)^2 + r^2}} \tag{2.2}$$

At all points on the surface defined by the equation $z = \frac{l}{2}$ the E_z component of the electric field is zero. If the unit charge were to follow a path along this surface from infinity to the radius $r_{1,1}$ then no axial force would be involved; all the electric forces would be radial. The work done on conductor 1 due to the charge on conductor 1 would be:

$$Vp_{1,1} = \frac{\rho}{2 \cdot \pi \cdot \varepsilon} \cdot \int_{r_{1,1}}^{\infty} \frac{l/2}{r \cdot \sqrt{(l/2)^2 + r^2}} \cdot dr(\mathrm{V}) \tag{2.3}$$

Performing the integration:

$$Vp_{1,1} = \frac{\rho}{2 \cdot \pi \cdot \varepsilon} \cdot \ln\left[\frac{l/2 + \sqrt{(l/2)^2 + {r_{1,1}}^2}}{r_{1,1}}\right] \tag{2.4}$$

In situations where $r_{1,1} << l$

$$Vp_{1,1} = \frac{\rho}{2 \cdot \pi \cdot \varepsilon} \cdot \ln\left(\frac{l}{r_{1,1}}\right) \tag{2.5}$$

The capacitance of the conductor is the ratio of charge to voltage. That is

$$Cp_{1,1} = \frac{\rho \cdot l}{Vp_{1,1}}(\mathrm{F}) \tag{2.6}$$

Substituting for the voltage:

$$Cp_{1,1} = \frac{2 \cdot \pi \cdot \varepsilon \cdot l}{\ln\left(\frac{l}{r_{1,1}}\right)} \tag{2.7}$$

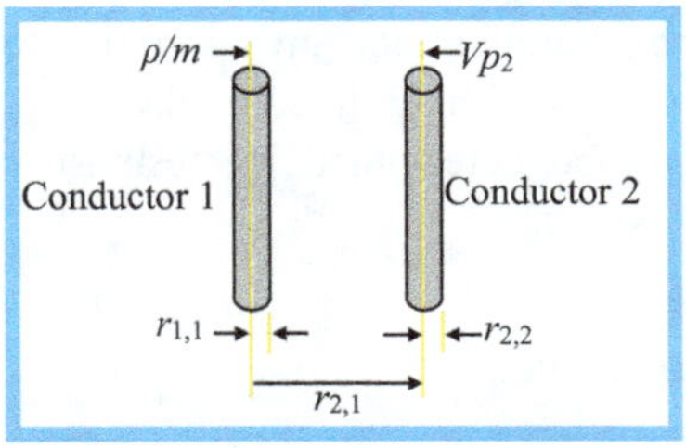

Figure 2.2 Voltage on conductor 2 due to charge on conductor 1

If a second conductor were to be placed parallel to conductor 1, as shown in Figure 2.2, then the energy level at the axis of conductor 2 due to the charge on conductor 1 would be:

$$Vp_{2,1} = \frac{\rho}{2 \cdot \pi \cdot \varepsilon} \cdot \ln\left(\frac{l}{r_{2,1}}\right) \tag{2.8}$$

And the primitive capacitance would be:

$$Cp_{2,1} = \frac{2 \cdot \pi \cdot \varepsilon \cdot l}{\ln\left(\frac{l}{r_{2,1}}\right)} \tag{2.9}$$

The general formula for primitive capacitance would be:

$$Cp_{i,j} = \frac{2 \cdot \pi \cdot \varepsilon \cdot l}{\ln\left(\frac{l}{r_{i,j}}\right)} \tag{2.10}$$

Where i and j are integers which identify the conductors. If $j = 1$ then the value is the primitive capacitance of conductor i due to the charge it carries.

This formulation depends on the assumption that the length is much greater than the radii and that the charge is uniformly distributed along the length. It has the practical advantage that it is simple. It works well in 'practice', in that it can be used as a building block in the construction of all circuit models. Moreover, any inaccuracies in the derivation are completely masked by the uncertainty in the value of the relative permittivity.

Capacitance values are a function of length, radii, and permittivity. Length and radii can be determined simply by measuring the physical parameters. Permittivity is a combination of two parameters:

$$\varepsilon = \varepsilon_o \cdot \varepsilon_r (\text{F/m}) \tag{2.11}$$

Where ε_0 is defined as the permittivity of free space, 8.854 pF/m, and ε_r is the relative permittivity.

In classical theory, the value of the relative permittivity is different for each type of material. When dealing with EMC problems, a variety of insulating materials are present in the vicinity of the conductor. Rather than taking the individual relative permittivity of each material and using data on the cross-section of the assembly to calculate the overall effect, an approximate 'overall' or 'effective' value can be used. In this book, ε_r is defined as the effective value of the relative permittivity. Its value can be determined by tests such as those described in Sections 7.4 and 7.5.

2.2 Primitive inductance

The derivation of primitive inductance starts off with basically the same configuration as with primitive capacitance. Figure 2.3 illustrates a short section of conductor of radius $r_{1,1}$ with a constant current Ip flowing in the z-direction.

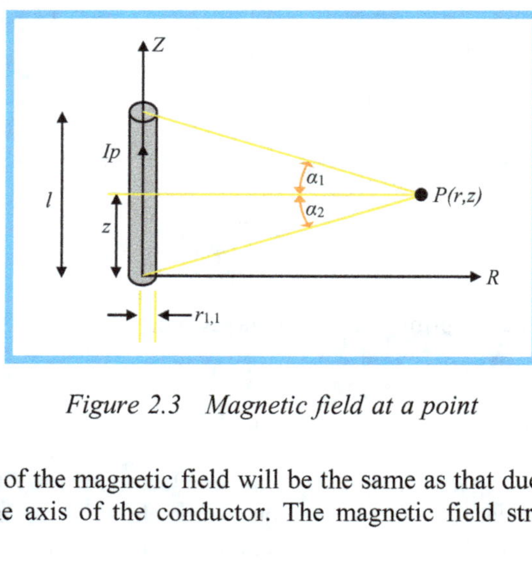

Figure 2.3 Magnetic field at a point

The distribution of the magnetic field will be the same as that due to a filamentary current along the axis of the conductor. The magnetic field strength H at point $P(r,z)$ is [14]:

$$H = \frac{Ip}{4 \cdot \pi \cdot r} \cdot [\sin(\alpha_1) + \sin(\alpha_2)](\mathrm{T}) \tag{2.12}$$

Converting this equation to one which uses the co-ordinates z and r gives:

$$H = \frac{Ip}{4 \cdot \pi \cdot r} \cdot \left[\frac{l - z}{\sqrt{(l - z)^2 + r^2}} + \frac{z}{\sqrt{z^2 + r^2}} \right] \tag{2.13}$$

To determine the inductance of this section of conductor, it is necessary to calculate the value of the total flux ϕ which passes through a rectangular strip extending from the surface of the conductor to a very large distance. This means that the flux density will need to be integrated over the defined area. Since,

$$B = \mu \cdot H(\mathrm{Wb/m^2}) \tag{2.14}$$

And

$$\phi = \int B \cdot ds(\mathrm{Wb}) \tag{2.15}$$

Where s is the surface. Then:

$$\begin{aligned}
\phi &= \frac{\mu \cdot Ip}{4 \cdot \pi} \cdot \int_{r_{1,1}}^{\infty} \int_{0}^{l} \frac{1}{r} \cdot \left(\frac{l-z}{\sqrt{(l-z)^2 + r^2}} + \frac{z}{\sqrt{z^2 + r^2}} \right) \cdot dz \cdot dr \\
&= \frac{\mu \cdot Ip}{4 \cdot \pi} \cdot \int_{r_{1,1}}^{\infty} \frac{\sqrt{l^2 + r^2} - r}{r} \cdot dr \\
&= \frac{\mu \cdot Ip \cdot l}{4 \cdot \pi} \cdot \left[\ln\left(\frac{l + \sqrt{l^2 + {r_{1,1}}^2}}{r_{1,1}} \right) + \left(\frac{r_{1,1} - \sqrt{l^2 + {r_{1,1}}^2}}{l} \right) \right]
\end{aligned}$$

The above derivation can be found in at least one textbook [13]. Inductance is defined as the ratio of magnetic flux to the current creating that flux. Since the flux ϕ is external to the conductor:

$$\text{Lexternal} = \frac{\phi}{Ip}(\mathrm{H}) \tag{2.16}$$

This leads to

$$\text{Lexternal} = \frac{\mu \cdot l}{4 \cdot \pi} \cdot \left[\ln\left(\frac{l + \sqrt{l^2 + {r_{1,1}}^2}}{r_{1,1}} \right) + \left(\frac{r_{1,1} - \sqrt{l^2 + {r_{1,1}}^2}}{l} \right) \right] \tag{2.17}$$

There are also internal flux linkages.

$$\text{Linternal} = \frac{\mu \cdot l}{2 \cdot \pi} \cdot \frac{1}{4}(\mathrm{H}) \tag{2.18}$$

Hence, the total inductance is the sum of *Lexternal* and *Linternal*

$$Lp_{1,1} = \frac{\mu \cdot l}{4 \cdot \pi} \cdot \left[\ln\left(\frac{l + \sqrt{l^2 + {r_{1,1}}^2}}{r_{1,1}} \right) + \left(\frac{r_{1,1} - \sqrt{l^2 + {r_{1,1}}^2}}{l} \right) + \frac{1}{4} \right] \tag{2.19}$$

If it is assumed that $l \gg r_{1,1}$ then

$$Lp_{1,1} = \frac{\mu \cdot l}{2 \cdot \pi} \cdot \left[\ln\left(\frac{2 \cdot l}{r_{1,1}}\right) - 1 + \frac{1}{4} \right]$$
$$= \frac{\mu \cdot l}{2 \cdot \pi} \cdot \left[\ln\left(\frac{l}{r_{1,1}}\right) + \ln 2 - 1 + \frac{1}{4} \right]$$
$$= \frac{\mu \cdot l}{2 \cdot \pi} \cdot \left[\ln\left(\frac{l}{r_{1,1}}\right) - 0.057 \right]$$

To a first approximation:

$$Lp_{1,1} = \frac{\mu \cdot l}{2 \cdot \pi} \cdot \ln\left(\frac{l}{r_{1,1}}\right) \tag{2.20}$$

Similar reasoning applies to a two-conductor assembly, as shown in Figure 2.4. It is assumed that there is a current *Ip* flowing in conductor 1. From (2.17), the inductance due to the flux linking conductor 2 due to current in conductor 1 is:

$$\text{Lmutual} = \frac{\mu \cdot l}{4 \cdot \pi} \cdot \left[\ln\left(\frac{l + \sqrt{l^2 + r_{2,1}{}^2}}{r_{2,1}}\right) + \left(\frac{r_{2,1} - \sqrt{l^2 + r_{2,1}{}^2}}{l}\right) \right] \tag{2.21}$$

When the pair of conductors is analysed as part of a complete loop, it must be assumed that the return path for the current *Ip* is along conductor 2. Magnetic flux due to current *Ip* in conductor 1 will link the region between infinity and the outer surface of conductor 2. It will also counteract the internal linkages of the return current *Ip* in conductor 2.

Hence, it is necessary to add the component *Linternal* to the value of the mutual inductance.

This leads to:

$$L_{2,1} = Lmutual + Linternal$$
$$= \frac{\mu \cdot l}{4 \cdot \pi} \cdot \left[\ln\left(\frac{l + \sqrt{l^2 + r_{2,1}{}^2}}{r_{2,1}}\right) + \left(\frac{r_{2,1} - \sqrt{l^2 + r_{2,1}{}^2}}{l}\right) + \frac{1}{4} \right] \tag{2.22}$$

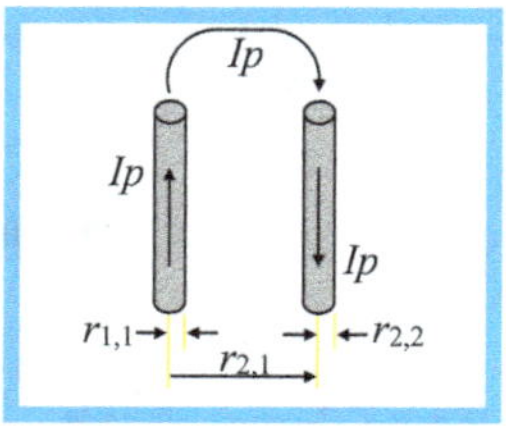

Figure 2.4 Current in a two-conductor assembly

If $l >> r_{2,1}$ then, to a first approximation:

$$Lp_{2,1} = \frac{\mu \cdot l}{2 \cdot \pi} \cdot \ln\left(\frac{l}{r_{2,1}}\right) \tag{2.23}$$

Since $r_{1,2} = r_{2,1}$ then $Lp_{1,2} = Lp_{2,1}$.

The general formula for primitive inductance can be defined as

$$Lp_{i,j} = \frac{\mu_o \cdot \mu_r \cdot l}{2 \cdot \pi} \cdot \ln\left(\frac{l}{r_{i,j}}\right) (\mathrm{H}) \tag{2.24}$$

where i and j are integers which identify the conductors.

As with the derivation of primitive capacitance, this formulation depends on the assumption that the length is much greater than any of the radii. It is also assumed that the current does not vary along the length. The fact that the formulation of (2.24) lacks precision is compensated by the fact that it is simple. From an engineering point of view, it is accurate enough for its intended purpose: to act as a building block for circuit models. The tests described in Sections 7.4 and 7.5 demonstrate that the value of the measured inductance was the same as that predicted by measurements of the geometry of the assembly-under-test.

It is also worth noting that any inaccuracies in the formulae for primitive inductance (and primitive capacitance) disappear when the loop parameters and circuit parameters are formulated. Section 2.4 explains why.

The permeability μ is a combination of two parameters:

$$\mu = \mu_o \cdot \mu_r (\mathrm{H/m}) \tag{2.25}$$

Where $\mu_o = 4 \cdot \pi \cdot 10^{-7}$ H/m. The parameter μ_r is a pure number, similar in concept to ε_r.

It can be defined here as the effective value of the relative permeability of the loop-under-review.

In this book, it is assumed that the value of μ_r is unity because in the majority of applications, there is no magnetic material in the cable assembly.

In those situations where magnetic material is present, the value of μ_r can be established by referring to data on the properties of materials, or by carrying out tests of the assembly-under-review.

Such tests would involve measurements of the frequency response of the current in a short-circuited loop. A circuit model of the assembly can be created by using data on the physical construction. This will produce a similar curve. The initial slope of both frequency response curves will be −20dB per decade. By modifying the value of μ_r in the model, both curves can be made to coincide over this range. The value of the relative permeability of the loop-under-test will be that of the model.

The test described in Section 7.5 illustrates the process used to measure ε_r, using an open-circuit line. If both ends of the line are short-circuited, the same process can be used to measure μ_r.

2.3 Duality of L and C

In general terms, the primitive components can be defined as:

$$Cp_{i,j} = \frac{2 \cdot \pi \cdot \varepsilon_o \cdot \varepsilon_o \cdot l}{\ln\left(\frac{l}{r_{i,j}}\right)} \text{(F)} \tag{2.26}$$

And

$$Lp_{i,j} = \frac{\mu_o \cdot \mu_r \cdot l}{2 \cdot \pi} \cdot \ln\left(\frac{l}{r_{i,j}}\right) \text{(H)} \tag{2.27}$$

where i and j are integers which identify the conductors. For example, $Lp_{i,j}$ gives the value of the primitive inductance of conductor i due to the current in conductor j. If $j = i$ then the value is the primitive inductance of conductor i due to the current it carries.

Since $r_{i,j} = r_{j,i}$ then $Lp_{i,j} = Lp_{j,i}$ and $Cp_{i,j} = Cp_{j,i}$.
Multiplying gives

$$Lp_{i,j} \cdot Cp_{i,j} = \mu_o \cdot \mu_r \cdot \varepsilon_o \cdot \varepsilon_r \cdot l^2 \tag{2.28}$$

From electromagnetic theory:

$$\sqrt{\mu_o \cdot \mu_r \cdot \varepsilon_o \cdot \varepsilon_r} = \frac{1}{v} \tag{2.29}$$

And

$$\sqrt{\mu_o \cdot \varepsilon_o} = \frac{1}{c} \tag{2.30}$$

where c is the velocity of light and v is the velocity of propagation of the electromagnetic field. Also

$$v = \lambda \cdot f \text{(m/s)} \tag{2.31}$$

where λ is the wavelength and f is the frequency. At the quarter-wave frequency, fq, of a monopole antenna, the current amplitude will peak. If the wavelength at this frequency is λq then the length of the cable will be:

$$l = \frac{\lambda q}{4} \text{(m)} \tag{2.32}$$

Taking the square root of (2.29) and using (2.28) to substitute for $\sqrt{\mu_o \cdot \mu_r \cdot \varepsilon_o \cdot \varepsilon_r}$ gives:

$$\sqrt{Lp_{i,j} \cdot Cp_{i,j}} = \frac{l}{v} \text{(s)} \tag{2.33}$$

The ratio $\frac{l}{v}$ is the propagation delay, the length of time it takes a transient pulse to propagate along the length of the conductor assembly. It can also be defined as the time constant *T*, as used in transient analysis. Section 6.3 derives its relationship to the inductance, capacitance, and characteristic impedance of the line.

Using (2.31) to substitute for v and (2.32) to substitute for l provides a relationship between the circuit components and the first resonant frequency:

$$\sqrt{Lp_{i,j} \cdot Cp_{i,j}} = \frac{1}{4 \cdot fq} \tag{2.34}$$

This means that, if the values of $Cp_{\mathrm{i,j}}$ and the frequency of quarter-wave resonance, fq, are known, then the value of $Lp_{\mathrm{i,j}}$ can be calculated.

Re-arranging (2.32) gives

$$\lambda q = 4 \cdot l$$

Using (2.31) to substitute for λq:

$$\frac{v}{fq} = 4 \cdot l$$

This leads to:

$$v = 4 \cdot l \cdot fq \tag{2.35}$$

From (2.29) and (2.30)

$$\frac{\sqrt{\mu_r \cdot \varepsilon_r}}{c} = \frac{1}{v}$$

This leads to:

$$\mu_r \cdot \varepsilon_r = \left(\frac{c}{v}\right)^2 \tag{2.36}$$

Since the conductors of most cables do not use magnetic material, it can usually be assumed that $\mu_r = 1$.

Equations (2.35) and (2.36) are extremely useful in establishing the value of the propagation velocity and the relative permittivity when the frequency of quarter-wave resonance is known.

The worksheet of Figure 7.40 makes use of this relationship to demonstrate that the antenna-mode current propagates at a higher velocity than the differential-mode current.

Knowledge of the relative permittivity of the cable allows the values of the associated capacitors to be calculated.

2.4 Loop parameters

Any cable can act either as an antenna or as a transmission line. Figure 2.5 illustrates the situation where the conductors are acting as an antenna. Currents Ip_1 and Ip_2 are assumed to flow in the same direction; in this case, from left to right.

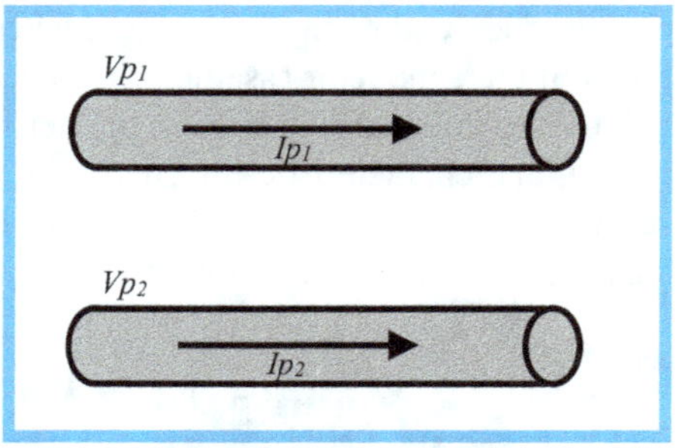

Figure 2.5 Primitive parameters

It is possible to deal with inductive and capacitive effects separately. Voltages Vp_1 and Vp_2 can initially be defined as the energy levels due to magnetic effects on Conductors 1 and 2. If it is assumed that currents and voltages are sinusoidal functions of time, then the relationship between them can be defined. Restricting consideration to magnetic effects, the primitive equations for two conductors are:

$$\begin{aligned} Vp_1 &= j \cdot \omega \cdot Lp_{1,1} \cdot Ip_1 + j \cdot \omega \cdot Lp_{1,2} \cdot Ip_2 \\ Vp_2 &= j \cdot \omega \cdot Lp_{2,1} \cdot Ip_1 + j \cdot \omega \cdot Lp_{2,2} \cdot Ip_2 \end{aligned} \tag{2.37}$$

Primitive equations define the behaviour of the conductor assembly as an antenna. When analysing the behaviour of the assembly as part of a circuit, it is necessary to deal with loop currents and loop voltages, as illustrated by Figure 2.6.

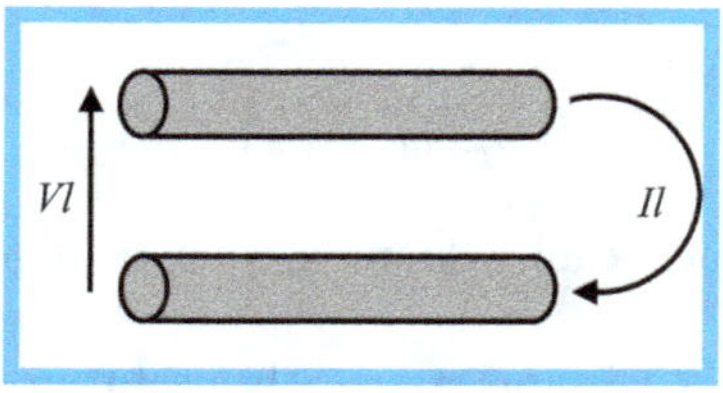

Figure 2.6 Loop parameters

The relationship between voltages is

$$Vl = Vp_1 - Vp_2 \tag{2.38}$$

The relationship between currents is

$$Ip_1 = Il = -Ip_2 \tag{2.39}$$

Using (2.39) to substitute loop currents for primitive currents in equation set (2.37) and then invoking (2.38) leads to:

$$Vl = j \cdot \omega \cdot \left(Lp_{1,1} - Lp_{1,2} - Lp_{2,1} + Lp_{2,2}\right) \cdot Il \tag{2.40}$$

This gives the value of the loop inductance of the conductor pair:

$$Ll = Lp_{1,1} - 2 \cdot Lp_{1,2} + Lp_{2,2} \tag{2.41}$$

This relationship takes into account the fact that $Lp_{1,2} = Lp_{2,1}$. Using (2.27) to substitute for the primitive inductors:

$$Ll = \frac{\mu_0 \cdot \mu_r \cdot l}{2 \cdot \pi} \cdot \ln\left(\frac{r_{1,2} \cdot r_{1,2}}{r_{1,1} \cdot r_{2,2}}\right) \tag{2.42}$$

The step between (2.41) and (2.42) is highly significant, in that the length parameter disappears from the logarithmic term. There is no longer any inaccuracy due to variation in the ratio between length and separation distance at any particular cross-section of the assembly-under-review. It does assume that the cross-section is uniform along the length l and that $l > 10 \cdot r$.

If $l \leq 10 \cdot r$, then propagation time needs to be considered and it becomes necessary to invoke the concepts of electromagnetic field theory. Even then, however, any inaccuracy in the predicted value of inductance does not negate the fact that loop inductance exists. The inaccuracy can be catered for by carrying out tests on the assembly-under-review.

Loop capacitance can be derived the same way. The primitive equations are:

$$\begin{aligned} Vp_1 &= \frac{1}{j \cdot \omega \cdot Cp_{1,1}} \cdot Ip_1 + \frac{1}{j \cdot \omega \cdot Cp_{1,2}} \cdot Ip_2 \\ Vp_2 &= \frac{1}{j \cdot \omega \cdot Cp_{2,1}} \cdot Ip_1 + \frac{1}{j \cdot \omega \cdot Cp_{2,2}} \cdot Ip_2 \end{aligned} \tag{2.43}$$

Using (2.39) to substitute loop currents for primitive currents in (2.43) leads to:

$$Vl = \frac{1}{j \cdot \omega} \cdot \left(\frac{1}{Cp_{1,1}} - \frac{1}{Cp_{1,2}} - \frac{1}{Cp_{2,1}} + \frac{1}{Cp_{2,2}}\right) \cdot Il \tag{2.44}$$

Hence

$$\frac{1}{Cl} = \frac{1}{Cp_{1,1}} - \frac{1}{Cp_{1,2}} - \frac{1}{Cp_{2,1}} + \frac{1}{Cp_{2,2}} \tag{2.45}$$

Using (2.26) to substitute for the primitive capacitors

$$\frac{1}{Cl} = \frac{1}{2 \cdot \pi \cdot \varepsilon_o \cdot \varepsilon_r \cdot l} \cdot \left[\ln \frac{r_{1,2} \cdot r_{1,2}}{r_{1,1} \cdot r_{2,2}}\right]$$

So

$$Cl = \frac{2 \cdot \pi \cdot \varepsilon_o \cdot \varepsilon_r \cdot l}{\ln\left(\frac{r_{1,2} \cdot r_{1,2}}{r_{1,1} \cdot r_{2,2}}\right)} \tag{2.46}$$

Equations (2.42) and (2.46) define the loop inductance and loop capacitance of a conductor pair. These are the values which could be measured with an LCR meter.

This derivation makes the assumption that all the current flowing in one conductor returns via the other. The nature of electromagnetic coupling ensures that this never happens in practice. Nevertheless, (2.42) and (2.46) can be regarded as providing a good approximation to the properties of an isolated twin-conductor cable. Section 5.2 develops these relationships to deal with the properties of such a cable as an antenna.

2.5 Circuit parameters

With EMC analysis, an important requirement is to maintain visibility of the voltage developed along each conductor, as well as the voltages appearing between conductor terminals. For the two-conductor configuration of Figure 2.5, there is little difficulty in meeting this requirement. It is assumed that all the current flowing in conductor 1 returns via conductor 2.

2.5.1 Inductance

The primitive equations for inductance define the voltage developed along each conductor in terms of the primitive currents. Reproducing (2.37):

$$\begin{aligned} Vp_1 &= j \cdot \omega \cdot Lp_{1,1} \cdot Ip_1 + j \cdot \omega \cdot Lp_{1,2} \cdot Ip_2 \\ Vp_2 &= j \cdot \omega \cdot Lp_{2,1} \cdot Ip_1 + j \cdot \omega \cdot Lp_{2,2} \cdot Ip_2 \end{aligned} \tag{2.37}$$

As far as inductance is concerned, the circuit model is shown in Figure 2.7.

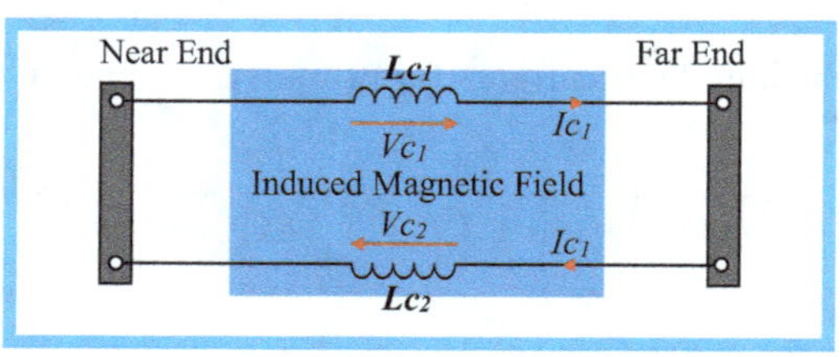

Figure 2.7 Circuit model for magnetic coupling

The circuit equations for inductive coupling, derived from Figure 2.7, are:

$$\begin{aligned} Vc_1 &= j\omega Lc_1 \cdot Ic_1 \\ Vc_2 &= j\omega Lc_2 \cdot Ic_1 \end{aligned} \tag{2.47}$$

Equations (2.37) and (2.47) were derived using different criteria. Equation (2.37) was derived from electromagnetic theory, while (2.47) came from circuit theory. Different assumptions were involved in their derivation. The equations can be correlated by defining the relationship between voltages and currents:

$$\begin{aligned} Vp_1 &= Vc_1 \\ -Vp_2 &= Vc_2 \end{aligned} \tag{2.48}$$

And

$$
\begin{aligned}
Ip_1 &= Ic_1 \\
Ip_2 &= -Ic_1
\end{aligned}
\tag{2.49}
$$

Making these substitutions allows the circuit parameters to be derived from the primitives:

$$
\begin{aligned}
Lc_1 &= Lp_{1,1} - Lp_{1,2} \\
Lc_2 &= Lp_{2,2} - Lp_{1,2}
\end{aligned}
\tag{2.50}
$$

Using (2.27) to relate primitive inductors to physical parameters gives formulae for the circuit inductors for a conductor pair:

$$
\begin{aligned}
Lc_1 &= \frac{\mu_o \cdot \mu_r \cdot l}{2 \cdot \pi} \cdot \ln\left(\frac{r_{1,2}}{r_{1,1}}\right) \\
Lc_2 &= \frac{\mu_o \cdot \mu_r \cdot l}{2 \cdot \pi} \cdot \ln\left(\frac{r_{1,2}}{r_{2,2}}\right)
\end{aligned}
\tag{2.51}
$$

2.5.2 Capacitance

Figure 2.8 illustrates the capacitive parameters associated with each conductor.

The primitive equations for capacitance define the voltage developed between each conductor and a theoretical surface at zero voltage in terms of the current flowing into the environment or out of the environment. Reproducing (2.43):

$$
\begin{aligned}
Vp_1 &= \frac{1}{j \cdot \omega \cdot Cp_{1,1}} \cdot Ip_1 + \frac{1}{j \cdot \omega \cdot Cp_{1,2}} \cdot Ip_2 \\
Vp_2 &= \frac{1}{j \cdot \omega \cdot Cp_{2,1}} \cdot Ip_1 + \frac{1}{j \cdot \omega \cdot Cp_{2,2}} \cdot Ip_2
\end{aligned}
\tag{2.43}
$$

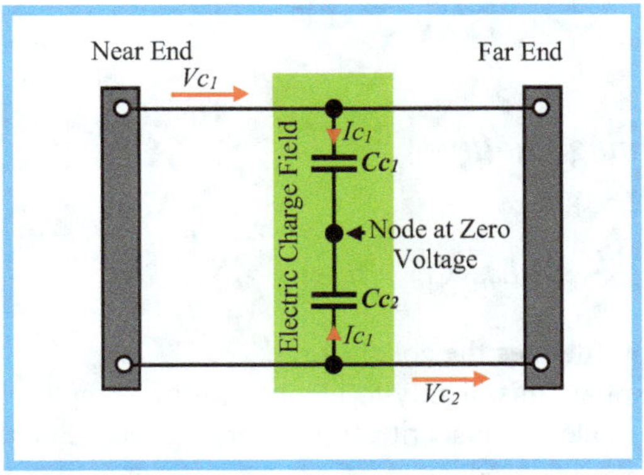

Figure 2.8 Circuit model of capacitive coupling

The circuit equations for capacitive coupling derived from Figure 2.8 are:

$$\begin{aligned} Vc_1 &= \frac{1}{j \cdot \omega \cdot Cc_1} \cdot Ic_1 \\ Vc_2 &= \frac{1}{j \cdot \omega \cdot Cc_2} \cdot Ic_1 \end{aligned} \tag{2.52}$$

Invoking (2.48) and (2.49) to substitute circuit parameters in the primitive equations leads to:

$$\begin{aligned} Vc_1 = Vp_1 &= \frac{1}{j \cdot \omega} \cdot \left[\frac{1}{Cp_{1,1}} - \frac{1}{Cp_{1,2}}\right] \cdot Ic_1 \\ Vc_2 = -Vp_2 &= \frac{1}{j \cdot \omega} \cdot \left[\frac{1}{Cp_{2,2}} - \frac{1}{Cp_{2,1}}\right] \cdot Ic_1 \end{aligned} \tag{2.53}$$

Correlating (2.52) and (2.53) allows circuit capacitors to be defined in terms of primitive capacitors:

$$\begin{aligned} \frac{1}{Cc_1} &= \frac{1}{Cp_{1,1}} - \frac{1}{Cp_{1,2}} \\ \frac{1}{Cc_2} &= \frac{1}{Cp_{2,2}} - \frac{1}{Cp_{1,2}} \end{aligned} \tag{2.54}$$

Using (2.26) to relate primitive capacitors to length and radii gives formulae for the circuit capacitors for a conductor pair:

$$\begin{aligned} Cc_1 &= \frac{2 \cdot \pi \cdot \varepsilon_o \cdot \varepsilon_r \cdot l}{\ln \frac{r_{1,2}}{r_{1,1}}} \\ Cc_2 &= \frac{2 \cdot \pi \cdot \varepsilon_o \cdot \varepsilon_r \cdot l}{\ln \frac{r_{1,2}}{r_{2,2}}} \end{aligned} \tag{2.55}$$

2.5.3 Maintaining duality

From (2.51) and (2.55):

$$Lc_i \cdot Cc_i = \mu_o \cdot \mu_r \cdot \varepsilon_o \cdot \varepsilon_r \cdot l^2 \tag{2.56}$$

where the subscript i defines the conductor.

Section 2.6 exploits this duality to identify further simplifications in the derivation of circuit models. This critical relationship between the inductive and capacitive properties of conductors is retained by avoiding the temptation to remove the zero-volt node (Figures 2.8, 2.19 and 5.10).

2.5.4 Resistance

The resistance of each conductor is a function of its cross-sectional area and its length. If conductor *i* is circular in section, the steady state resistance is:

$$Rss_i = \frac{\rho \cdot l}{\pi \cdot \left(r_{i,i}\right)^2} \tag{2.57}$$

Where ρ is the resistivity of the conducting material. Skin effect makes resistance a function of frequency. Textbooks on electromagnetic theory derive the formula [15]:

$$Rskin_i = \frac{l}{2 \cdot r_{i,i}} \cdot \sqrt{\frac{\mu \cdot f}{\pi \cdot \sigma}} \tag{2.58}$$

Where σ is the conductivity. Conductivity and resistivity are related:

$$\rho = \frac{1}{\sigma} \tag{2.59}$$

At high frequencies, resistance increases as the square root of the frequency; that is, at 10dB per decade. The crossover point occurs when *Rskin* is equal to *Rss*. From (2.57), (2.58) and (2.59):

$$\frac{l}{2 \cdot r_{i,i}} \cdot \sqrt{\frac{\mu \cdot \rho \cdot Fx}{\pi}} = \frac{\rho \cdot l}{\pi \cdot r_{i,i}^2}$$

Where *Fx* is the crossover frequency. This leads to the formula:

$$Fx = \frac{4 \cdot \rho}{\mu \cdot \pi} \cdot \frac{1}{r_{i,i}{}^2} \tag{2.60}$$

Hence, the general formula for conductor resistance is:

$$Rc_i = Rss_i \cdot \sqrt{1 + \frac{f}{Fx_i}} \tag{2.61}$$

The graph of Figure 2.9 illustrates the variation of resistance with frequency. The curves for *Rc* and *Rskin* were derived from (2.61) and (2.58), respectively. These give the relationships for a 15m length of 1mm diameter copper conductor. In this case, the cross-over frequency is 69kHz.

This leads to a third representation of a twin-conductor cable (Figure 2.10). Unlike inductance and capacitance, resistance is independent of the current in other conductors. This means that:

$$Rc_i = Rp_i \tag{2.62}$$

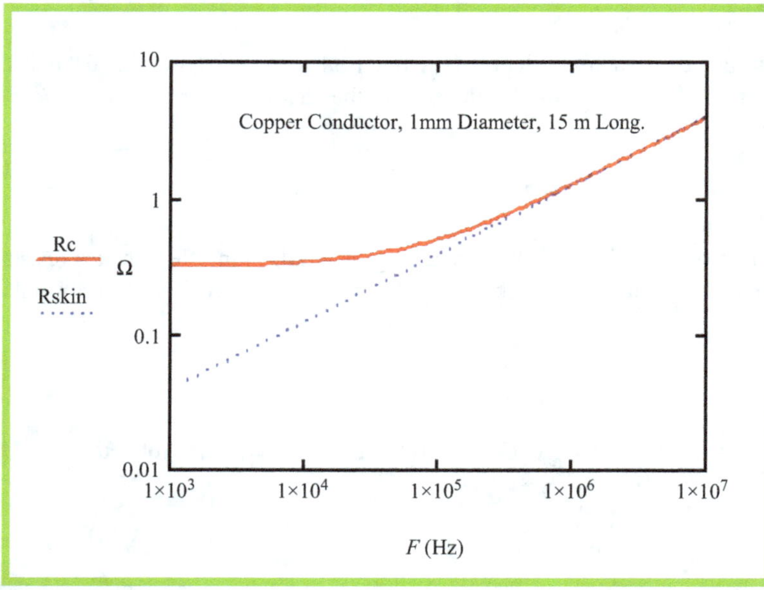

Figure 2.9 Relationship between resistance and frequency

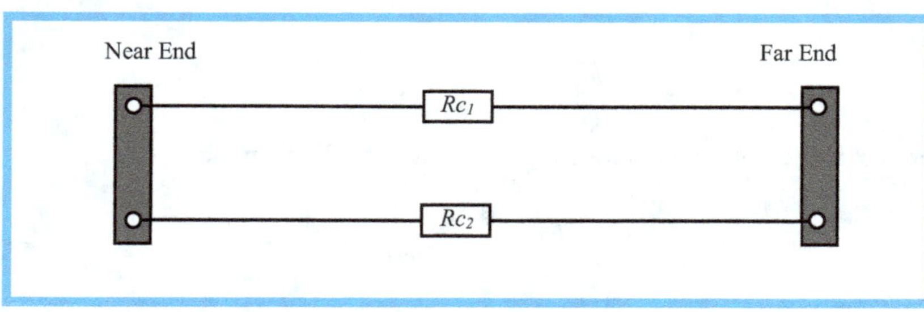

Figure 2.10 Circuit model for resistive coupling

2.5.5 Basic assumption

A basic assumption made in the derivation of the above formulae for inductance and capacitance of the conductors was that the current in the return conductor is equal in value but opposite in direction to that in the send conductor. This is never the case for a twin-conductor; some current radiates out into the environment. However, when this is taken into account, the formulae for conductor inductance and capacitance remain unaltered. Section 5.2 on the virtual conductor provides the more comprehensive derivation.

2.6 Twin-conductor model

Since the phenomenon causing interference is an electromagnetic field, there is no point in trying to separate out the electric field effects and the magnetic field effects. Resistive effects also have a significant role in the coupling mechanism.

Any form of analysis of interference must take into account the combined effects of resistance, inductance and capacitance. Since every conductor possesses these properties, it is certainly not valid to assume that a zero-impedance path for current exists anywhere in the system.

Since the simplest configuration capable of efficiently carrying an electronic signal from one location to another is the twin-core cable, the best place to start is to analyse the properties of such a cable. Figure 2.11 departs from the conventional approach by recognising the fact that both conductors have the same set of properties.

The most significant property of this model is the duality of the inductors and capacitors. If a voltage source is connected to the near end and the terminals at the far end are short circuited, then, as far as reactive parameters are concerned, the configuration behaves as a bridge network.

From (2.56)

$$Lc_1 \cdot Cc_1 = Lc_2 \cdot Cc_2$$

Hence:

$$\frac{Lc_1}{Lc_2} = \frac{Cc_2}{Cc_1} \tag{2.63}$$

From the circuit model of Figure 2.12:

$$V_1 = \frac{I_1 - I_2}{j \cdot \omega \cdot C_1} \tag{2.64}$$

$$V_2 = \frac{I_1 - I_2}{j \cdot \omega \cdot C_2} \tag{2.65}$$

$$V_3 = j \cdot \omega \cdot \frac{Lc_1}{2} \cdot I_2 \tag{2.66}$$

$$V_4 = j \cdot \omega \cdot \frac{Lc_2}{2} \cdot I_2 \tag{2.67}$$

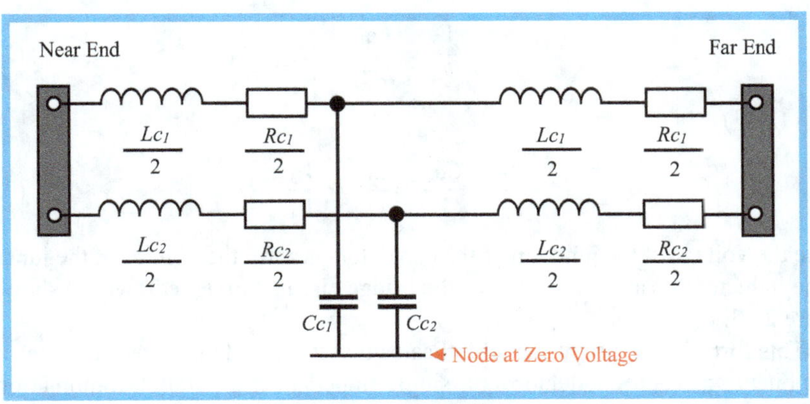

Figure 2.11 Circuit model of conductor pair

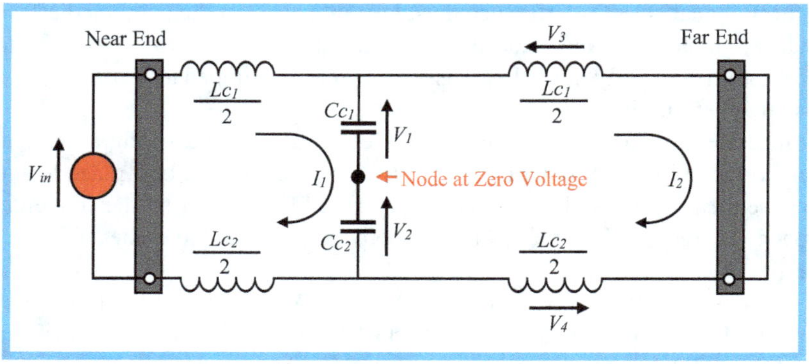

Figure 2.12 Bridge network

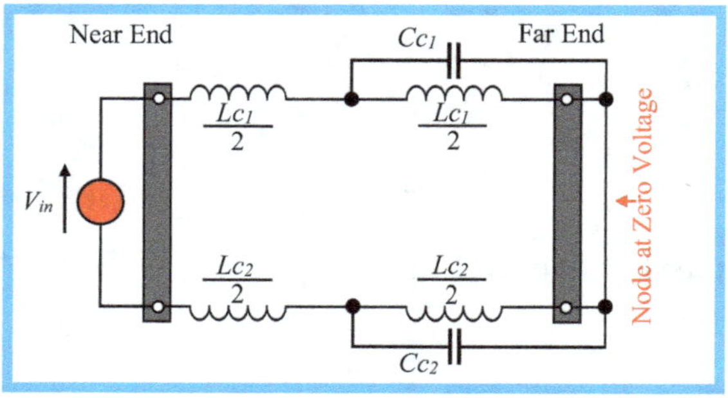

Figure 2.13 Equivalent circuit

Hence

$$\frac{V_1}{V_2} = \frac{Cc_2}{Cc_1} \tag{2.68}$$

And

$$\frac{V_3}{V_4} = \frac{Lc_1}{Lc_2} \tag{2.69}$$

From (2.63)

$$\frac{V_1}{V_2} = \frac{Cc_1}{Cc_2} = \frac{Lc_2}{Lc_1} = \frac{V_3}{V_4} \tag{2.70}$$

Since the voltage at the junction of the capacitors is zero, the voltage at the junction of the inductors' is also zero. Hence, the bridge circuit can be redrawn, as shown in Figure 2.13.

This form of the circuit model indicates that it is valid to represent the reactive components of each conductor as a single impedance: a parallel combination of inductance and capacitance in series with an inductance. Such a representation is extremely useful in simplifying the analysis of the transfer admittance in Section 2.9 and in deriving parameter values for the virtual conductor in Section 5.2.

2.7 Three-conductor model

The process established in the previous sections can be simplified by using *Z*-parameters, where '*Z*' can represent the impedance presented by any individual component, be it a capacitance, an inductance, or a resistance. Primitive impedances can be identified as *Zp*, loop impedances by *Zl* and circuit impedances by *Zc*. Figure 2.14 illustrates a three-conductor assembly.

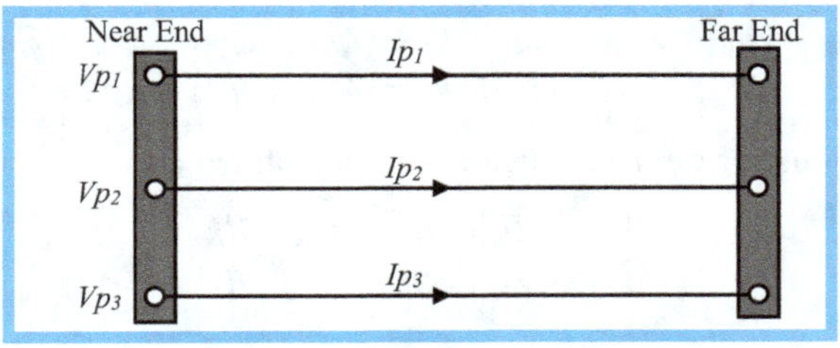

Figure 2.14 Three-conductor assembly

The primitive equations for three conductors are:

$$\begin{aligned} Vp_1 &= Zp_{1,1} \cdot Ip_1 + Zp_{1,2} \cdot Ip_2 + Zp_{1,3} \cdot Ip_3 \\ Vp_2 &= Zp_{2,1} \cdot Ip_1 + Zp_{2,2} \cdot Ip_2 + Zp_{2,3} \cdot Ip_3 \\ Vp_3 &= Zp_{3,1} \cdot Ip_1 + Zp_{3,2} \cdot Ip_2 + Zp_{3,3} \cdot Ip_3 \end{aligned} \tag{2.71}$$

Figure 2.15 illustrates a configuration in which the cable assembly is terminated by short circuits at both ends, with voltage sources Vl_1 and Vl_2 inserted in the loops at the near end.

Comparing Figures 2.14 and 2.15 allows the relationships between primitive parameters and loop parameters to be defined. Voltages are related by:

$$\begin{aligned} Vl_1 &= Vp_1 - Vp_2 \\ Vl_2 &= Vp_2 - Vp_3 \end{aligned} \tag{2.72}$$

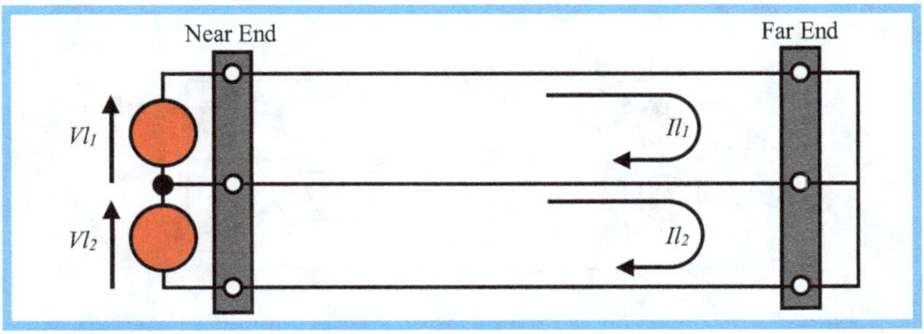

Figure 2.15 Loop voltages and currents

Currents are related by:

$$\begin{aligned} Ip_1 &= Il_1 \\ Ip_2 &= Il_2 - Il_1 \\ Ip_3 &= -Il_2 \end{aligned} \tag{2.73}$$

Substituting loop currents for primitive currents in (2.71) gives:

$$\begin{aligned} Vp_1 &= Zp_{1,1} \cdot Il_1 + Zp_{1,2} \cdot (Il_2 - Il_1) - Zp_{1,3} \cdot Il_2 \\ Vp_2 &= Zp_{2,1} \cdot Il_1 + Zp_{2,2} \cdot (Il_2 - Il_1) - Zp_{2,3} \cdot Il_2 \\ Vp_3 &= Zp_{3,1} \cdot Il_1 + Zp_{3,2} \cdot (Il_2 - Il_1) - Zp_{3,3} \cdot Il_2 \end{aligned} \tag{2.74}$$

Subtracting adjacent rows of (2.74) gives:

$$\begin{aligned} Vp_1 - Vp_2 &= (Zp_{1,1} - Zp_{1,2} - Zp_{2,1} + Zp_{2,2}) \cdot Il_1 + (Zp_{1,2} - Zp_{1,3} - Zp_{2,2} + Zp_{2,3}) \cdot Il_2 \\ Vp_2 - Vp_3 &= (Zp_{2,1} - Zp_{2,2} - Zp_{3,1} + Zp_{3,2}) \cdot Il_1 + (Zp_{2,2} - Zp_{2,3} - Zp_{3,2} + Zp_{3,3}) \cdot Il_2 \end{aligned} \tag{2.75}$$

The loop equations for three conductors can be defined as:

$$\begin{aligned} Vl_1 &= Zl_{1,1} \cdot Il_1 + Zl_{1,2} \cdot Il_2 \\ Vl_2 &= Zl_{2,1} \cdot Il_1 + Zl_{2,2} \cdot Il_2 \end{aligned} \tag{2.76}$$

Where:

$$\begin{aligned} Zl_{1,1} &= Zp_{1,1} - Zp_{1,2} - Zp_{2,1} + Zp_{2,2} \\ Zl_{1,2} &= Zp_{1,2} - Zp_{1,3} - Zp_{2,2} + Zp_{2,3} \\ Zl_{2,1} &= Zp_{2,1} - Zp_{2,2} - Zp_{3,1} + Zp_{3,2} \\ Zl_{2,2} &= Zp_{2,2} - Zp_{2,3} - Zp_{3,2} + Zp_{3,3} \end{aligned} \tag{2.77}$$

Since $Zp_{i,j} = Zp_{j,i}$, then

$$Zl_{1,2} = Zl_{2,1} \tag{2.78}$$

The circuit model which can simulate the behaviour of the loop equations is shown in Figure 2.16. The circuit equations for three conductors are:

$$\begin{aligned} Vc_1 &= (Zc_1 + Zc_2) \cdot Ic_1 - Zc_2 \cdot Ic_2 \\ Vc_2 &= -Zc_2 \cdot Ic_1 + (Zc_2 + Zc_3) \cdot Ic_2 \end{aligned} \tag{2.79}$$

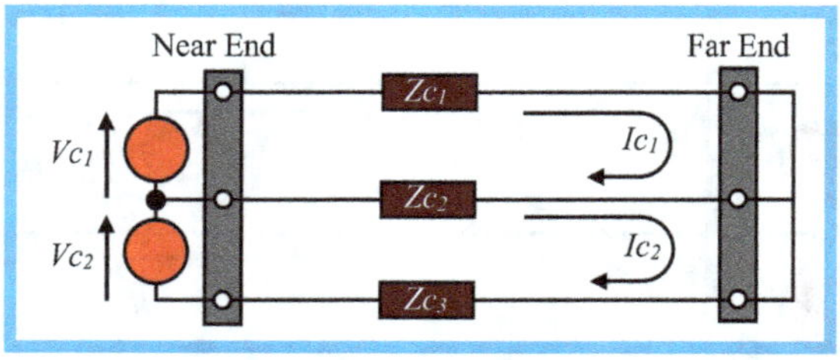

Figure 2.16 Circuit impedances for three-conductor assembly

Comparison between (2.76) and (2.79) shows clearly that there is a one-to-one correlation between loop parameters and circuit equation. Hence, loop impedances can be related to circuit impedances:

$$\begin{aligned} Zl_{1,1} &= Zc_1 + Zc_2 \\ Zl_{1,2} &= -Zc_2 \\ Zl_{2,2} &= Zc_2 + Zc_3 \end{aligned}$$

Defining circuit impedances in terms of loop impedances gives:

$$\begin{aligned} Zc_1 &= Zl_{1,1} + Zl_{1,2} \\ Zc_2 &= -Zl_{1,2} \\ Zc_3 &= Zl_{2,2} + Zl_{1,2} \end{aligned} \tag{2.80}$$

Using equation set (2.77) to substitute primitive impedances for loop impedances gives the circuit impedances for three conductors in terms of the primitive impedances:

$$\begin{aligned} Zc_1 &= Zp_{1,1} - Zp_{2,1} - Zp_{1,3} + Zp_{2,3} \\ Zc_2 &= Zp_{2,2} - Zp_{1,2} - Zp_{2,3} + Zp_{1,3} \\ Zc_3 &= Zp_{3,3} - Zp_{3,1} - Zp_{2,3} + Zp_{2,1} \end{aligned} \tag{2.81}$$

If the radial dimensions of the conductor assembly are defined as shown in Figure 2.17, then circuit components can be defined in terms of spatial parameters.

If the impedances are assumed to be purely inductive, that is $Zp_{i,j} = j \cdot \omega \cdot Lp_{i,j}$, then (2.27) can be used to define the circuit inductors for three conductors:

$$\begin{aligned} Lc_1 &= \frac{\mu_o \cdot \mu_r \cdot l}{2 \cdot \pi} \cdot \ln \frac{r_{1,2} \cdot r_{1,3}}{r_{1,1} \cdot r_{2,3}} \\ Lc_2 &= \frac{\mu_o \cdot \mu_r \cdot l}{2 \cdot \pi} \cdot \ln \frac{r_{1,2} \cdot r_{2,3}}{r_{2,2} \cdot r_{1,3}} \\ Lc_3 &= \frac{\mu_o \cdot \mu_r \cdot l}{2 \cdot \pi} \cdot \ln \frac{r_{1,3} \cdot r_{2,3}}{r_{3,3} \cdot r_{1,2}} \end{aligned} \tag{2.82}$$

Where l is the length of the assembly.

This establishes the inductance values for the assembly depicted in Figure 2.17.

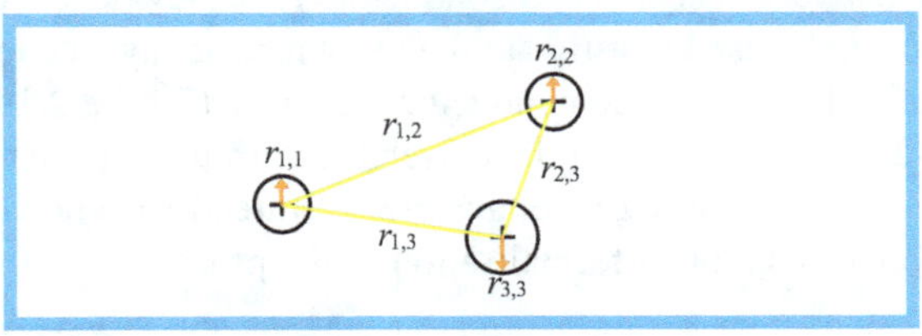

Figure 2.17 Cross-section of conductor assembly

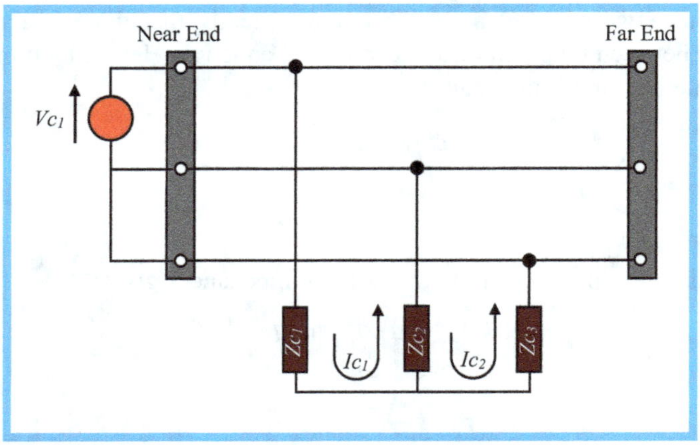

Figure 2.18 Circuit model of capacitive coupling

If Figure 2.16 were to be redrawn as shown in Figure 2.18, then the same process could be used to derive capacitor values. Equation (2.81) applies to both figures.

This time, however, the impedances are assumed to be purely capacitive. The impedance parameters would then be related to capacitors using:

$$Zp_{i,j} = \frac{1}{j \cdot \omega \cdot Cp_{i,j}}$$

Making this substitution in (2.81) and then invoking (2.26) leads to formulae for the circuit capacitors for three conductors:

$$\begin{aligned} Cc_1 &= \frac{2 \cdot \pi \cdot \varepsilon_o \cdot \varepsilon_r \cdot l}{\ln \dfrac{r_{1,2} \cdot r_{1,3}}{r_{1,1} \cdot r_{2,3}}} \\ Cc_1 &= \frac{2 \cdot \pi \cdot \varepsilon_o \cdot \varepsilon_r \cdot l}{\ln \dfrac{r_{1,2} \cdot r_{2,3}}{r_{2,2} \cdot r_{1,3}}} \\ Cc_1 &= \frac{2 \cdot \pi \cdot \varepsilon_o \cdot \varepsilon_r \cdot l}{\ln \dfrac{r_{1,3} \cdot r_{2,3}}{r_{3,3} \cdot r_{1,2}}} \end{aligned} \tag{2.83}$$

The relationship between resistors and conductors is self-evident. Equations (2.57), (2.60) and (2.61) can be used to estimate the resistance at any frequency. This leads to the triple-T model for the three-conductor assembly (Figure 2.19).

Equations (2.82) and (2.83) can be correlated with the formulae developed for phase inductance [7] and phase capacitance [8] of three-phase power lines. In fact, the derivation above is simply a formalisation of the process described in that basic textbook.

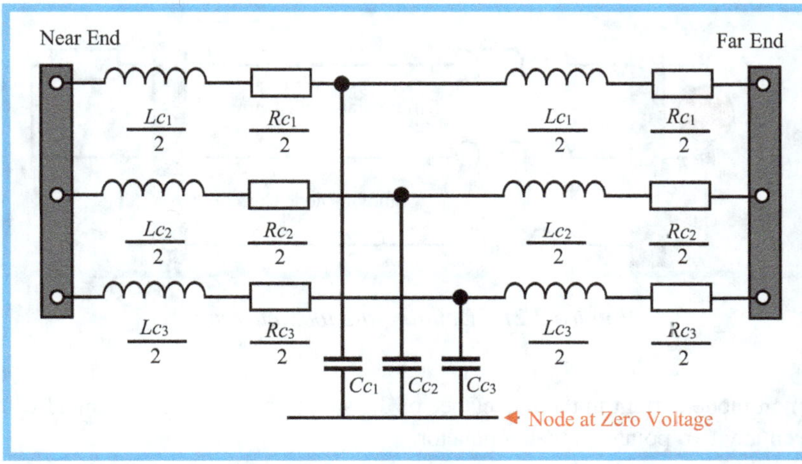

Figure 2.19 Triple-T circuit model of three-conductor assembly

2.8 Optimum coupling

The formulae derived in Section 2.7 provide useful guidance as to how to reduce interference during the design process. If the cross-sectional view of the assembly of Figure 2.17 is compared with the circuit model in the light of (2.82) and (2.83), then some fundamental relationships can be established.

First and foremost, mesh analysis, rather than nodal analysis, has been used to establish the formulae. There are two interdependent loops. In Figure 2.20, conductors 1 and 2 are assigned to carry the signal from the near end to the far end. This signal link can be defined as the differential-mode loop. In this loop, all the current flowing down the signal conductor is assumed to return via the return conductor. Figure 2.21 illustrates this.

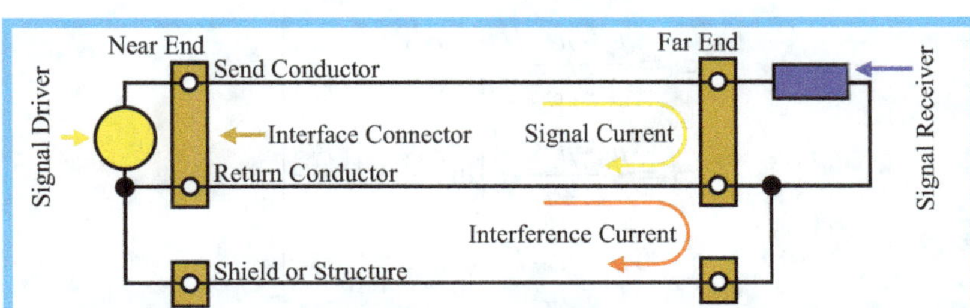

Figure 2.20 Defining the function of conductors

Inevitably, the differential-mode current in the return conductor develops a voltage across its inductance. This creates a current flow in the loop formed by the return conductor and the structure – the common-mode loop, as shown in Figure 2.21.

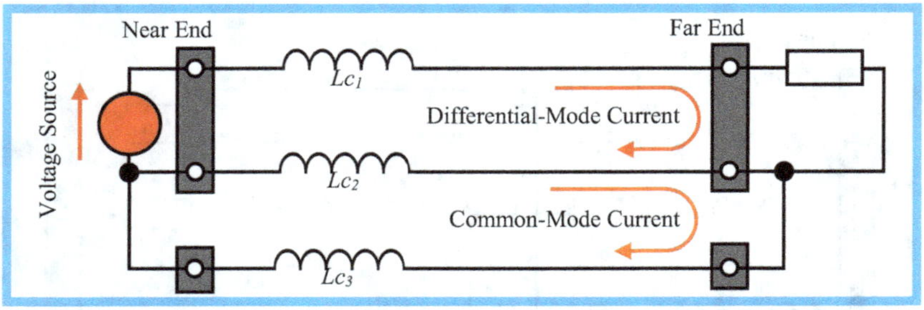

Figure 2.21 Defining the loop currents

Common-mode current in the inductance of the structure creates an unwanted voltage between any two points on that conductor.

Similarly, any voltage developed along the structure by an external source will create a current in the common-mode loop. In turn, this will induce an unwanted signal in the differential-mode loop.

In either case, any current in the common-mode loop is undesirable, since it constitutes interference. Whether the level of interference is acceptable or not is a matter of system design.

As far as EMC is concerned, the objective would be to maximise the differential-mode current and minimise the common-mode current.

Low-frequency simulation of this configuration is modelled by the circuit of Figure 2.21. As far as EMC is concerned, the objective would be to maximise the differential-mode current and minimise the common-mode current.

The physical design should ensure that the impedance of the differential-mode loop is as low as possible. That is, the inductors Lc_1 and Lc_2 should be as small as possible. Conversely, the value of Lc_3 should be as high as possible.

If the three conductors are circular in cross-section as illustrated in Figure 2.22, then the inductor values would be as defined by (2.82). These are re-written below:

$$
\begin{aligned}
Lc_1 &= \frac{\mu_o \cdot \mu_r \cdot l}{2 \cdot \pi} \cdot \left[\ln \frac{r_{1,2}}{r_{1,1}} + \ln \frac{r_{1,3}}{r_{2,3}} \right] \\
Lc_2 &= \frac{\mu_o \cdot \mu_r \cdot l}{2 \cdot \pi} \cdot \left[\ln \frac{r_{1,2}}{r_{2,2}} + \ln \frac{r_{2,3}}{r_{1,3}} \right] \\
Lc_3 &= \frac{\mu_o \cdot \mu_r \cdot l}{2 \cdot \pi} \cdot \left[\ln \frac{r_{1,3}}{r_{1,2}} + \ln \frac{r_{2,3}}{r_{3,3}} \right]
\end{aligned}
\tag{2.84}
$$

Given this set of equations, it can be seen that reducing r_{12} would reduce the value of Lc_1 and Lc_2, while increasing the value of Lc_3. That is, altering just one physical dimension will affect the value of all three inductors. Reducing the spacing between the signal and return conductors will have a significant effect in improving EMC.

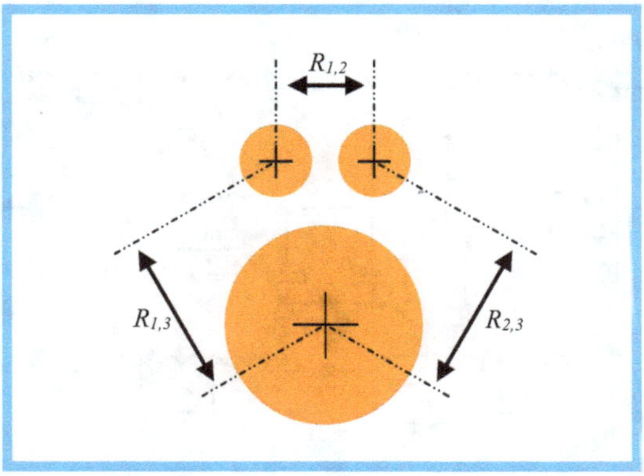

Figure 2.22 Cross-section of conductor assembly

Further improvement can be achieved by locating conductors 1 and 2 an equal distance away from conductor 3.

If

$$r_{1,3} = r_{2,3}$$

then

$$\ln\frac{r_{1,3}}{r_{2,3}} = \ln\frac{r_{2,3}}{r_{1,3}} = 0$$

That is, if the conductors carrying the differential-mode current are equidistant from the structure, then these logarithmic factors can be deleted from (2.84). This means that Lc_1 and Lc_2 can be reduced further.

If, in addition, the radii of the signal and return conductors were the same, then

$$\frac{r_{1,2}}{r_{1,1}} = \frac{r_{1,2}}{r_{2,2}}$$

giving the relationship $Lc_1 = Lc_2$.

This is a desirable characteristic in configurations where balanced drivers and balanced receivers are utilised.

When capacitive coupling is brought into the picture, the circuit model changes to that shown in Figure 2.23.

If Lc_1 and Lc_2 are reduced in value, then Cc_1 and Cc_2 are increased. If the value of Lc_3 increases, then the value of Cc_3 reduces. It can be seen from Figure 2.23 that these changes to capacitance values will act in the same way as inductance value changes; more current will flow in the signal loop and less will flow in the common-mode loop.

This means that, when the separation between signal and return conductors is reduced, all the reactive components will change in a way that enhances EMC.

Hence, the basic requirement for optimum coupling is that the signal and return conductors be held as closely together as possible along the entire route from signal driver to signal receiver.

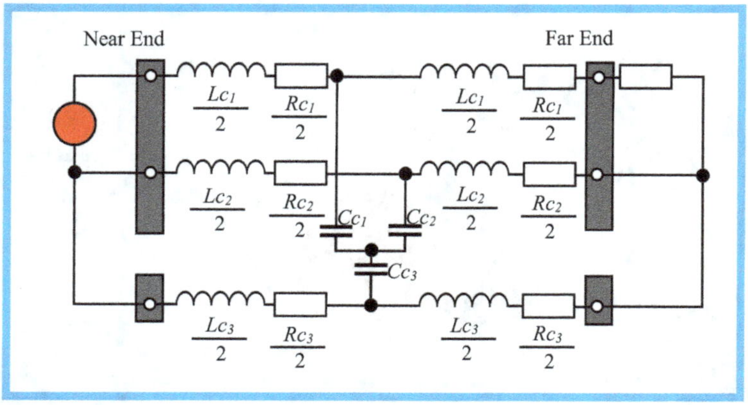

Figure 2.23 Adding capacitors and resistors to model

2.9 Transfer admittance

It was shown in Section 2.6 that the reactive components of the model will form a bridge circuit. The same natural balance exists in the three-conductor model of Figure 2.24. If the terminations at the far end are short-circuited, then the voltage at the junction of the three inductors will be the same as the voltage at the junction of the three capacitors.

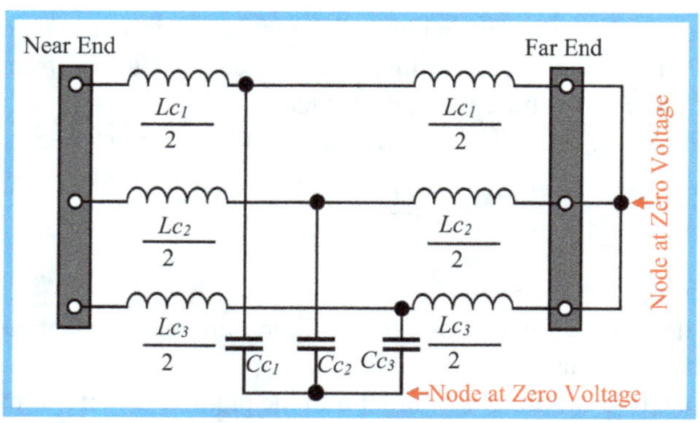

Figure 2.24 Voltage balance in three-conductor assembly

If the voltages at these two nodes are at the same potential, it is valid to join them together. Hence, it is valid to redraw the model to give the diagram of Figure 2.25.

At any particular frequency, this model can be further simplified to that shown in Figure 2.26, where *Z1*, *Z2* and *Z3* represent the impedances associated with each conductor.

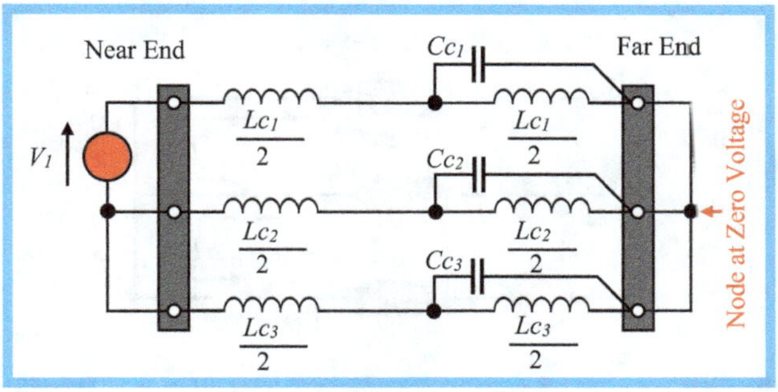

Figure 2.25 Equivalent circuit of configuration of Figure 2.24

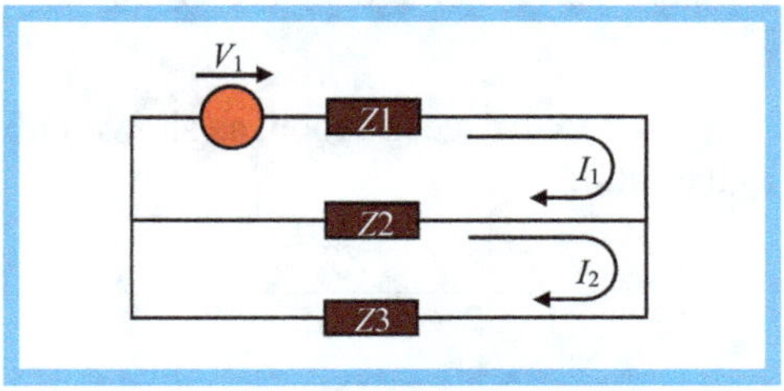

Figure 2.26 Simplified model

The circuit equations for Figure 2.26 are:

$$V_1 = (Z1 + Z2) \cdot I_1 - Z2 \cdot I_2 \tag{2.85}$$

$$0 = -Z2 \cdot I_1 + (Z2 + Z3) \cdot I_2 \tag{2.86}$$

From (2.86)

$$I_1 = \frac{Z2 + Z3}{Z2} \cdot I_2 \tag{2.87}$$

Substituting for I_1 in (2.85) gives:

$$V_1 = \frac{(Z1 + Z2) \cdot (Z2 + Z3)}{Z2} \cdot I_2 - Z2 \cdot I_2$$

This expands to:

$$V_1 = \frac{Z1 \cdot Z2 + Z1 \cdot Z3 + Z2 \cdot Z2 + Z2 \cdot Z3 - Z2 \cdot Z2}{Z2} \cdot I_2$$

Giving

$$\frac{I_2}{V_1} = \frac{Z2}{Z1 \cdot Z2 + Z1 \cdot Z3 + Z2 \cdot Z3} \tag{2.88}$$

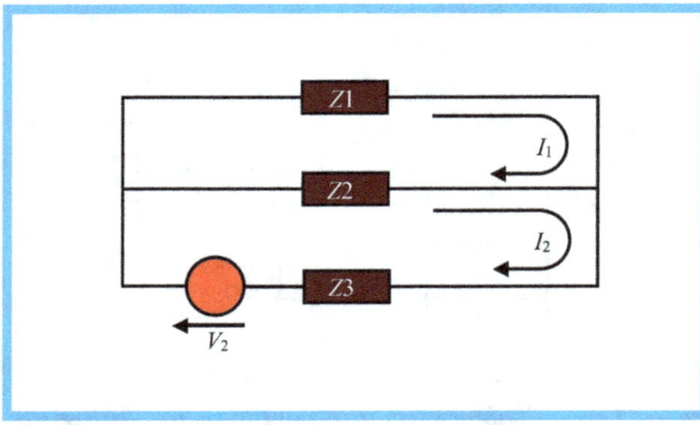

Figure 2.27 Voltage source in loop 2

If the voltage source is placed in the second loop, as shown in Figure 2.27, then the same process leads to the ratio:

$$\frac{I_1}{V_2} = \frac{Z2}{Z1 \cdot Z2 + Z1 \cdot Z3 + Z2 \cdot Z3} \tag{2.89}$$

This identifies the fact that there is a duality in any circuit network. Comparing Figures 2.26 and 2.27:

$$YT = \frac{I_1}{V_2} = \frac{I_2}{V_1} \tag{2.90}$$

Since the ratio of current to voltage is termed admittance, and since the ratio is between parameters in different loops, then the parameters of (2.88) and (2.89) are described as *Transfer Admittance, YT*.

The transfer admittance provides a direct measure of the ratio of unwanted signal in the victim loop to the source voltage in the culprit loop. It is the basic parameter which enables the interference characteristic of any assembly to be defined.

Figure 2.26 is effectively a circuit model of the conducted emission test carried out by engineers at an EMC Test House. It allows the ratio of common-mode current to injection voltage to be simulated over a range of frequencies. Figure 2.27 simulates the setup used in a conducted susceptibility test. It defines the ratio of unwanted signal current to a source voltage in the common-mode loop.

This reasoning leads to a definition of transfer admittance:

Transfer Admittance is the ratio of the current in the victim loop to the source voltage in the culprit loop when there are no other voltage sources.

Most significantly, transfer admittance has the same value for conducted emission as for conducted susceptibility. This means that a conducted susceptibility test on a cable assembly can be used to predict the response of a conducted emission test, and vice versa. It also identifies a useful check which can be made on the integrity of any program used to analyse any circuit model.

Interchange the locations of source voltage and monitored current and re-run the program. If the results from both runs are not identical, then there is an error somewhere.

In practice, there are many factors which tend to compromise this analysis of transfer admittance. Most notable is the fact that the test methods and test equipment used with the two types of tests are different. Even so, it is fair to say that a cable configuration which exhibits high susceptibility at a particular frequency will also create a high level of emission at that frequency.

2.10 Co-axial coupling

The special property of a co-axial cable is that the separation between the axis of the inner conductor and any point on the outer conductor is the radius of the outer conductor. The fact that the inner and outer conductors share the same axis also means that the separation between the inner conductor and the external conductor is exactly the same as the separation between the outer conductor and the external conductor. Figure 2.28 shows a typical cross-section.

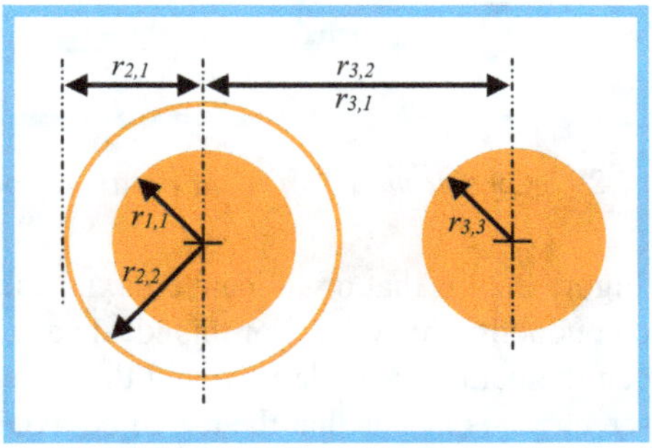

Figure 2.28 Co-axial cable with external conductor

Hence

$$r_{1,2} = r_{2,1} = r_{2,2} \tag{2.91}$$

And

$$r_{1,3} = r_{3,1} = r_{2,3} = r_{3,2} \tag{2.92}$$

The circuit model for a three-conductor cable is shown in Figure 2.19 and the formula for inductance values is given by (2.82). When the relationships of (2.91) and (2.93) are incorporated, the values of the inductors become:

$$\begin{aligned} Lc_1 &= \frac{\mu_o \cdot \mu_r \cdot l}{2 \cdot \pi} \cdot \ln \frac{r_{2,2}}{r_{1,1}} \\ Lc_2 &= 0 \\ Lc_3 &= \frac{\mu_o \cdot \mu_r \cdot l}{2 \cdot \pi} \cdot \ln \frac{r_{3,2} \cdot r_{2,3}}{r_{3,3} \cdot r_{2,2}} \end{aligned} \tag{2.93}$$

The value of Lc_1 is the familiar formula for the loop inductance of co-axial cable. The value of Lc_2 turns out to be zero, while the value of Lc_3 is the loop inductance of conductors 2 and 3 acting as a conductor pair. Since Lc_2 is zero, this inductance disappears from the circuit model. Values for the capacitors of the circuit model are related to the inductors by (2.28). This means that the theoretical value of Cc_2 is infinite. In effect, it becomes a short circuit. When these modifications are carried out on the model of Figure 2.19, it changes to the circuit model of co-axial coupling shown in Figure 2.29.

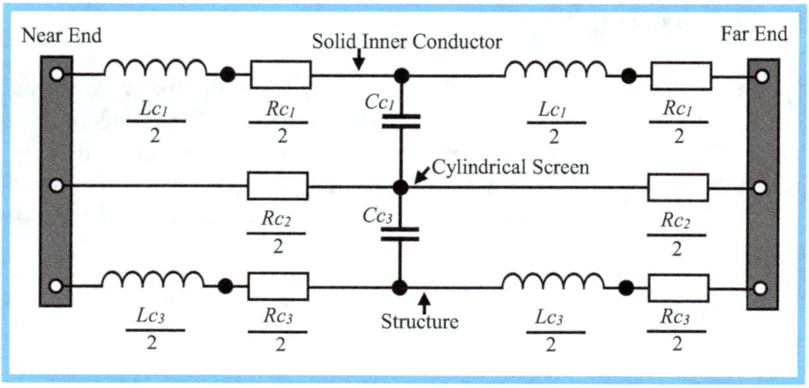

Figure 2.29 Circuit model of ideal co-axial coupling

The circuit model of Figure 2.29 is that of an ideal co-axial cable, where the screen is constructed of solid conducting material. For this ideal case, the only component at the interface between the common-mode loop and the differential-mode loop is the resistance of the screen. This means that the use of co-axial cable for the signal link will ensure that the value of the transfer admittance is extremely small. Put another way, the common-mode rejection will be extremely high.

However, it is usually impractical to use such a cable. In the vast majority of assemblies, the outer shield of the screened cables is made from thin wires interwound to form a flexible braid. Gaps in the braid allow penetration of the external field and emission of the internal field. At high frequencies, the shielding effectiveness of the screen deteriorates.

A more realistic model for a braided co-axial cable is shown in Figure 2.30. The magnetic field coupling between inner and outer loops is simulated by Lc_2, while Cc_2 represents the effect of the electric field coupling. Effectively, the model reverts to that shown in Figure 2.19. There is a significant difference. The values of the reactive parameters associated with the screen conductor are much less than those of the separate return conductor derived in Section 2.7.

A rule-of-thumb estimate for the value of these parameters would be to assume that:

$$Lc_2 = \frac{Lc_1}{10} \text{ and } Cc_2 = \frac{Cc_1}{10}$$

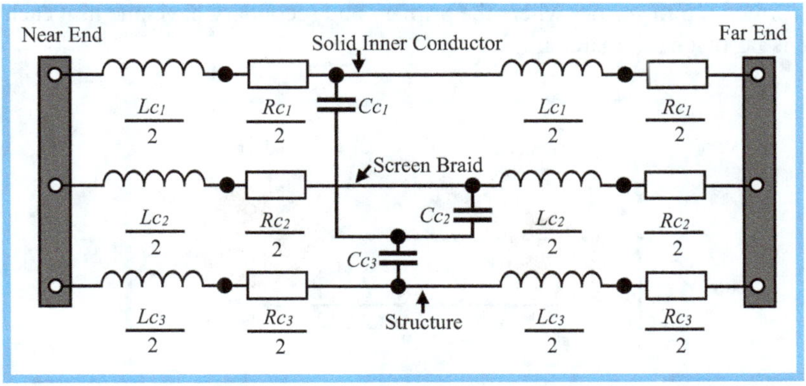

Figure 2.30 Circuit model of braided co-axial cable

The best way of establishing more accurate values is to carry out tests on a representative assembly. Such tests usually result in a parameter described as the transfer impedance and are defined in terms of a frequency response relating impedance to frequency. The relationship between transfer admittance and transfer impedance indicates that *Z2* in Figure 2.26 can be defined as the transfer impedance of the assembly-under-review.

Some manufacturers provide data on the transfer impedance in the form of a frequency response curve for a 1 m length. It should not be too difficult to create a model which replicates this curve and to derive values for the resistance, inductance and capacitance.

In any event, it can be said that co-axial cable will provide much better EMC than the twin-conductor cable, since the transfer impedance is much less. It may be that a shielded twisted pair could provide even better immunity to interference. However, it is more difficult to derive parameter values for such a configuration. This topic is dealt with in detail in the next chapter.

2.11 The ground plane

The concept of the conducting plane can be found in every textbook on electromagnetic theory. It exploits the fact that, on a non-conducting plane surface midway between two charged conductors, the electric field is orthogonal to that surface. If this surface was conducting, then the electric field distribution would not change. Figure 2.31 illustrates a cross-section of two conductors over a ground plane.

Conversely, a flat conducting surface can be represented by a non-conducting surface midway between the two conductors and a second pair of identical conductors. That is, the configuration of Figure 2.31 can be replaced by that of Figure 2.32.

If these four conductors are connected as two separate circuit loops, then the picture changes to that shown by Figure 2.33. Here, the coupling is between two isolated loops. As far as magnetic coupling is concerned, this is essentially a transformer; a transformer where the primary and secondary have one turn each and there is no magnetic material.

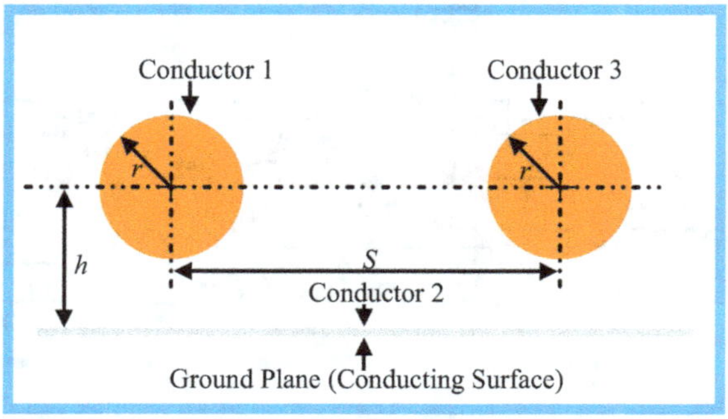

Figure 2.31 Two conductors over a ground plane

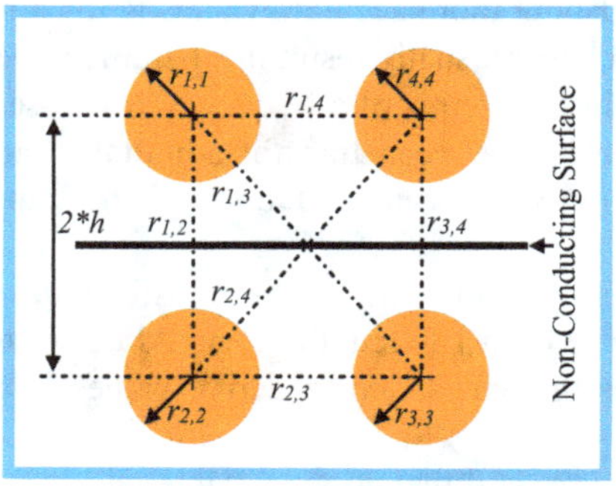

Figure 2.32 Simulating the effect of the plane

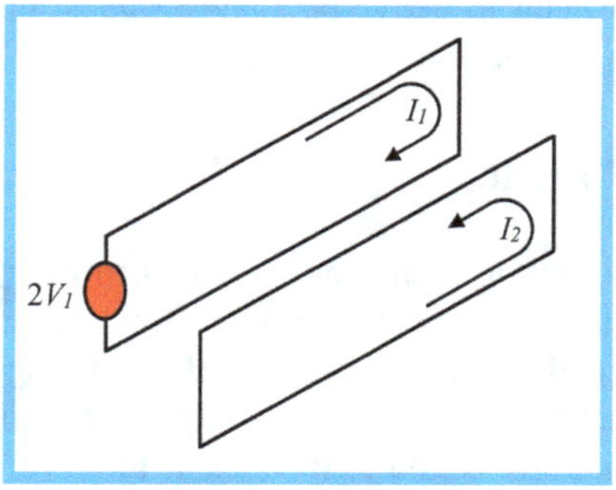

Figure 2.33 Transformer coupling

The circuit model for this configuration is the familiar transformer model of Figure 2.34. Here, Lc_1 represents the leakage inductance of the primary, Lc_2 represents the mutual inductance, and Lc_3 represents the leakage inductance of the secondary.

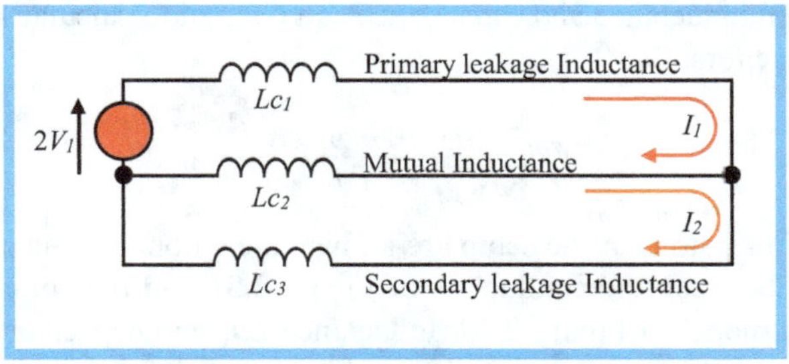

Figure 2.34 Circuit model of transformer coupling

Following the procedure outlined in Section 2.7, the three inductors can be defined for a length l of the assembly:

$$\begin{aligned} Lc_1 &= \frac{\mu_o \cdot \mu_r \cdot l}{2 \cdot \pi} \cdot \ln \frac{r_{1,2} \cdot r_{2,1} \cdot r_{1,4} \cdot r_{2,3}}{r_{1,1} \cdot r_{2,2} \cdot r_{1,3} \cdot r_{2,4}} \\ Lc_2 &= \frac{\mu_o \cdot \mu_r \cdot l}{2 \cdot \pi} \cdot \ln \frac{r_{1,3} \cdot r_{2,4}}{r_{1,4} \cdot r_{2,3}} \\ Lc_3 &= \frac{\mu_o \cdot \mu_r \cdot l}{2 \cdot \pi} \cdot \ln \frac{r_{3,2} \cdot r_{4,1} \cdot r_{3,4} \cdot r_{4,3}}{r_{3,1} \cdot r_{4,2} \cdot r_{3,3} \cdot r_{4,4}} \end{aligned} \tag{2.94}$$

The next step is to use the transformer coupling model to derive a model which simulates the coupling between the conductors of Figure 2.31. The relationships between the radial parameters of Figure 2.32 and the dimensions of Figure 2.31 are:

$$\begin{aligned} r_{1,2} &= r_{2,1} = r_{3,4} = r_{4,3} = 2 \cdot h \\ r_{1,3} &= r_{3,1} = r_{2,4} = r_{4,2} = \sqrt{s^2 + 4 \cdot h^2} \\ r_{1,4} &= r_{4,1} = r_{2,3} = r_{3,2} = s \\ r_{1,1} &= r_{2,2} = r_{3,3} = r_{4,4} = r \end{aligned} \tag{2.95}$$

The voltage source is defined as $2 \cdot V_1$ in Figure 2.33 because the voltage between any conductor and its image is twice that which would exist between the conductor and the ground plane. Since the formulae of (2.94) have been derived for a voltage source of $2 \cdot V_1$, the inductor values are twice those associated with conductors over a plane.

Dividing each of the inductive parameters of (2.94) by two and invoking the relationships of (2.95) leads to:

$$\begin{aligned} Ld_1 &= \frac{Lc_1}{2} = \frac{\mu_o \cdot \mu_r \cdot l}{2 \cdot \pi} \cdot \ln \frac{2 \cdot h \cdot s}{r \cdot \sqrt{s^2 + 4 \cdot h^2}} \\ Ld_2 &= \frac{Lc_2}{2} = \frac{\mu_o \cdot \mu_r \cdot l}{2 \cdot \pi} \cdot \ln \frac{\sqrt{s^2 + 4 \cdot h^2}}{s} \\ Ld_3 &= \frac{Lc_3}{2} = \frac{\mu_o \cdot \mu_r \cdot l}{2 \cdot \pi} \cdot \ln \frac{2 \cdot h \cdot s}{r \cdot \sqrt{s^2 + 4 \cdot h^2}} \end{aligned} \tag{2.96}$$

Duality between inductance and capacitance can be used to calculate values for the associated capacitors:

$$Cd_i = \frac{\mu_0 \cdot \mu_r \cdot \varepsilon_0 \cdot \varepsilon_r \cdot l^2}{Ld_i} \tag{2.97}$$

If the spatial dimensions of the setup are known, then a one-to-one correlation can be established between the conductors of Figure 2.31 and the components of the triple-T circuit model of Figure 2.19. Inductance Ld_1 and capacitance Cd_1 can be assigned to conductor 1, while inductance Ld_3 and capacitance Cd_3 can be assigned to conductor 3.

The most significant feature of (2.96) is that a value can be assigned to Ld_2. This is the inductance of the ground plane. Similarly, the capacitance of the ground plane is Cd_2.

The existence of Lc_2 means that any transient current in the plane will create an end-to-end voltage along the surface. If a voltage exists along the surface, then that surface cannot possibly be equipotential. To assume that the conducting plane is an equipotential surface is to ignore the lessons learnt from electromagnetic theory.

The relationship between separation distance and inductance values can be determined by inspection of (2.96). As the separation between conductors 1 and 2 reduces, the values of Ld_1 and Ld_3 reduce while the value of Ld_2 increases. When the spacing is at a minimum, the assembly behaves like a transformer. Conversely, as the spacing increases, the coupling between the two loops decreases, and the value of Ld_2 reduces.

The reasoning applied to inductor values can also be applied to capacitor values. As the spacing between the conductors increases, capacitive coupling reduces.

This means that, if the ground plane is acting as a return conductor for signals in conductors 1 and 3 (as with a printed circuit board), then interference coupling can be reduced by separating conductors 1 and 3 as much as possible and by reducing the separation between these conductors and the ground plane. The reasoning applies to both inductive effects and capacitive effects. The action of the ground plane in printed circuit boards is analysed further in Sections 8.2 and 9.3.

If the ground plane is used to represent the properties of the structure when it is acting as a shield, then it is desirable for conductors 1 and 3 to be as close together as possible. This will increase the coupling between the send and return conductors and enhance the balance between the currents in those conductors, while reducing the amplitude of common-mode current in the ground conductor.

Chapter 3
Other cross-sections

It is possible to develop a process to determine the properties of conductor assemblies of virtually any cross-section. The starting point is a technique devised by researchers at Culham to predict induced voltages in aircraft cables [9].

In this technique, the assembly-under-review is represented by an array of parallel conductors. It is assumed that the conductors at each end are short-circuited. So, the end-to-end voltage of each conductor is the same. Since the voltage along the length of one conductor of this array is determined by the currents in all the conductors, a set of primitive equations can be defined. Solving this set of equations allows the current in each conductor to be calculated. When the currents are known, it is possible to calculate the magnetic potential of any point in the vicinity. This allows the magnetic field pattern in the region to be determined.

The composite conductor can be defined as a set of elemental conductors, aligned in parallel, which enables the distribution of currents or voltages in the actual conductor to be simulated.

An elemental conductor can be defined as a conductor which represents a small segment of the surface of a composite conductor.

In the method described here, the primitive equations are set up and the currents in the elemental conductors are calculated, but the focus remains on the behaviour of those currents.

Section 3.1 illustrates the process by representing the cross-section of a circular tube as an array of elemental conductors. The current in each element is calculated. Adding all these currents together gives a figure for the total current in the section. The ratio of the voltage to the total current gives a figure for the partial inductance of the composite conductor.

A circular section was chosen for this first illustration because it was easy to check the accuracy of the result.

Since the process necessarily involves the use of a computer program, Mathcad software was used to illustrate the details of the computation. Since mathematical notation is used throughout, the program is much easier to follow than one written in, say, the Java language. A few special features of Mathcad are described in Appendix A.1

Section 3.2 develops the method to deal with the differential-mode current in two conductors to derive a value for the loop inductance. Duality between inductance and capacitance is then used to determine the value of the loop capacitance. This value is the same as that obtained when the 'method of images' is invoked. Such a check effectively validates the technique.

Section 3.3 develops the technique one stage further to enable a circuit model to be created for a three-conductor assembly. The example chosen is that of a screened pair cable.

The program defines the geometry of the elemental conductors of the three composite conductors, creates an array of primitive inductors, and converts this to an array of loop inductors. If it is assumed that a sinusoidal voltage of 1 V is applied between the screen and the inner pair, the loop currents in the elemental conductors are calculated.

Then, the primitive current in every elemental conductor is calculated. That is, current flow from left to right is deemed to be positive current. Negative current flows from right to left. Summing the currents in the appropriate elements allows the current in each composite to be calculated.

Knowledge of currents and impedance values allows a three-by-three matrix of partial voltages to be determined. From this, it is possible to calculate the values for a three-by-three array of *partial inductors*. Then, the loop inductance values for the composite conductors are calculated. This is a two-by-two matrix.

Using this information, the value of each of the three inductors of the circuit model can be calculated. Invoking the duality between inductance and capacitance gives the value of each of the capacitors. This allows a circuit model to be defined for a 10 m length of twin-core, screened cable.

The Mathcad worksheet, which carries out all the necessary computations, is reproduced. This can be hand copied into a new worksheet by the reader. Alternatively, the worksheet file can be downloaded from the website designemc.info. MATLAB® files, which replicate the calculations in every worksheet, are also available for download. Modifying this program, virtually any cross-section can be modelled. The only changes would be a re-definition of the geometry of the cross-section and of the length of the cable.

Intermediate results from the calculation are used to create a bubble chart which defines the current distribution in the cable. This is more meaningful to a circuit designer than a diagram of the magnetic field distribution. It also illustrates the fact that the skin effect is not uniform. Differential-mode current concentrates on adjacent surfaces.

Figure 3.18 illustrates the fact that common-mode current tends to flow on surfaces of the signal-carrying conductors that are as far apart as possible. This is because the common-mode current in the wire pair is flowing in the same direction. Since the common-mode current returns via the outer shield, it concentrates on that part of the shield which is closest to the current concentration in the wire pair.

Current in any conductor causes a magnetic field which has both internal and external linkages with that conductor, and the formula derived for primitive inductance takes account of both. By representing the surface of the conductor as an array of elemental conductors, this treatment enables the effect of internal linkages to be included, even though they are concentrated on that surface.

3.1 Single composite conductor

To derive values corresponding to the primitive inductance and capacitance of any non-circular conductor, the first step is to represent the cross-section as an array of parallel conductors. The process is best illustrated by using it to represent a circular section conductor. This allows the resultant values to be confirmed by comparing them with known values.

Figure 3.1(a) illustrates a section of the conductor to be analysed, while Figure 3.1(b) shows the method of simulation. Each elemental conductor represents a small segment, s, of the surface. Hence:

$$2 \cdot \pi \cdot r = s = \theta \cdot Rad$$

Giving a general formula for the radius of each elemental conductor:

$$r = \frac{\theta \cdot Rrad}{2 \cdot \pi} \tag{3.1}$$

In this particular case, the conductor being simulated is circular. So, there are n elemental conductors equally spaced around the periphery, and this leads to the relationship:

$$n \cdot \theta = 2 \cdot \pi \tag{3.2}$$

Substituting for $2 \cdot \pi$ in (3.1):

$$r = \frac{Rad}{n} \tag{3.3}$$

Since these conductors each have a defined radius, they cannot be termed 'filamentary'.

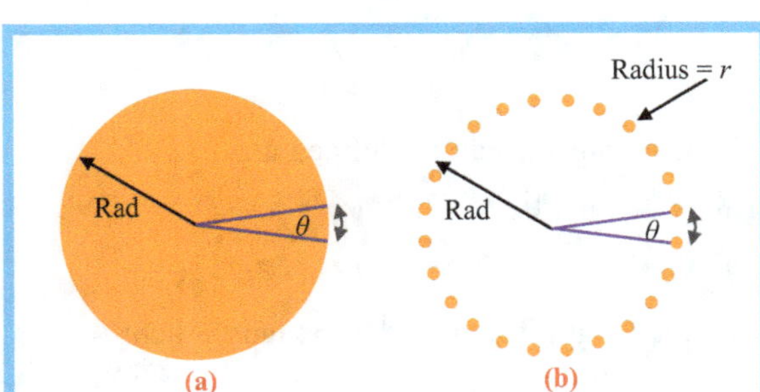

Figure 3.1 (a) Circular section conductor and (b) array of elemental conductors

If it is assumed that a current Ip_i is flowing in elemental conductor i, then a set of primitive equations can be set up. The voltage on each conductor will be a function of the current in every other conductor, as well as the current it carries itself. In the case of the three-conductor assembly of Section 2.7, there were three equations. For an assembly of *n* conductors, there are *n* equations.

$$Vp_i = \sum_{j=1}^{n} Zp_{i,j} \cdot Ip_j \tag{3.4}$$

Defining the primitive equations for elemental conductors in terms of vector algebra

$$\mathbf{Vp} = \mathbf{Zp} \cdot \mathbf{Ip} \tag{3.5}$$

Where **Vp** and **Ip** are vectors of *n* elements, and **Zp** is a square matrix of n^2 elements.

If the impedances are assumed to be inductive and the terminals at each end are shorted together, then the voltage along the length of each elemental conductor will be the same.

Impedance and inductance are related by

$$\mathbf{Zp} = j \cdot \omega \cdot \mathbf{Lp}$$

Whatever value is chosen for ω, it will remain the same throughout this computation process. If it is chosen to be unity, then

$$\mathbf{Zp} = |\mathbf{Lp}| \quad \text{where} \quad \omega = 1$$

Since the voltage along the length of each elemental conductor is a fixed value, it is possible to define every component of the **Vp** vector. This being so, there is enough information to calculate the value of every element of the **Ip** vector. With personal computers, the task is completed in a fraction of a second. The relevant function in Mathcad is:

$$\mathbf{Ip} = \text{lsolve}\left(\mathbf{Zp}, \mathbf{Vp}\right) \tag{3.6}$$

The total current in the composite conductor is the sum of the currents in the elemental conductors. That is:

$$Iq = \sum_{i=1}^{n} Ip_i \tag{3.7}$$

The impedance of the composite can be defined as:

$$Zq = \frac{Vq}{Iq} \tag{3.8}$$

If the voltage along the length is assumed to be unity, then:

$$Zq = \frac{1}{|Iq|} \quad \text{where} \quad Vq = 1$$

If the value of ω is set at unity:

$$Lq = \frac{1}{|Iq|} \quad \text{where} \quad Vq = 1 \quad \text{and} \quad \omega = 1 \tag{3.9}$$

The parameter Lq can be correlated with the primitive inductance Lp defined by (2.20) in that it defines the inductance of a composite conductor when that conductor is acting as an antenna. To avoid confusion between the two terms, it is useful to describe Lq as a *partial inductance*.

It follows that Vq, Iq, and Zq can be described as partial voltage, partial current, and partial impedance.

The set of equations developed above is amenable to the creation of a computer program to calculate the partial inductance of virtually any cross-section. Two stages are involved:

- Defining the co-ordinates of the elemental conductors.
- Calculating the values of the electrical parameters.

Figure 3.2 is a copy of a Mathcad worksheet which carries out the computations for the first stage. In this case, the process is quite simple. For a more complex cross-section,

Worksheet 3.1 page 1

$\text{Rad} := 10 \cdot 10^{-3}$	Radius of composite conductor, m
$l := 10$	Length of composite conductor, m
$n := 24$	Number of elements
$i := 1 .. n$	Control variable
$\theta_i := \frac{2 \cdot \pi}{n} \cdot (i - 0.5)$	Angle at which each element is located, radian
$x_i := \text{Rad} \cdot \cos(\theta_i)$	x co-ordinate of each elemental conductor, m
$y_i := \text{Rad} \cdot \sin(\theta_i)$	y co-ordinate of each elemental conductor, m
$r_i := \frac{\text{Rad}}{n}$	Radius of each elemental conductor, m

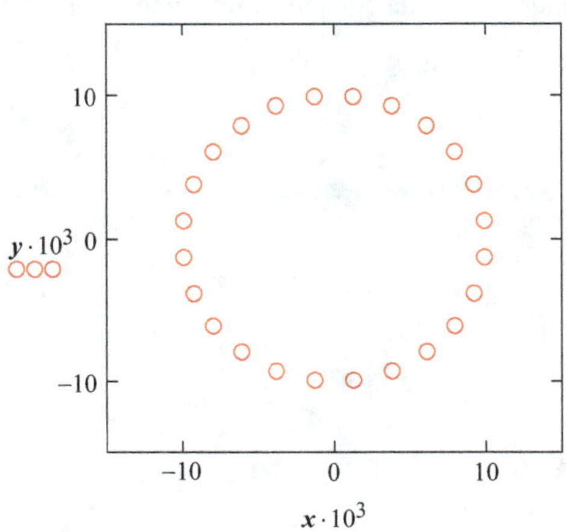

Plot of co-ordinates

Figure 3.2 Defining the physical parameters of the composite conductor

the definition of the co-ordinates would probably take the form of a three-column table of values.

Figure 3.4 invokes (3.4)–(3.9) to determine the value of Lq, the partial inductance of the composite conductor. It then carries out a check to confirm the resultant value is the same as the primitive inductance of a conductor with the same radius as the composite. This check provides a level of confidence that the process is valid.

Finally, the value of the *partial capacitance* is calculated, using the relationship between inductance and capacitance defined by (2.28).

3.2 The composite pair

The concept of an array of elemental conductors can be extended to the modelling of a pair of conductors of any cross-section. The following illustration assumes circular sections.

Figure 3.3 depicts the cross-section of a pair of composite conductors, while Figure 3.5 illustrates how the primitive currents can be related to the loop currents. Only one loop can have a voltage applied, that is between the composite pair. All the other loop voltages must be zero.

Following the approach of Section 2.7, the first step is to create the primitive equations for the composite pair. Using vector notation:

$$\mathbf{Vp} = \mathbf{Zp} \cdot \mathbf{Ip} \tag{3.10}$$

If it is assumed that there are $n1$ elemental conductors in composite 1 and $n2$ elemental conductors in composite 2, then the voltage source will be in loop $n1$. The total number of conductors is N. Where

$$N = n1 + n2 \tag{3.11}$$

Inspection of (2.77) reveals a clear correlation between primitive impedances and loop impedances. The general formula for loop impedance can be defined as:

$$Zl_{i,j} = Zp_{i,j} - Zp_{i,j+1} - Zp_{i+1,j} + Zp_{i+1,j+1} \tag{3.12}$$

Figure 3.3 Cross-section of composite pair

Worksheet 3.1 page 2

$\mu_o := 4 \cdot \pi \cdot 10^{-7}$ H/m $\qquad \mu_r := 1$

$K := \dfrac{\mu_o \cdot \mu_r \cdot l}{2 \cdot \pi} = 2 \times 10^{-6}$ H $\qquad$ K is a constant for this computation

$$Zp := \begin{array}{|l} \text{for } i \in 1..n \\ \quad \text{for } j \in 1..n \\ \quad \quad \begin{array}{|l} h \leftarrow x_j - x_i \\ v \leftarrow y_j - y_i \\ \text{rad} \leftarrow \sqrt{h^2 + v^2} \\ \text{rad} \leftarrow r_i \text{ if rad} = 0 \\ \text{Lp}_{i,j} \leftarrow K \cdot \ln\left(\dfrac{l}{\text{rad}}\right) \end{array} \\ \text{Lp} \end{array}$$

Zp = array of n*n inductance values, H

Rad = separation between pair of elements
= radius of element if separation = 0

See Equation (2.27)

$\text{Vp}_i := 1$ $\qquad$ Input voltage, V

$\text{Ip} := \text{lsolve}(\text{Zp}, \text{Vp})$ $\qquad$ Output current, A

$\text{Iq} := \sum_{i=1}^{n} \text{Iq}_i = 7.238 \times 10^4$ $\qquad$ Sum of currents in elements

$\omega := 1$ radian/s $\qquad \text{Lq} := \dfrac{\text{Vp}_1 \cdot \omega}{\text{Iq}} \qquad \text{Lq} = 1.382 \times 10^{-5}$ H

Check: $\qquad \text{Lp} := 2 \cdot 10^{-7} \cdot l \cdot \ln\left(\dfrac{l}{\text{Rad}}\right) \qquad \text{Lp} = 1.382 \times 10^{-5}$ H

$\varepsilon_o := 8.854 \cdot 10^{-12}$ F/m $\qquad \varepsilon_r := 1$

$\text{Cq} := \dfrac{\varepsilon_o \cdot \varepsilon_r \cdot \mu_o \cdot \mu_r \cdot l^2}{\text{Lq}} \qquad \text{Cq} = 8.053 \times 10^{-11}$ F

Figure 3.4 Partial inductance and capacitance of a single composite conductor

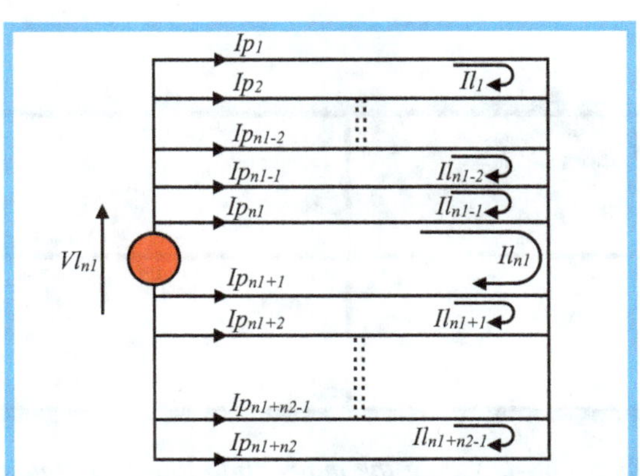

Figure 3.5 Primitive currents and loop currents

Using this relationship, it is possible to create a matrix of loop impedances **Zl**. The loop equations can then be set up. Again, in vector notation:

$$\mathbf{Vl} = \mathbf{Zl} \cdot \mathbf{Il} \tag{3.13}$$

It is clear from Figure 3.5 that the number of elements in the **Vl** and **Il** vectors is N-1

Voltages in every loop can be defined as:

$$\begin{aligned} Vl_i &= 1 \quad if \quad i = n1 \\ Vl_i &= 0 \quad if \quad i \neq n1 \end{aligned} \tag{3.14}$$

Since all the voltages and impedances have been defined for the loop equations, it is possible to determine the loop currents. The relevant function in Mathcad is:

$$\mathbf{Il} = \text{lsolve}\,(\mathbf{Zl}, \mathbf{Vl}) \tag{3.15}$$

Once the loop currents have been determined, it is a simple matter to calculate the values of the primitive currents. The relationship can be derived from Figure 3.5.

$$\begin{aligned} Ip_1 &= Il_1 \\ Ip_i &= Il_i \quad \text{for} \quad i = 2 \text{ to } N-1 \\ Ip_N &= -Il_{N-1} \end{aligned} \tag{3.16}$$

It now becomes possible to calculate every voltage component in every elemental conductor.

$$v_{i,j} = Zp_{i,j} \cdot Ip_j \tag{3.17}$$

If the variables are set out in tabular form, the result would be an array of N by N values.

Since there are only two composite conductors, the objective is to reduce this to an array of 2 by 2 values. This can be done by dividing the array into four sections, each representing the contribution made by the current in a composite conductor.

Figure 3.6 provides a picture of the resultant sub-matrices.

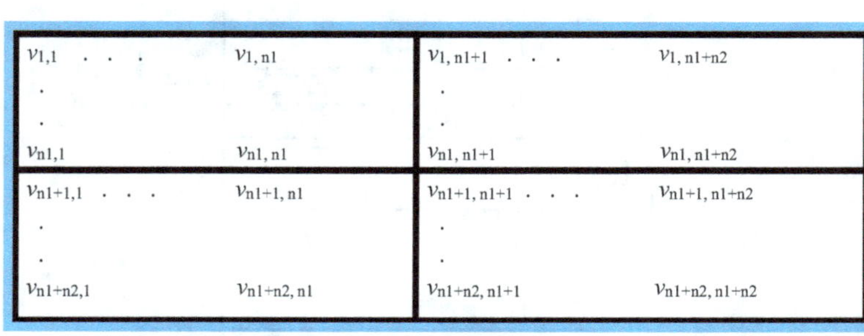

Figure 3.6 Dividing the matrix into four sub-matrices

Since there are $n1$ elemental conductors in composite 1, the average value of the contributions of current in these conductors to the voltage experienced by composite 1 is:

$$vq_{1,1} = \frac{1}{n1} \cdot \sum_{i=1}^{n1} \sum_{j=1}^{n1} v_{i,j} \tag{3.18}$$

Similar reasoning applies to the other three sections of Figure 3.6, giving:

$$vq_{1,2} = \frac{1}{n1} \cdot \sum_{i=1}^{n1} \sum_{j=n1+1}^{n1+n2} v_{i,j} \tag{3.19}$$

$$vq_{2,1} = \frac{1}{n2} \cdot \sum_{i=n1+1}^{n1+n2} \sum_{j=1}^{n1} v_{i,j} \tag{3.20}$$

$$vq_{2,2} = \frac{1}{n2} \cdot \sum_{i=n1+1}^{n1+n2} \sum_{j=n1+1}^{n1+n2} v_{i,j} \tag{3.21}$$

Calculating the average value of the current in each conductor is much simpler:

$$Iq_1 = \sum_{i=1}^{n1} Ip_i \tag{3.22}$$

$$Iq_2 = \sum_{i=n1+1}^{n1+n2} Ip_i \tag{3.23}$$

The four voltage components can be set out in the form of a small array of Figure 3.7. Where:

$vq_{1,1}$ is the voltage in composite 1 due to Iq_1
$vq_{1,2}$ is the voltage in composite 1 due to Iq_2
$vq_{2,1}$ is the voltage in composite 2 due to Iq_1
$vq_{2,2}$ is the voltage in composite 2 due to Iq_2

Partial impedance values can be derived by using the relationship:

$$Zq_{h,k} = \frac{vq_{h,k}}{Iq_k} \tag{3.24}$$

The integers h and k are used to identify the two composite conductors.

$vq_{1,1}$	$vq_{1,2}$
$vq_{2,1}$	$vq_{2,2}$

Figure 3.7 Voltage components derived from Figure 3.6

The number of equations has now been reduced from N to 2

$$\begin{aligned} Vq_1 &= Zq_{1,1} \cdot Iq_1 + Zq_{1,2} \cdot Iq_2 \\ Vq_2 &= Zq_{2,1} \cdot Iq_1 + Zq_{2,2} \cdot Iq_2 \end{aligned} \tag{3.25}$$

Comparing this equation with (2.37) or (2.43) indicates that the parameters Vq, Iq, and Zq can be treated as primitives.

As indicated in Section 3.1, a clear distinction between the properties of elemental conductors and those of composite conductors can be achieved by invoking the term 'partial' to describe the parameters involved.

Hence, (3.25) can be described as the *partial equations* for the composite pair. If the impedances $Zq_{i,j}$ are assumed to be due to inductive effects, then the relationship of (2.50) can be used to determine the circuit inductance associated with each conductor.

$$\begin{aligned} Zc_1 &= j \cdot \omega \cdot Lc_1 = j \cdot \omega \cdot \left(Lq_{1,1} - Lq_{1,2}\right) \\ Zc_2 &= j \cdot \omega \cdot Lc_2 = j \cdot \omega \cdot \left(Lq_{2,2} - Lq_{2,1}\right) \end{aligned}$$

Hence:

$$\begin{aligned} Lc_1 &= Lq_{1,1} - Lq_{1,2} \\ Lc_2 &= Lq_{2,2} - Lq_{2,1} \end{aligned} \tag{3.26}$$

In summary, the process above has established a relationship between the primitive parameters of the elemental conductors and the inductances of a circuit model of the composite pair. Since the primitive parameters of the elemental conductors can be related to the parameters of radius and length, it becomes possible to derive the inductive components from knowledge of the structure of the assembly. Circuit capacitances can be derived by invoking the duality between inductance and capacitance described in Section 2.3.

A three-page worksheet derived from this set of equations is illustrated in Figures 3.8–3.10. It is assumed that the diameter of the conductors is 2 mm, that the centres are spaced 4 mm apart, and that the length of the cable is 1 m.

Page 1 of the worksheet derives the co-ordinates of the elemental conductors (Figure 3.8).
Page 2 calculates the value of the current in each elemental conductor (Figure 3.9).
Page 3 processes this data to derive component values for the circuit model (Figure 3.10).

It is useful to print out the intermediate results in the computation. If there are any errors in the program, then the results would become implausible. For example, the sum of the currents in the **Iq** vector should always be zero.

The **Vq** vector indicates that the input voltage is split equally between the two conductors. This is intuitively correct, since the two conductors are defined to be identical. Vq_2 is negative because Iq_2 is negative.

Worksheet 3.2 Page 1

$l := 1$		Length of assembly, m
$X1 := 0$	$X2 := 4 \cdot 10^{-3}$	Co-ordinates of centre of composite conductors 1 and 2, m
$Y1 := 0$	$Y2 := 0$	
$R1 := 1 \cdot 10^{-3}$	$R2 := 1 \cdot 10^{-3}$	Radius of composite conductors 1 and 2, m.
$n_1 := 12$	$n2 := 12$	Number of elemental conductors in composite 1 and 2
$N := n_1 + n_2$		Total number of elemental conductors

Defining elemental conductors for composite 1:

$i := 1 .. n_1$ Range variable

$$\theta_i := \frac{2 \cdot \pi}{n_1} \cdot (i - 0.5)$$

$$x_i := X1 + R1 \cdot \cos(\theta_i) \qquad y_i := Y1 + R1 \cdot \sin(\theta_i) \qquad r_i := \frac{R1}{n_1}$$

Defining elemental conductors for composite 2:

$i := n_1 + 1 .. N$ Extending the range variable

$$\theta_i := \frac{2 \cdot \pi}{n_2} \cdot (i - n1 - 0.5)$$

$$x_i := X2 + R2 \cdot \cos(\theta_i) \qquad y_i := Y2 + R2 \cdot \sin(\theta_i) \qquad r_i := \frac{R2}{n_2}$$

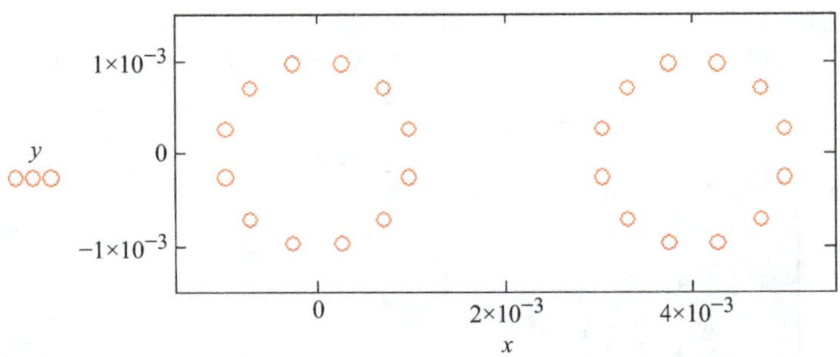

Figure 3.8 Deriving the co-ordinates of the elemental conductors

Figure 3.11 is a circuit model which effectively summarises the results of the computations.

It is possible to check this result by comparing it with that derived from a textbook on electromagnetic theory [16]. Figure 3.12 is a copy of the final three steps in the Mathcad worksheet. The fact that the results agree to three decimal places provides a high degree of confidence in the method. The fact that the value of the loop capacitance is the last parameter to be computed indicates that all the preceding results are also correct.

However, it is worth bearing in mind that the relative permittivity and the relative permeability are both assumed to be unity. It is best to confirm the component values using electrical tests. Alternatively, estimated values for these parameters can be defined at the start of the program.

Worksheet 3.2 page 2

$\mu_o := 4 \cdot \pi \cdot 10^{-7}$ H/m $\qquad \mu_r := 1 \qquad K := \dfrac{\mu_o \cdot \mu_r \cdot l}{2 \cdot \pi} = 2 \times 10^{-7}$ H

```
Zp :=  | for i ∈ 1..N                         See Figure 3.4
       |    for j ∈ 1..N
       |       | h ← x_j − x_i
       |       | v ← y_j − y_i
       |       | rad ← √(h² + v²)
       |       | rad ← r_i  if rad = 0
       |       | Lp_i,j ← K·ln(l / rad)
       | Lp

Zloop := | for i ∈ 1..N − 1                   Equation (3.12)
         |    for j ∈ 1..N − 1
         |       Lloop_i,j ← Zp_i,j − Zp_i,j+1 −Zp_i,j+1 + Zp_i+1,j+1
         | Lloop

Vloop := | for i ∈ 1..N − 1                   Equation (3.14)
         |    | V_i ← 0
         |    | V_i ← 1  if i = n_1
         | V

Iloop := lsolve (Zloop, Vloop)                Equation (3.15)

Ip := | I_1 ← Iloop_1                         Equation (3.16)
      | for i ∈ 2..N − 1
      |    I_i ← Iloop_i − Iloop_i−1
      | I_N ← −Iloop_N−1
      | I
```

Figure 3.9 Calculating the currents in the elemental conductors

A noteworthy feature of this configuration is the distribution of current in the conductors, illustrated by the bubble plot of Figure 3.13. The diameter of each circle is proportional to the amplitude of the current in that conductor. Current in the left-hand conductor is assumed to be flowing down 'into' the page, whilst current in the right-hand conductor is assumed to flow up 'out of' the page.

This illustration shows that the current is concentrated in the two facing surfaces. This asymmetrical distribution means that conductor resistance will probably increase more rapidly with frequency than the increase predicted in Section 2.5. However, tests indicate that this aspect of the response can be catered for by assuming a slightly higher value for the steady-state resistance than that given by (2.57) and by retaining the use of (2.61).

Worksheet 3.2 page 3

$\text{Start} := \begin{pmatrix} 1 \\ n_1 + 1 \end{pmatrix}$ $\quad \text{End} := \begin{pmatrix} n_1 \\ n_1 + n_2 \end{pmatrix}$ See Figure 3.6

$h := 1..2 \quad k := 1..2$ Pointers to sub-matrices

$$\text{vq}_{h,k} := \begin{array}{|l} v \leftarrow 0 \\ \text{for } i \in \text{Start}_h..\text{End}_h \\ \quad \text{for } j \in \text{Start}_k..\text{End}_k \\ \quad\quad v \leftarrow v + \text{Zp}_{i,j} \cdot \text{Ip}_j \\ \dfrac{v}{n_h} \end{array}$$

Equations (3.17) to (3.21)

$$\text{vq} = \begin{pmatrix} 2.622 & -2.122 \\ 2.122 & -2.622 \end{pmatrix} V$$

$$\text{Vq}_h := \begin{array}{|l} v \leftarrow 0 \\ \text{for } k \in 1..2 \\ \quad v \leftarrow v + \text{vq}_{h,k} \\ v \end{array}$$

Check: voltages along conductors

$$\text{Vq} = \begin{pmatrix} 0.5 \\ -0.5 \end{pmatrix} V$$

$$\text{Iq}_h := \begin{array}{|l} I \leftarrow 0 \\ \text{for } i \in \text{Start}_h..\text{End}_h \\ \quad I \leftarrow I + \text{Ip}i \\ I \end{array}$$

Equations (3.22) and (3.23)

$$\text{Iq} = \begin{pmatrix} 1.898 \times 10^{6} \\ -1.898 \times 10^{6} \end{pmatrix} A$$

$$\text{Lq}_{h,k} := \frac{\text{vq}_{h,k}}{\text{Iq}_k}$$ Equation (3.24)

$$\text{Lq} = \begin{pmatrix} 1.382 \times 10^{-6} & 1.118 \times 10^{-6} \\ 1.118 \times 10^{-6} & 1.382 \times 10^{-6} \end{pmatrix}$$

$\text{Lc}_1 := \text{Lq}_{1,1} - \text{Lq}_{1,2}$

$\text{Lc}_2 := \text{Lq}_{2,2} - \text{Lq}_{2,1}$ Equation (2.50)

$$\frac{\text{Lc}}{2} = \begin{pmatrix} 1.317 \times 10^{-7} \\ 1.317 \times 10^{-7} \end{pmatrix} H$$

$\varepsilon_o := 8.854 \cdot 10^{-12} \quad \varepsilon_r := 1$

$$\text{Cc} := \frac{\mu_o \cdot \mu_r \cdot \varepsilon_o \cdot \varepsilon_r \cdot l^2}{\text{Lc}}$$ Equation (2.28)

$$\text{Cc} = \begin{pmatrix} 4.224 \times 10^{-11} \\ 4.224 \times 10^{-11} \end{pmatrix} F$$

Figure 3.10 Calculating circuit components for composite conductors

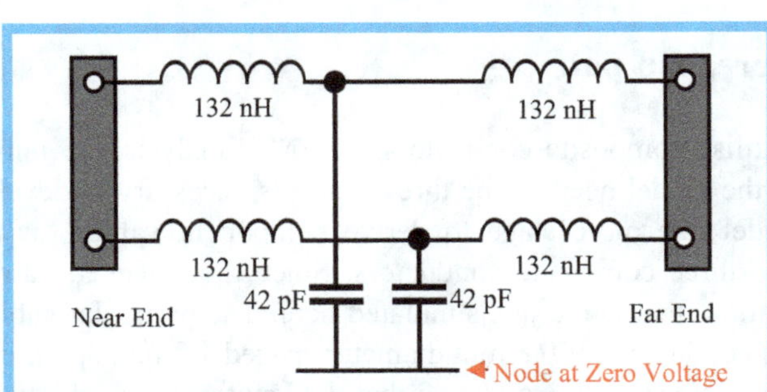

Figure 3.11 Circuit components for the composite pair

Worksheet 3.2 page 4

$$\text{Cloop} := \frac{\text{Cc}_1 \cdot \text{Cc}_2}{\text{Cc}_1 + \text{Cc}_2} \qquad \text{Cloop} = 2.112 \times 10^{-11}$$

$$b := \frac{X2}{2} \qquad r := \text{R1}$$

"Electromagnetic concepts and applications
GG Skitek & SV Marshall, pages 208 to 209.
Capacitance between two cylindrical conductors."

$$\text{Ctheory} := \frac{\pi \cdot \varepsilon_0 \cdot \varepsilon_r \cdot l}{\ln\left(\frac{b + \sqrt{b^2 - r^2}}{r}\right)} \qquad \text{Ctheory} = 2.112 \times 10^{-11}$$

Figure 3.12 Comparison of result with textbook theory

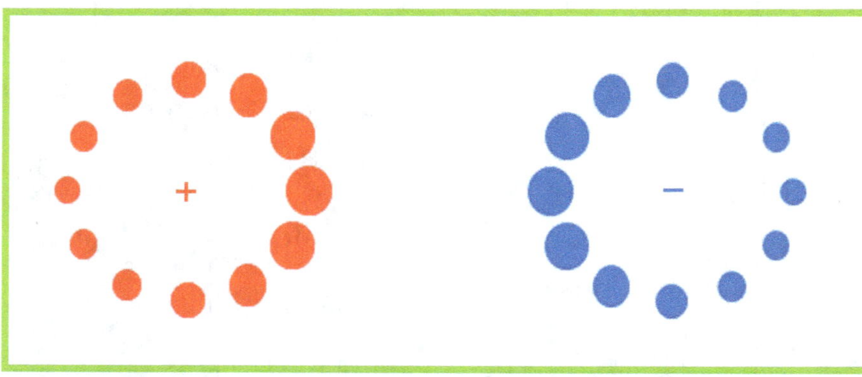

Figure 3.13 Bubble plot for composite pair, showing distribution of current

If it were assumed that the two conductors were isolated at the far end, then the plot would show the distribution of charge on the conductors, where the left-hand conductor is positively charged and the right-hand conductor is assumed to hold negative charge.

3.3 The screened pair

To properly utilise composite conductors for EMC analysis, the number of such conductors in the model needs to be three. So, it is necessary to develop the twin-conductor model one more stage to derive component values for an assembly comprising the three composite conductors. Since the screened pair is a widely used configuration, that is the one simulated here. The particular cable chosen for simulation has conductors of 0.8 mm diameter spaced 1.2 mm apart, enclosed by a screen of 3 mm diameter. It is assumed that the length of the cable is 10 m.

Worksheet 3.3 page 1

$l := 10\ \text{m}$

$X1 := 0\ \text{m}$	$X2 := 0.6 \cdot 10^{-3}\ \text{m}$	$X3 := -0.6 \cdot 10^{-3}\ \text{m}$
$Y1 := 0\ \text{m}$	$Y2 := 0\ \text{m}$	$Y3 := 0\ \text{m}$
$R1 := 1.5 \cdot 10^{-3}\ \Omega$	$R2 := 0.4 \cdot 10^{-3}\ \Omega$	$R3 := 0.4 \cdot 10^{-3}\ \Omega$
$n_1 := 30$	$n_2 := 12$	$n_3 := 12$
$N := n_1 + n_2 + n_3$		
$i := 1..n_1$	$\theta_i := \frac{2 \cdot \pi}{n_1} \cdot (i - 0.5)\ \text{radian}$	$r_i := \frac{R1}{n_1}$
$x_i := X1 + R1 \cdot \cos(\theta_i)$	$y_i := Y1 + R1 \cdot \sin(\theta_i)$	
$i := n_1 + 1..n_1 + n_2$	$\theta_i := \frac{2 \cdot \pi}{n_2} \cdot (i - n_1 - 0.5)$	$r_i := \frac{R2}{n_2}$
$x_i := X2 + R2 \cdot \cos(\theta_i)$	$y_i := Y2 + R2 \cdot \sin(\theta_i)$	
$i := n_1 + n_2 + 1..N$	$\theta_i := \frac{2 \cdot \pi}{n_2} \cdot (i - n_1 - n_2 - 0.5)$	$r_i := \frac{R2}{n_3}$
$x_i := X3 + R3 \cdot \cos(\theta_i)$	$y_i := Y3 + R3 \cdot \sin(\theta_i)$	

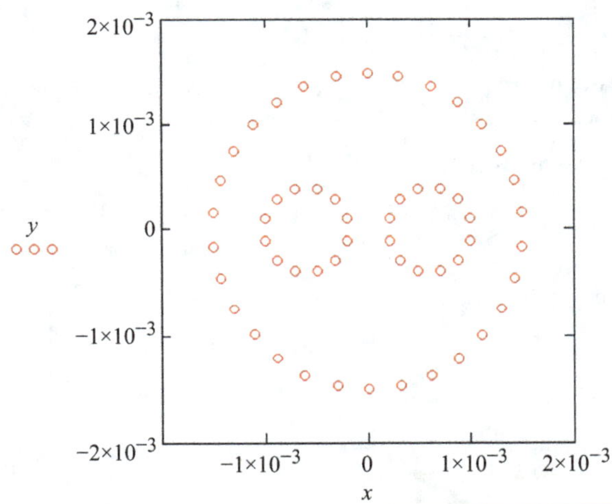

Figure 3.14 Definition of elemental conductors

Figure 3.14 is a copy of the first page of a four-page worksheet. It illustrates the calculations used to define the co-ordinates of the conductors. This is only a small development of the process illustrated by the worksheet of Figure 3.8. Three conductors are defined, rather than two. Here, the screen is defined as composite conductor 1, whilst the signal and return conductors are identified as conductors 2 and 3.

To analyse the common-mode current flow, it is assumed that all three conductors are shorted together at the far end, that the signal and return conductors are shorted together at the near end, and that a sinusoidal voltage source of 1 V is applied between the screen and signal conductors at the near end. The task is to determine the three inductive parameters and use them to calculate the capacitor values.

Worksheet 3.3 page3 (Worksheet 3.3 page 2 is identical to Figure 3.9)

$$n = \begin{pmatrix} 30 \\ 12 \\ 12 \end{pmatrix}$$

Number of elemental conductors in each composite.

$$\text{Start} := \begin{pmatrix} 1 \\ n_1 + 1 \\ n_1 + n_2 + 1 \end{pmatrix} \quad \text{End} := \begin{pmatrix} n_1 \\ n_1 + n_2 \\ N \end{pmatrix}$$

Pointers to nine sub-matrices.

$h := 1..3 \qquad k := 1..3$

Range variables

$$\text{vq}_{h,k} := \left| \begin{array}{l} v \leftarrow 0 \\ \text{for } i \in \text{Start}_h .. \text{End}_h \\ \quad \text{for } j \in \text{Start}_k .. \text{End}_k \\ \quad\quad v \leftarrow v + \text{Zp}_{i,j} \cdot \text{Ip}_j \\ \dfrac{v}{n_h} \end{array} \right.$$

Voltage components of sub-matrices:

$$\text{vq} = \begin{pmatrix} 12.667 & -6.334 & -6.334 \\ 12.694 & -7.284 & -6.41 \\ 12.694 & -6.41 & -7.284 \end{pmatrix}$$

$$\text{Vq}_h := \left| \begin{array}{l} v \leftarrow 0 \\ \text{for } k \in 1..3 \\ \quad v \leftarrow v + \text{vq}_{h,k} \\ v \end{array} \right.$$

Voltage along composite conductors:

$$\text{Vq} = \begin{pmatrix} -5.874 \times 10^{-8} \\ -1 \\ -1 \end{pmatrix}$$

$$\text{Iq}_h := \left| \begin{array}{l} I \leftarrow 0 \\ \text{for } i \in \text{Start}_h .. \text{End}_h \\ \quad I \leftarrow I + \text{Ip}_i \\ I \end{array} \right.$$

Current in composite conductors:

$$\text{Iq} = \begin{pmatrix} 7.193 \times 10^{5} \\ -3.597 \times 10^{5} \\ -3.597 \times 10^{5} \end{pmatrix}$$

$\text{Itotal} := \text{Iq}_1 + \text{Iq}_2 + \text{Iq}_3 = 0$

Sum of currents in conductors

Figure 3.15 Computing values for voltages and currents in composite conductors

Calculating the current in each elemental conductor is carried out on the second page of the worksheet. Since this page is identical to that of Figure 3.9, there is no need to replicate it.

Once the currents in all the elemental conductors have been defined, the next stage is to compute the values of the currents and voltages in the composite conductors, and this set of subroutines is illustrated in Figure 3.15.

This process is very similar to that of Figure 3.10, the main difference being that the number of composites has increased from two to three. Two sets of intermediate results are worthy of note: the vectors for partial voltage *Vq* and partial current *Iq*.

The value of Vq_1 illustrates the fact that the sum of the voltages induced in the screen is effectively zero. This is essentially the same as a statement that all internally generated electromagnetic fields are contained within the confines of the screen.

The net effect of currents in the three conductors is to balance out the voltages induced in the screen. As far as external circuits are concerned, the surface of the screen is at zero voltage. The screen behaves in exactly the same way as the outer conductor of a co-axial cable.

Worksheet 3.3 page 4

$$Lq_{h,k} := \frac{vq_{h,k}}{Iq_k}$$

Deriving primitive inductance values from current and voltage data. See Equation (3.24)

$$Lq = \begin{pmatrix} 1.761\times10^{-5} & 1.761\times10^{-5} & 1.761\times10^{-5} \\ 1.765\times10^{-5} & 2.025\times10^{-5} & 1.782\times10^{-5} \\ 1.765\times10^{-5} & 1.782\times10^{-5} & 2.025\times10^{-5} \end{pmatrix}$$

Deriving loop inductance values for three-conductor assembly. See Equation (3.12)

```
L_loop := | for h ∈ 1..2
          |   for k ∈ 1..2
          |     L_h,k ← Lq_h,k − Lq_h,k+1 − Lq_h+1,k + Lq_h+1,k+1
          | L
```

$$L\_loop = \begin{pmatrix} 2.606\times10^{-6} & -2.432\times10^{-6} \\ -2.432\times10^{-6} & 4.864\times10^{-6} \end{pmatrix}$$

$$Lc_1 := L\_loop_{1,1} + L\_loop_{1,2}$$
$$Lc_2 := -L\_loop_{1,2}$$
$$Lc_3 := L\_loop_{2,2} + L\_loop_{2,1}$$

Deriving circuit inductors for three-conductor assembly. See Equation (2.80).

$$\frac{Lc}{2} = \begin{pmatrix} 8.711\times10^{-8} \\ 1.216\times10^{-6} \\ 1.216\times10^{-6} \end{pmatrix} H$$

$$\varepsilon_o := 8.854\cdot10^{-12}\ \text{F/m}$$
$$\varepsilon_r := 1$$
$$Cc := \frac{\mu_o\cdot\mu_r\cdot\varepsilon_o\cdot\varepsilon_r\cdot l^2}{Lc}$$

Deriving capacitor values. See Equation (2.28)

$$Cc := \begin{pmatrix} 6.386\times10^{-9} \\ 4.575\times10^{-10} \\ 4.575\times10^{-10} \end{pmatrix} F$$

Figure 3.16 Calculating values of circuit components

The voltages on conductors 2 and 3 are both 1 V in amplitude, since these conductors are shorted together and the voltage source is set at 1 V. The sign is negative because the currents Iq_1 and Iq_2 are both negative.

An important check to make is to add all the partial currents in the composite conductors together. If the sum of the currents is not zero, then there can be no confidence in any subsequent calculations. It is to be expected that the returning current is equally split between composites 2 and 3, so this is also a useful check.

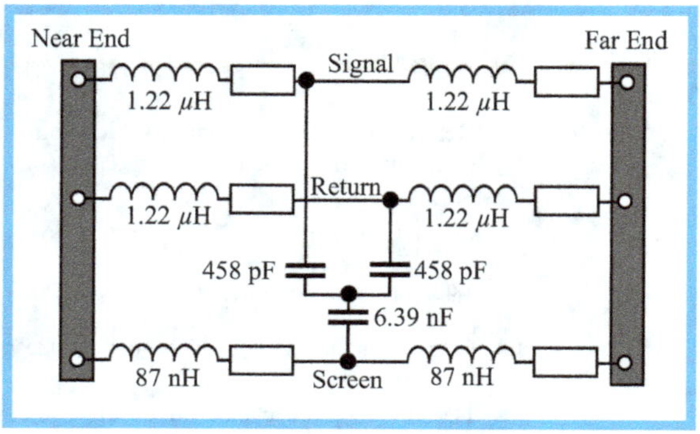

Figure 3.17 Representative circuit model of the screened pair example

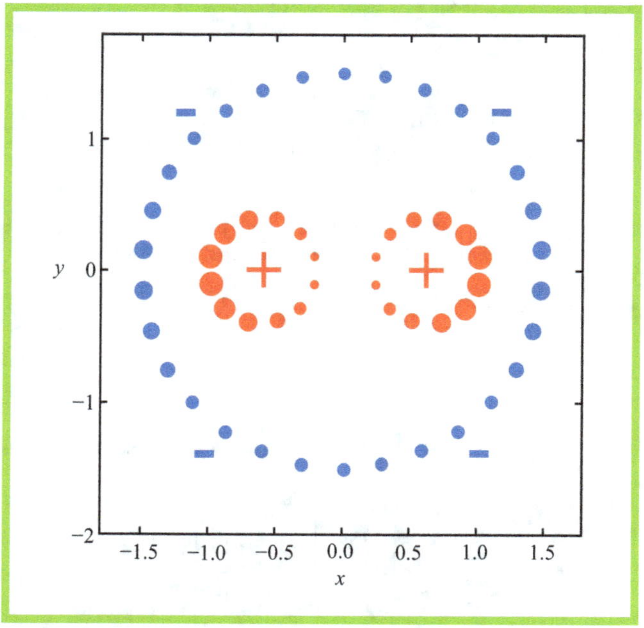

Figure 3.18 Bubble plot for screened pair, showing distribution of current

Figure 3.16 illustrates the final set of calculations in the worksheet. The partial inductance values are derived from current and voltage data, and these values are used to create the matrix of loop inductors. For this particular cross-section, the loop matrix *L_loop* is symmetrical, so there is no problem in proceeding to the final two stages: the determination of circuit inductance and circuit capacitance values. Since half-values of inductors are used in the circuit model, these are the values printed out. This leads to the model of Figure 3.17.

The only parameters left to define are the resistors, and this can be done by using general-purpose test equipment to measure conductor resistance, or by invoking (2.57).

If the cross-section of the cable assembly had been asymmetrical, the loop matrix would also have been asymmetrical, leading to a problem in determining a circuit model. Due to the nature of circuit theory, the impedance matrices derived from mesh equations are always symmetrical. It would not have been possible to achieve a one-to-one correlation between an asymmetrical matrix of loop impedances and a symmetrical matrix of circuit impedances. This does not preclude the creation of circuit models, but it does mean that the models for susceptibility and emission would be different.

A useful check on the plausibility of the computations is to produce a bubble plot of the distribution of current in the assembly. This is essentially the same as the cross-section illustrated in Figure 3.14, but with the radius at each location proportional to the amplitude of the current rather than the actual radius.

Figure 3.18 illustrates the resultant plot. It is assumed that the current in the outer screen is flowing out of the page, whilst the current flow in the inner conductors is into the page. Since the signal and return conductors are shorted together at each end, the picture that emerges is one showing the distribution of common-mode current. In contrast to Figure 3.13, the current is concentrated on the outer surfaces of the wire pair.

Chapter 4
Transmission line models

It is often required to extend the model to frequencies well beyond the one-tenth wavelength limit of the lumped component model. Such a task can be carried out by invoking the concepts of transmission line theory. This involves the need for circuit components to be defined in terms of Ω/m, F/m, H/m, and S/m.

Using this approach, transmission line theory derives a pair of hybrid equations relating current and voltage at the sending end of the line to the current and voltage at the far end.

Although at least one book on Electromagnetic Theory derives these equations [17], some books do not. So, a derivation is provided in Appendix C.

In Section 4.1, it is postulated that a T-network circuit model can be used to replicate the relationships of the hybrid equations. A pair of loop equations is derived from this model.

It is assumed that a one-to-one relationship exists between the currents and voltages of the two sets of equations. This allows formulae to be developed which define the impedances of the T-network model. These impedances can be described as distributed parameters since they are based on the assumption that R, L and C are distributed along the length of the cable.

Since this model represents a defined length of cable, the impedances of the branches can be defined in terms of the lumped parameters; that is, in terms of resistors, capacitors, inductors, and frequency, as used in textbooks on circuit theory. A one-to-one correlation is thus established between the components of the distributed-parameter model and the lumped-parameter model. The distributed-parameter model is defined by Figure 4.2 and the relationships are defined by (4.21)–(4.24). The key feature of this set of relationships is that it eliminates the need for the circuit designer to use 'per unit length' parameters.

There is still a fundamental limitation. The formulation assumes that action and reaction are instantaneous between adjacent conductors. An electromagnetic wave takes a finite time to cross the gap between these conductors. Section 4.1 provides a way of estimating the maximum usable frequency.

Section 4.2 develops this model to represent a three-conductor line by representing each conductor as a T-network. Although the impedance values of each branch are not simple functions of inductance, capacitance, and resistance, the distributed-parameter model is still amenable to analysis using circuit theory. This means that it should be possible to develop SPICE software to implement this approach.

In Section 4.3, a Mathcad program is used to simulate the coupling between two conductors over a ground plane. This is done for three situations, where the terminations at the far end are open-circuit, short circuit, and loaded by the characteristic impedance. The resultant graphs are extremely informative in comparing the frequency response characteristics of both the grounded and floating configurations.

Section 4.4 shows how the circuit model can be used in conjunction with conducted susceptibility tests.

4.1 Single-T model

The twin-T model of Figure 2.11 can be simplified to the T-network of Figure 4.1

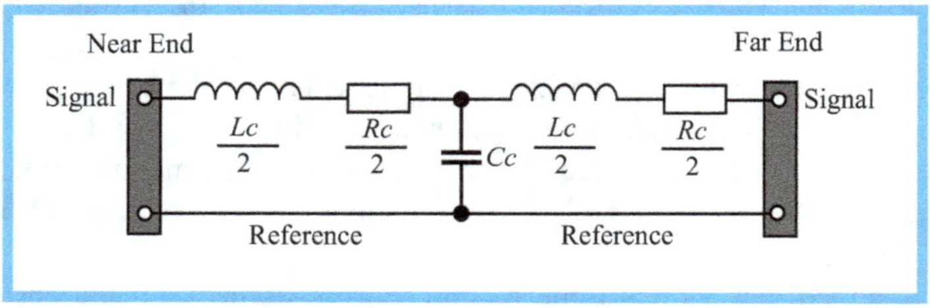

Figure 4.1 Single-T model using lumped parameters

Component values are related using

$$Rc = Rc_1 + Rc_2 \tag{4.1}$$

$$Lc = Lc_1 + Lc_2 \tag{4.2}$$

$$Cc = \frac{Cc_1 \cdot Cc_2}{Cc_1 + Cc_2} \tag{4.3}$$

Where Lc_1 and Lc_2 are defined by (2.51) and where Cc_1 and Cc_2 are as defined by (2.54).

This simplification to the model does not mean that the reference terminals at the near and far ends are at the same voltage. It merely reduces the number of parameters that mathematics has to handle.

A two-conductor transmission line can be simulated by connecting a large number of such models in series. However, there is a better way. The model can be constructed using distributed parameters, as shown in Figure 4.2.

Here, *Vs* and *Is* are the voltage and current at the sending end, whilst *Vr* and *Ir* are the voltage and current at the receiving end.

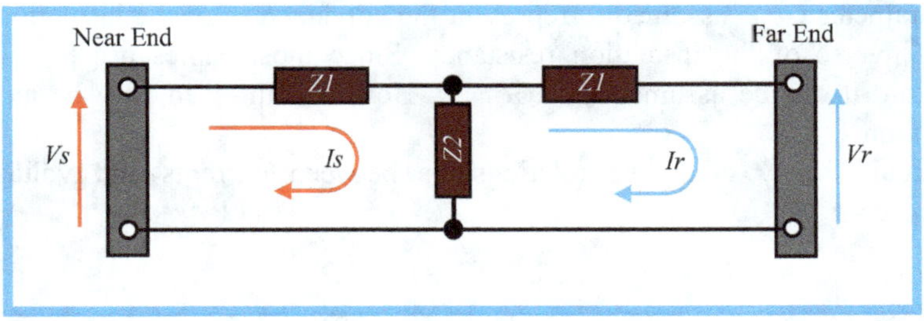

Figure 4.2 Single-T model, using distributed parameters

Appendix C shows how a pair of hybrid equations relating these four parameters can be derived from fundamental concepts:

$$Vs = Vr \cdot \cosh(\gamma \cdot l) + Zo \cdot Ir \cdot \sinh(\gamma \cdot l) \tag{4.4}$$

$$Is = \frac{Vr}{Zo} \cdot \sinh(\gamma \cdot l) + Ir \cdot \cosh(\gamma \cdot l) \tag{4.5}$$

The circuitry at each interface can be anything from a short circuit to an open circuit; it can be resistive, capacitive, inductive, or any combination of all three. It can also be another transmission line. Appendix C is well worth reading, since it recognises the fact that the equations relate to conducting loops. It does not assume the existence of a zero-volt surface.

The propagation constant γ is a function of distributed parameters: resistance per metre, inductance per metre, conductance per metre, and capacitance per metre:

$$\gamma = \sqrt{\left(\frac{Rc}{l} + \frac{j \cdot \omega \cdot Lc}{l}\right) \cdot \left(\frac{Gc}{l} + \frac{j \cdot \omega \cdot Cc}{l}\right)}$$

Simplifying this gives:

$$\gamma = \frac{1}{l} \cdot \sqrt{(Rc + j \cdot \omega \cdot Lc) \cdot (Gc + j \cdot \omega \cdot Cc)} \tag{4.6}$$

The characteristic impedance Zo is also a function of distributed parameters:

$$Zo = \sqrt{\frac{Rc + j \cdot \omega \cdot Lc}{l} \cdot \frac{l}{Gc + j \cdot \omega \cdot Cc}}$$

This can also be simplified:

$$Zo = \sqrt{\frac{Rc + j \cdot \omega \cdot Lc}{Gc + j \cdot \omega \cdot Cc}} \tag{4.7}$$

The parameter Gc is assumed to represent the conductance of the insulation. That is, the inverse of the insulation resistance. Since most cables use high-quality insulation, it can be assumed that $Gc = 0$. However, this can only be an initial assumption.

Circuit theory gives the relationships between currents and voltages of Figure 4.2 as:

$$Vs = (Z1 + Z2) \cdot Is - Z2 \cdot Ir \tag{4.8}$$

$$Vr = Z2 \cdot Is - (Z1 + Z2) \cdot Ir \tag{4.9}$$

Re-arranging (4.9) gives:

$$Is = \frac{Vr}{Z2} + \left(1 + \frac{Z1}{Z2}\right) \cdot Ir \tag{4.10}$$

Substituting for Is in (4.8) gives:

$$\begin{aligned} Vs &= (Z1 + Z2) \cdot \left[\frac{Vr}{Z2} + \left(1 + \frac{Z1}{Z2}\right) \cdot Ir\right] - Z2 \cdot Ir \\ &= \left(1 + \frac{Z1}{Z2}\right) \cdot Vr + \left[Z1 + Z2 + \frac{Z1^2}{Z2} + Z1 - Z2\right] \cdot Ir \end{aligned}$$

Simplifying further:

$$Vs = \left(1 + \frac{Z1}{Z2}\right) \cdot Vr + Z1 \cdot \left(2 + \frac{Z1}{Z2}\right) \cdot Ir \tag{4.11}$$

Equations (4.11) and (4.10) form a pair of equations which can be correlated with (4.4) and (4.5).

Comparing (4.4) with (4.11) gives:

$$1 + \frac{Z1}{Z2} = \cosh(\gamma \cdot l) \tag{4.12}$$

And

$$Z1 \cdot \left(2 + \frac{Z1}{Z2}\right) = Zo \cdot \sinh(\gamma \cdot l) \tag{4.13}$$

Comparing (4.10) with (4.5) gives:

$$\frac{1}{Z2} = \frac{1}{Zo} \cdot \sinh(\gamma \cdot l) \tag{4.14}$$

And also replicates (4.12)

$$1 + \frac{Z1}{Z2} = \cosh(\gamma \cdot l)$$

To make subsequent equations less unwieldy, let

$$\theta = \gamma \cdot l \tag{4.15}$$

The parameter θ can be described as a phase variable since it effectively determines the phase shift between the input and output signals of a transmission line.

From (4.12) and (4.15)

$$\frac{Z1}{Z2} = \cosh\ \theta - 1$$

Substituting for $\frac{Z1}{Z2}$ in (4.13) gives

$$Z1 \cdot (1 + \cosh\ \theta) = Zo \cdot \sinh\ \theta$$

This leads to

$$Z1 = \frac{\sinh\ \theta}{1 + \cosh\ \theta} \cdot Zo$$

Invoking the half-angle formula

$$Z1 = \frac{2 \cdot \sinh\ \frac{\theta}{2} \cdot \cosh\ \frac{\theta}{2}}{\cosh^2\left(\frac{\theta}{2}\right) + \sinh^2\left(\frac{\theta}{2}\right) + \cosh^2\left(\frac{\theta}{2}\right) - \sinh^2\left(\frac{\theta}{2}\right)} \cdot Zo$$

This can be simplified to

$$Z1 = Zo \cdot \tanh\left(\frac{\theta}{2}\right) \tag{4.16}$$

Re-arranging (4.14)

$$Z2 = Zo \cdot \operatorname{cosech} \theta \tag{4.17}$$

Equations (4.16) and (4.17) relate the Z-parameters of Figure 4.2 to the characteristic impedance and the phase variable of the transmission line.

Finally, from (4.6) and (4.15):

$$\theta = \sqrt{(Rc + j \cdot \omega \cdot Lc) \cdot (Gc + j \cdot \omega \cdot Cc)} \tag{4.18}$$

Using (4.7) and (4.18), Zo and θ can be calculated. These can then be used in (4.16) and (4.17) to obtain values for the components of Figure 4.2.

This means that the lumped parameter model of Figure 4.1 can easily be transformed into one which uses distributed parameters. The transformation would be:

$$\begin{aligned} &\frac{1}{2} \cdot (Rc + j \cdot \omega \cdot Lc) \rightarrow Zo \cdot \tanh\left(\frac{\theta}{2}\right) \\ &\frac{1}{(Gc + j \cdot \omega \cdot Cc)} \rightarrow Zo \cdot \operatorname{cosech} \theta \end{aligned} \tag{4.19}$$

Invoking this transformation allows the circuit model of Figure 4.2 to simulate the response of a transmission line over the same range of frequencies at which the hybrid equations are valid.

The characteristic impedance *Zo* is independent of the length of the line. Given knowledge of the values of *Rc, Lc, Gc,* and *Cc*, the phase variable θ is also independent of the line length. The variable *l* does not appear in (4.18). Although it is necessary to invoke the concept of 'distributed parameters' to derive the hybrid equations, it is not necessary to use the concept when calculating impedance values for the model of Figure 4.2.

An electromagnetic wave takes a finite time to cross the gap between the conductors of any cable form. Just as SPICE modelling is limited by the electrical length of the assembly, modelling using distributed parameters is limited by the radial separation of the adjacent conductors. It can be reasoned that, if *rmax* is the maximum separation between the centres of two conductors in the cross-section of the assembly-under-review, then the maximum frequency at which the distributed-parameter model is usable is *fmax*, where:

$$fmax = \frac{vmin}{10 \cdot rmax} \tag{4.20}$$

and *vmin* is the propagation velocity of the electromagnetic wave in the insulating medium with the highest relative permittivity in the section-under-review.

That is, if the velocity of propagation is 200 m/μs and the maximum spacing between conductors is 10 mm, then the maximum frequency at which this modelling is usable would be about 2 GHz.

When distributed parameters are used, there is no theoretical limit to the length of the cable which can be modelled. It is assumed that the cross-section of the cable is constant and that the lengths of the conductors are identical.

4.2 Triple-T model

The next step is to create a distributed-parameter model for the three-conductor transmission line. Figure 4.3 illustrates the starting point; the lumped parameter model developed in section 2.7.

Representing each conductor by the T-network of Section 4.1 leads to Figure 4.4.

The relationships between components of the two models are:

$$Zo_i = \sqrt{\frac{Rc_i + j \cdot \omega \cdot Lc_i}{Gc_i + j \cdot \omega \cdot Cc_i}} \tag{4.21}$$

$$\theta_i = \sqrt{(Rc_i + j \cdot \omega \cdot Lc_i) \cdot (Gc_i + j \cdot \omega \cdot Cc_i)} \tag{4.22}$$

$$Z_{1,i} = Zo_i \cdot \tanh\left(\frac{\theta_i}{2}\right) \tag{4.23}$$

$$Z_{2,i} = Zo \cdot \text{cosech}(\theta_i) \tag{4.24}$$

These four equations can be collected together to compute the values of a pair of Z-parameters from a set of *R*, *L*, *C*, and *G* values. Indices are used to retain correlation between the parameters and the conductors.

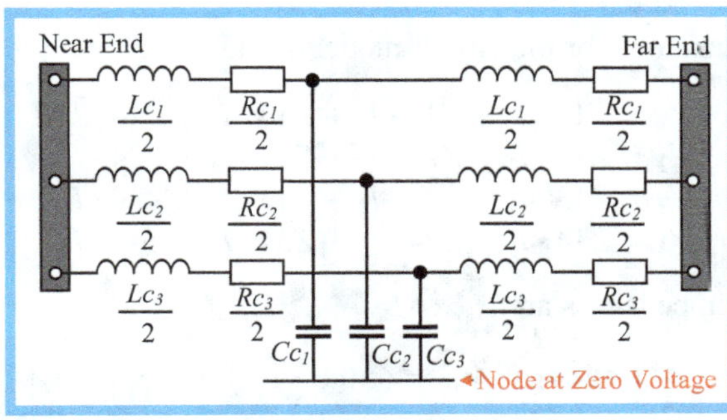

Figure 4.3 Triple-T network of lumped parameters

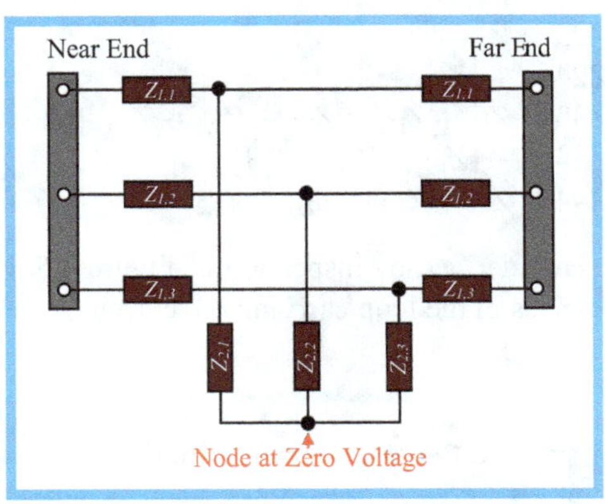

Figure 4.4 Triple-T network of distributed parameters

Given the definitions above, the result of the computation would be an array of two rows by three columns. The first row would hold values correlating with horizontal branches of Figure 4.4. The second row would hold values for the vertical branches. Each column would be associated with a conductor. The conductance parameter G_i is included in the formulae, which allows losses in the insulation to be simulated. It is fair to make an initial assumption that all conductance values are zero.

One feature of the Triple-T model is that it only includes components representing the cable assembly.

It is necessary for the components at the cable terminations to be included before any simulation is possible. This results in the full model of Figure 4.5.

Impedances within the unit at the near end are identified as Zn_i, whilst those within the equipment at the far end are designated Zf_i.

Four separate loops can be identified. Currents in each loop have been defined, as well as all possible voltage sources. It is assumed that conductors 1 and 2 carry the differential-mode signal current. So, voltage sources V_1 and V_3 are located within the equipment units. Voltage sources V_2 and V_4 represent possible interference sources in the common-mode loop.

The four loop equations for the circuit model of Figure 4.5 are

$$\begin{aligned}V_1 &= Z11 \cdot I_1 + Z12 \cdot I_2 + Z13 \cdot I_3 + Z14 \cdot I_4 \\ V_2 &= Z12 \cdot I_1 + Z22 \cdot I_2 + Z23 \cdot I_3 + Z24 \cdot I_4 \\ V_3 &= Z13 \cdot I_1 + Z23 \cdot I_2 + Z33 \cdot I_3 + Z34 \cdot I_4 \\ V_4 &= Z14 \cdot I_1 + Z24 \cdot I_2 + Z34 \cdot I_3 + Z44 \cdot I_4\end{aligned} \tag{4.25}$$

Where the loop impedances are:

$$\begin{aligned}Z11 &= Zn_1 + Z_{1,1} + Z_{2,1} + Z_{2,2} + Z_{1,2} + Zn_2 \\ Z12 &= -(Z_{1,2} + Z_{2,2} + Zn_2) \\ Z13 &= -(Z_{2,1} + Z_{2,2}) \\ Z14 &= Z_{2,2} \\ Z22 &= Zn_2 + Z_{1,2} + Z_{2,2} + Z_{2,3} + Z_{1,3} + Zn_3 \\ Z23 &= Z_{2,2} \\ Z24 &= -(Z_{2,2} + Z_{2,3}) \\ Z33 &= Zf_2 + Z_{1,2} + Z_{2,2} + Z_{2,1} + Z_{1,1} + Zf_1 \\ Z34 &= -(Z_{1,2} + Z_{2,2} + Zf_2) \\ Z44 &= Zf_3 + Z_{1,3} + Z_{2,3} + Z_{2,2} + Z_{1,2} + Zf_2\end{aligned} \tag{4.26}$$

The loop impedances are derived by inspection of Figure 4.5. For example, $Z11$ is the sum of the impedances in the loop carrying the current I_1, whilst $Z34$ is the sum

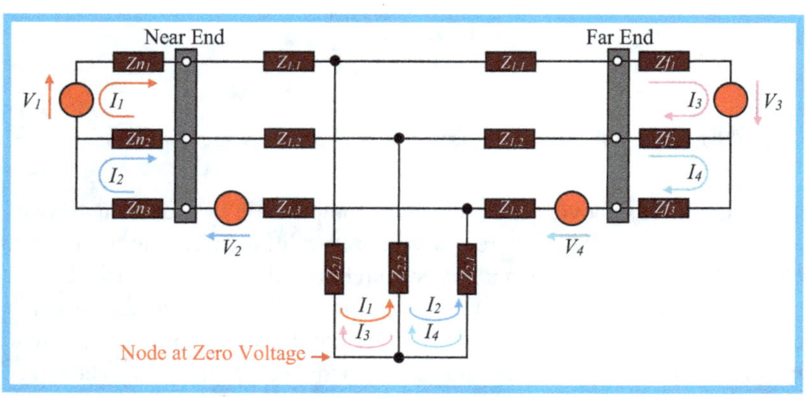

Figure 4.5 Full circuit network using lumped and distributed parameters

of the impedances which share currents I_3 and I_4. The negative sign indicates that these two currents flow in opposite directions.

Equations (4.21)–(4.26) provide a traceable relationship between conventional circuit components and the array of loop impedances. If all the component values are defined and all the source voltages are defined, then vector algebra can be used to derive all the currents.

It then becomes a routine matter to calculate the voltage between any two points or the current in any component, at any particular frequency.

Since there is now a one-to-one correlation between the branches of the lumped-parameter and distributed-parameter networks, it becomes possible to define the circuit model in terms of the lumped parameters, but analyse it using the loop impedances of (4.26). This leads to the general circuit model of Figure 4.6. As it is possible to replace the circuitry at the near and far ends with components representing the interface circuits of the equipment-under-review.

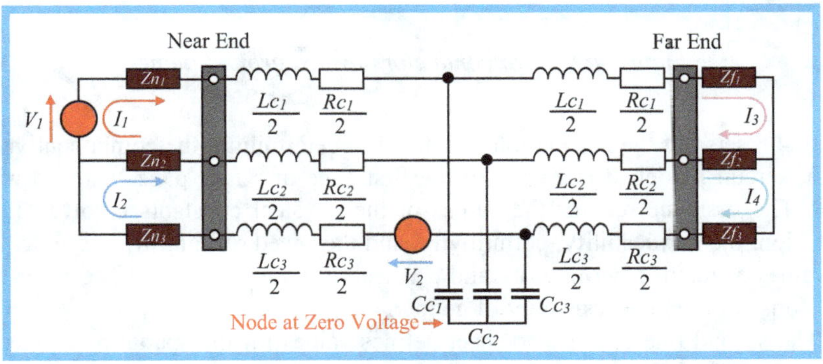

Figure 4.6 General circuit model of a three-conductor signal link

In this model, the values for the components of the triple-T network can be derived from geometrical data of the cable and structure, whilst values for the components of the interface circuitry at the near end and far end of the cable are defined by the designer.

It is also possible to assign values to all the components of the model by carrying out electrical tests on an actual assembly. Chapter 7 describes how this can be done.

4.3 Cross-coupling

Enough analytical tools now exist to allow an initial assessment to be made of a wide range of intra-system interference problems. Perhaps the simplest illustration of the process lies in the analysis of the cross-coupling between two conductors over a ground plane.

Figure 4.7 illustrates the configuration. It is assumed that the conductors are both 1 mm in diameter and that each is spaced 1 mm above the plane. The separation between them is set at 4 mm and the length of the assembly is assumed to be 1 metre.

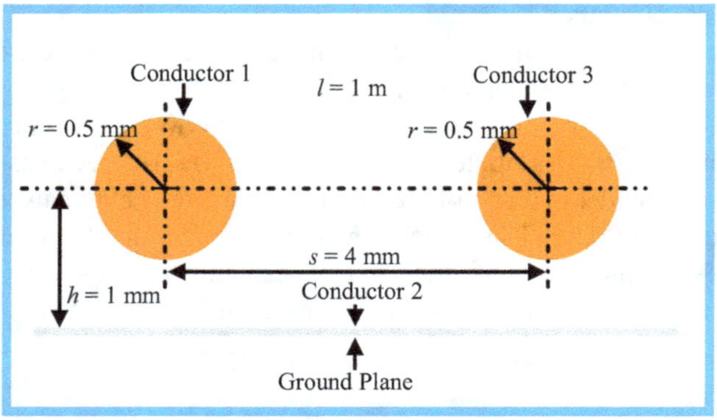

Figure 4.7 Two conductors over a ground plane

Figure 4.8 sets out the calculations involved in determining the component values for the circuit model. It is a copy of the first page of a four-page Mathcad worksheet. The first line records the values of the physical constants involved in the derivation: the permeability, permittivity, and the speed of light. It is assumed that the values of relative permittivity and permeability are both unity. The other physical constant is ρ, the resistivity of copper.

The second line of the worksheet defines values for the spatial parameters of the configuration. These correlate with the parameters of Figure 2.31.

Steady-state resistance values *Rss* for the two circular-section conductors are determined using (2.57). A guess value of 5 mΩ is assigned to the ground plane. Above a certain frequency, skin effect will cause an increase in the resistance of the conductors. The crossover frequency *Fx* is determined using (2.60) and the relationship between *Rc* and *f* is defined by (2.61).

Inductance values for the circuit model are derived from (2.96). Since the two wires are equidistant from the plane, Lc_3 is equal to Lc_1. The three values are stored in the vector *Lc*. Capacitor values are obtained quite simply by invoking (2.33).

Another significant parameter is *Fq*, the frequency at which quarter-wave resonance occurs. This is determined using (2.34). Knowledge of this frequency allows subsequent analysis to include maximum values of the peaks in the response curves.

Since half-values of resistance and inductance are used in the circuit model of Figure 4.3, these are the values printed out in the worksheet. Adding the values of the three capacitors provides sufficient information to draw the representative circuit model for the conductor assembly.

Before the response can be analysed, it is necessary to include values for the interface components of Figure 4.6. Invoking worst-case conditions, it is assumed that the components at the near end are short-circuits and that the wire terminals at the far end are open circuits.

Following normal practice used in SPICE analysis, the open circuits are represented by 10 MΩ resistors. To allow for the option of simulating losses in the insulation, conductance values *Gc* are also defined. In this model, all three are assumed to be zero.

Data derived from this first page of the worksheet can be illustrated by the circuit model of Figure 4.9. It is assumed that the assembly of Figure 4.7 is configured as a pair of transmission lines, with the ground plane acting as the return conductor for both signals. It is also assumed that there is only one voltage source between conductor 1 and the ground plane. This model represents the situation

Worksheet 4.3 page 1

$\mu_o := 4 \cdot \pi \cdot 10^{-7}$ H/m $\qquad \varepsilon_o := 8.854 \cdot 10^{-12}$ F/m $\qquad c := 2.998 \cdot 10^{8}$ m/s

$\rho := 1.7 \cdot 10^{-8}$ Ω/m $\qquad l := 1$ m $\qquad h := 1 \cdot 10^{-3}$ m/s

$s := 4 \cdot 10^{-3}$ m $\qquad r := 0.5 \cdot 10^{-3}$ m $\qquad$ See Figure 4.7

$Rss_1 := \dfrac{\rho \cdot l}{\pi \cdot r^2} \qquad Rss_3 := Rss_1 \qquad Rss_2 := 0.005\ \Omega$ $\qquad$ Equation (2.57)

$$Fx := \frac{4 \cdot \rho}{\mu_o \cdot \pi \cdot r^2} = 6.89 \times 10^4 \text{ Hz} \qquad \text{Equation (2.60)}$$

$$Lc_1 := \frac{\mu_o \cdot l}{2 \cdot \pi} \cdot \ln\left(\frac{2 \cdot h \cdot s}{r \cdot \sqrt{s^2 + 4 \cdot h^2}}\right) H$$

$$Lc_2 := \frac{\mu_o \cdot l}{2 \cdot \pi} \cdot \ln\left(\frac{\sqrt{s^2 + 4 \cdot h^2}}{s}\right) H \qquad \text{Equation (2.96)}$$

$$Lc_3 := Lc_1$$

$$Cc := \frac{1}{Lc} \cdot \left(\frac{l}{c}\right)^2 F \qquad \text{Equation (2.33)}$$

$$Fq := \frac{1}{4 \cdot \sqrt{Lc_1 \cdot Cc_1}} = 7.495 \times 10^7 \text{ Hz} \qquad \text{Equation (2.34)}$$

Component values for circuit model of Figure 4.7:

$$\frac{Rss}{2} = \begin{pmatrix} 0.011 \\ 2.5 \times 10^{-3} \\ 0.011 \end{pmatrix} \qquad \frac{Lc}{2} = \begin{pmatrix} 1.275 \times 10^{-7} \\ 1.116 \times 10^{-8} \\ 1.275 \times 10^{-7} \end{pmatrix} \qquad Cc = \begin{pmatrix} 4.364 \times 10^{-11} \\ 4.986 \times 10^{-10} \\ 4.364 \times 10^{-11} \end{pmatrix}$$

$$Zn := \begin{pmatrix} 0 \\ 0 \\ 0 \end{pmatrix} \Omega \qquad Zf := \begin{pmatrix} 10^7 \\ 0 \\ 10^7 \end{pmatrix} \Omega \qquad Gc := \begin{pmatrix} 0 \\ 0 \\ 0 \end{pmatrix} S$$

Figure 4.8 Deriving circuit parameters

where the culprit circuit uses Conductor 1, whilst the victim circuit uses Conductor 3.

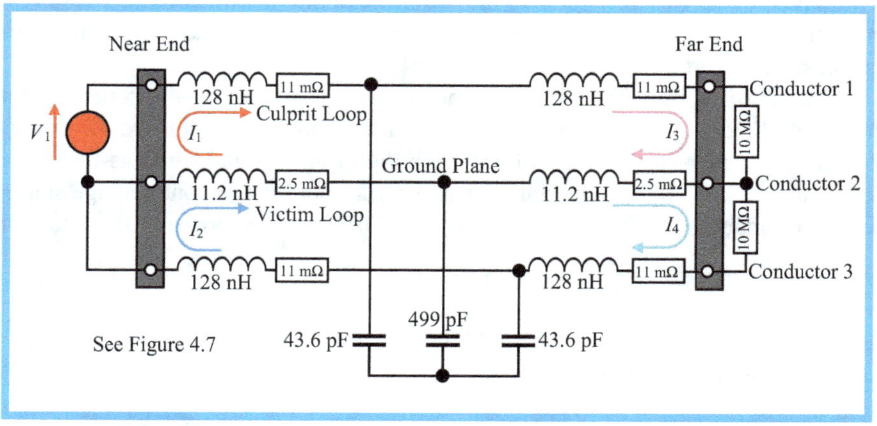

Figure 4.9 Representative circuit model of cross-coupling example

The purpose of the analysis is to simulate the current that would flow in the victim loop over a range of frequencies. Since the far end is open-circuit, only the current at the near end will be of any significance. Hence, the objective is to simulate the current I_2 when voltage V_1 is applied. If V_1 is held at a constant value of one volt, the result will be the frequency response of the transfer admittance.

Figure 4.10 sets out the two program functions derived in Section 4.2. It constitutes page 2 of the Mathcad worksheet.

The function *Zbranch(f)* supplies one input variable: the frequency *f*. It calculates the values of the two distributed impedances associated with each conductor, and assembles them in the output array *Z*. As indicated in Section 4.2, this contains two rows of three columns.

A feature worth noting is the fact that the value of each resistor is calculated at each frequency. This allows skin effect to be incorporated into the analysis.

In addition to supplying frequency as an input variable, the function *Zloop(f,Zf)* also includes the vector containing component values at the far interface. This allows any of these components to be varied, subsequent to the definition of the function.

The first program step in the *Zloop* function is to call up the *Zbranch(f)* function and place the results in the local variable *Z*. This allows component values of individual branches to be accessed as required when determining values for the loop impedances. These impedance values are then placed in the four-by-four array at the end of the subroutine.

Creating a graph of the response involves the definition of two vectors: the frequency and the output variable, and this is the task performed by page 3 of the worksheet. Figure 4.11 illustrates the process.

Worksheet 4.3 page 2

$$\text{Zbranch}(f) := \left| \begin{array}{l} \omega \leftarrow 2\cdot\pi\cdot f \\ \text{for } i \in 1..3 \\ \quad \left| \begin{array}{ll} \text{Rc}_i \leftarrow \text{Rss}_i\cdot\sqrt{1+\dfrac{f}{\text{Fx}}} & \text{Equation (2.61)} \\ \theta \leftarrow \sqrt{(\text{Rc}_i+j\cdot\omega\cdot\text{Lc}_i)\cdot(\text{Gc}_i+j\cdot\omega\cdot\text{Cc}_i)} & \text{Equation (4.22)} \\ \text{Zo} \leftarrow \sqrt{\dfrac{\text{Rc}_i+j\cdot\omega\cdot\text{Lc}_i}{\text{Gc}_i+j\cdot\omega\cdot\text{Cc}_i}} & \text{Equation (4.21)} \\ Z_{1,i} \leftarrow \text{Zo}\cdot\tanh\left(\dfrac{\theta}{2}\right) & \text{Equation (4.23)} \\ Z_{2,i} \leftarrow \text{Zo}\cdot\text{csch}(\theta) & \text{Equation (4.24)} \end{array} \right. \\ Z \end{array} \right.$$

$$\text{Zloop}(f, \text{Zf}) := \left| \begin{array}{l} Z \leftarrow \text{Zbranch}(f) \qquad \text{Equation (4.26)} \\ Z11 \leftarrow Zn_1 + Z_{1,1} + Z_{2,1} + Z_{2,2} + Z_{1,2} + Zn_2 \\ Z12 \leftarrow -(Z_{1,2} + Z_{2,2} + Zn_2) \\ Z13 \leftarrow -(Z_{2,1} + Z_{2,2}) \\ Z14 \leftarrow Z_{2,2} \\ Z22 \leftarrow Zn_2 + Z_{1,2} + Z_{2,2} + Z_{2,3} + Z_{1,3} + Zn_3 \\ Z23 \leftarrow Z_{2,2} \\ Z24 \leftarrow -(Z_{2,2} + Z_{2,3}) \\ Z33 \leftarrow Zf_2 + Z_{1,2} + Z_{2,2} + Z_{2,1} + Z_{1,1} + Zf_1 \\ Z34 \leftarrow -(Z_{1,2} + Z_{2,2} + Zf_2) \\ Z44 \leftarrow Zf_3 + Z_{1,3} + Z_{2,3} + Z_{2,2} + Z_{1,2} + Zf_2 \\ \begin{pmatrix} Z11 & Z12 & Z13 & Z14 \\ Z12 & Z22 & Z23 & Z24 \\ Z13 & Z23 & Z33 & Z34 \\ Z14 & Z24 & Z34 & Z44 \end{pmatrix} \end{array} \right.$$

Figure 4.10 Calculating branch impedances and loop impedances

A set of spot frequencies is defined in the first line. The integer n is used to set the number of spot frequencies in the range between zero and the frequency of quarter-wave resonance, Fq. To cover the range up to a full wavelength, the control variable s is set to make four times this number of steps. This allows each spot frequency, F_s, to be a multiple of the quarter-wave frequency. Using this method of defining the spot frequencies ensures that all the resonant frequencies are selected and that the amplitude of every peak and every trough in the response is calculated.

Worksheet 4.3 page 3

$n := 100 \qquad s := 1 .. 4 \cdot n \qquad F_s := s \cdot \frac{\text{Fq}}{n}$ Defining the spot frequencies

$$V := \begin{pmatrix} 1 \\ 0 \\ 0 \\ 0 \end{pmatrix} V$$ Defining the voltage source

$$\text{Yoc}_s := \left| \begin{array}{l} f \leftarrow F_s \\ Z \leftarrow \text{Zloop}(f, \text{Zf}) \\ I \leftarrow \text{lsolve}(Z, V) \\ |I_2| \end{array} \right.$$

Main program

Defining the current at the near end of the victim line when the far end is open-circuit. See Figure 4.9.

Figure 4.11 Calculating the frequency response of the open-circuit transfer admittance

In the model of Figure 4.9, the source voltage generator is located in loop 1. Setting the amplitude of loop 1 at 1 Volt and the other three sources at zero voltage allows the voltage vector V to be defined.

Since all the inputs have now been defined, it becomes possible to create the main program, and this is illustrated on the third line of the worksheet of Figure 4.11.

The first program step is to select a spot frequency from the vector F and assign it to the local variable f. The second line calls up the function *Zloop(f,Zf)*. Since *Zf* has already been defined (see Figure 4.8), both variables are 'visible' to the function. The output is stored in the local variable Z.

The next step is to calculate the values of all four loop currents and place them in the vector I.

The final step in the main subroutine is to select I_2, the current at the near end of the victim loop, and determine its amplitude. Dividing this current by the input voltage V_1 gives the value of the transfer admittance.

Since the value of V_1 is unity, the values of current and transfer admittance are identical. Finally, the output is stored in the relevant location in the vector *Yoc*; the transfer admittance when the line terminations are open-circuit. The graph of this function is shown in Figure 4.12.

Although normal convention is to display such a function using a logarithmic scale for both parameters, in this case a linear scale is used for frequency. This highlights the symmetrical nature of the response. Symmetry is evident in the fact that the response between 150 MHz and 300 MHz is almost the same as between zero and 150 MHz; almost the same, but not identical. Skin effect causes the resistance to increase, reducing the amplitude of the second peak and increasing the amplitude of the second trough.

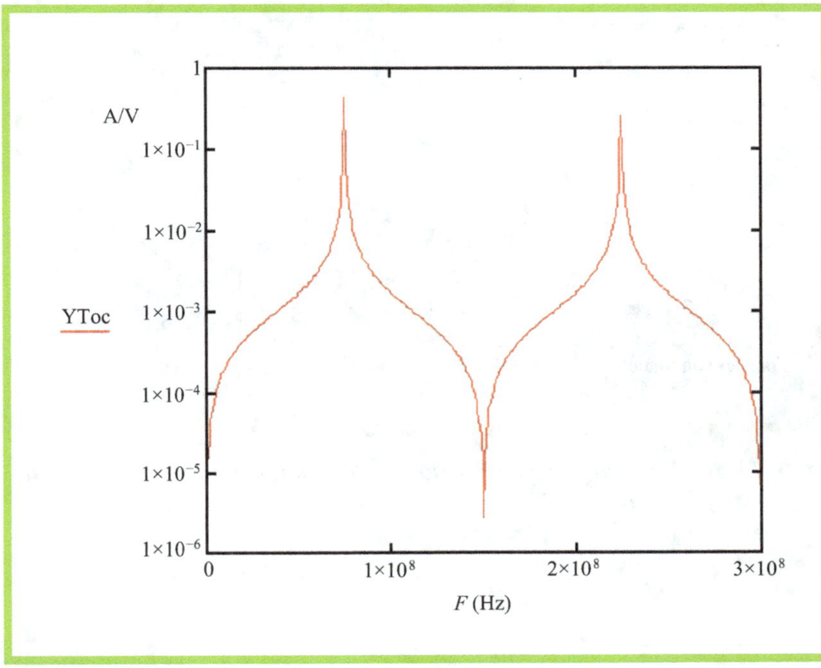

Figure 4.12 Transfer admittance with open-circuit terminations

By resetting all the components of the impedance vector *Zf* to zero and re-running the main program, the vector *Ysc* can be created to define the response of the assembly when the terminations are short-circuited. The response of the assembly when it is critically damped can also be determined. The program steps are illustrated in Figure 4.13.

Figure 4.14 provides a further illustration of the symmetry of the response, by comparing the open-circuit response with the response when the terminations at the far end are short-circuited. Transfer admittance under short circuit conditions, *Ysc*, is illustrated by the dotted curve.

When each conductor is terminated by its characteristic impedance, the response is perfectly flat over the entire frequency range. This is illustrated by the *Ycrit* curve, the dashed line.

The bottom half of the graph, below the *Ycrit* curve, is almost the mirror image of the top half.

Many lessons can be derived from this set of curves, not least among them the fact that the only configuration which is free from peaks in the transfer admittance is also the one which provides the highest efficiency in terms of signal transmission.

In practice, the peaks will be much lower and much more rounded than those illustrated, since this model does not take account of losses due to radiated emission. Chapter 5 develops the use of models to cater for this effect.

Worksheet 4.3 page 4

$$Zf := \begin{pmatrix} 0 \\ 0 \\ 0 \end{pmatrix}$$

$$Ysc_s := \begin{vmatrix} f \leftarrow F_s \\ Z \leftarrow \mathrm{Zloop}\,(f, \mathrm{Zf}) \\ I \leftarrow \mathrm{lsolve}\,(Z, V) \\ \left|I_2\right| \end{vmatrix}$$

Transfer admittance with short circuit terminations

$$Zf := \sqrt{\frac{Lc}{Cc}} \qquad Zf = \begin{pmatrix} 76.432 \\ 6.69 \\ 76.432 \end{pmatrix}$$

$$Ycrit_s := \begin{vmatrix} f \leftarrow F_s \\ Z \leftarrow \mathrm{Zloop}\,(f, \mathrm{Zf}) \\ I \leftarrow \mathrm{lsolve}\,(Z, V) \\ \left|I_2\right| \end{vmatrix}$$

Transfer admittance with critical damping

Figure 4.13 Calculating frequency response for short-circuited and critically damped lines

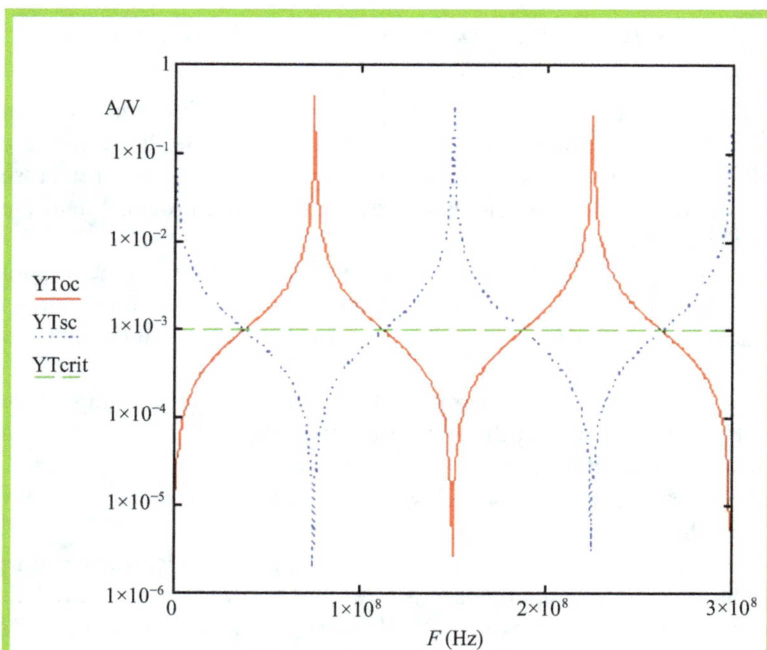

Figure 4.14 Transfer admittance for open-circuit, short-circuit, and critically damped lines

4.4 Bench test models

Bench testing is a vital part of the product development process. It provides assurance that the design requirements are being met. Correlating the observed performance with the predicted performance allows the assumptions used in the theoretical analysis to be verified, or modified in the light of more accurate information.

It allows corrections to be made to the design to improve its performance. This is true of requirements such as system function, response time, reliability, size, and cost.

It can also apply to the requirements of electromagnetic compatibility. Since the EMC Test Houses have already acquired a wide experience in meeting the regulatory requirements, it is a logical exercise to tap into that experience to identify the relevant test equipment and test methods.

Two problems immediately emerge. The equipment is extremely expensive, and the test procedures require the services of highly skilled engineers. It would be far too costly for a small organisation to attempt to emulate this approach.

However, the product designer also has one advantage; a big advantage. It is possible to design and use interface equipment which allows direct visibility of the signals being processed. This is not an option available to the Test Houses, since any special modification of the equipment-under-test would render the test results invalid.

Another advantage is the fact that General Purpose Test Equipment is already at hand to carry out functional tests of the assembly-under-review. This could also be used for EMC testing. Also, the frequency range of the tests can be tailored to the range over which the equipment-under-test normally functions. There is no need to cover the range covered by the Test Houses.

The simplest set of bench tests would be to analyse the performance of the interconnecting cables, specifically to measure the frequency response of the transfer admittance. The results could be compared directly with those obtained by circuit modelling. But first, the representative circuit model needs to be created.

Figure 4.15 illustrates a setup which can be used to measure the conducted emission of any wiring configuration. In this case, it is assumed that the assembly-under-review is a 10 m length of a conductor pair routed 5 mm above a ground plane.

An input signal can be applied to the near-end terminals via a splitter box. This allows the input voltage to be monitored on channel 1 of the oscilloscope. Channel 2 is used to monitor the common-mode current. The 100 Ω load at the far end of the cable is floating.

The ratio of output to input gives the value of the transfer admittance at the frequency of the signal generator output. Repeating the measurements over a range of spot frequencies enables the frequency response characteristic of the conducted emission to be determined.

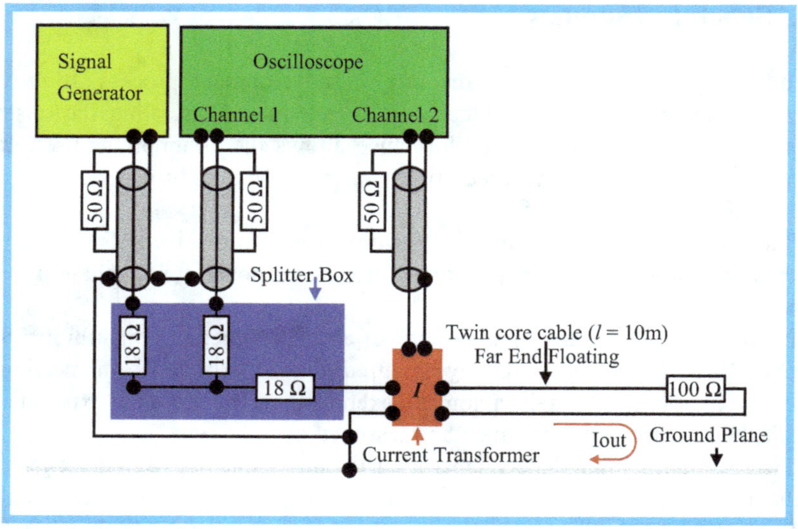

Figure 4.15 Setup for conducted emission test (floating configuration)

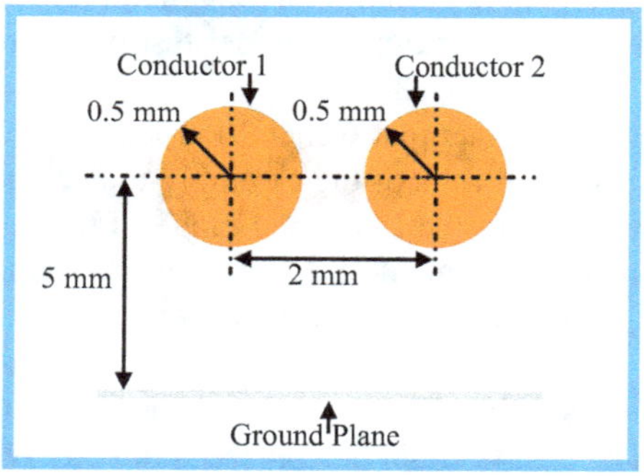

Figure 4.16 Cross-sectional view of cable assembly

If the physical geometry of the signal link is as shown in Figure 4.16, then the circuit model can be derived by invoking the calculations recorded on the work-sheet of Figure 4.8.

As with SPICE modelling, it is necessary to simulate the open circuit as a high-value resistance; in this case 10 MΩ.

The resultant model is shown in Figure 4.17. It is worth noting that the source voltage is in the differential-mode loop whilst the output current is in the common-mode loop.

Conducted susceptibility measurements would involve a setup similar to that illustrated in Figure 4.18. Here, the input voltage is applied to the common-mode loop via an injection transformer. Since any transformer has an output impedance, the applied voltage can be reduced as the load current increases.

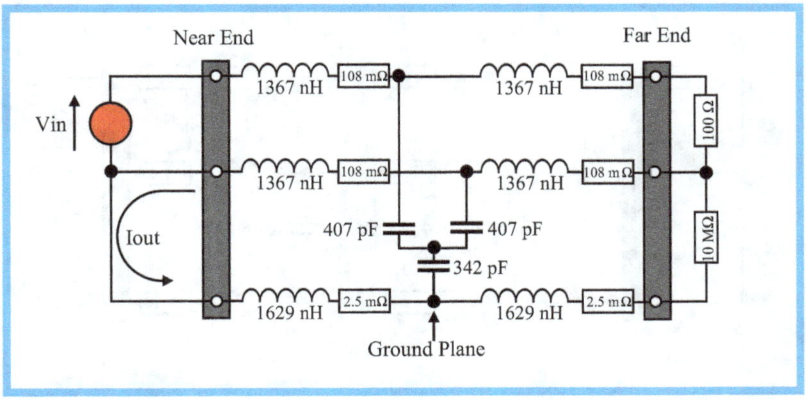

Figure 4.17 Representative circuit model of conducted emission test setup

This effect is catered for by using a separate turn on the transformer to monitor the actual voltage applied. One channel of the oscilloscope provides the means of measurement. This enables the amplitude of the input voltage to be determined.

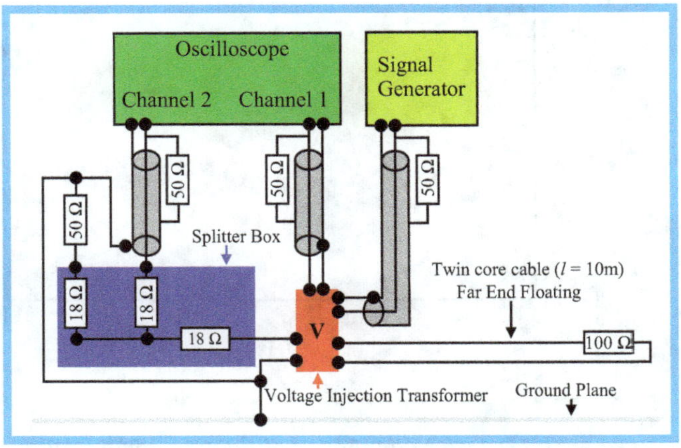

Figure 4.18 Setup for conducted susceptibility test

The second channel of the oscilloscope allows the resultant current in the differential-mode loop to be measured. Again, the ratio of output to input provides a value for the transfer admittance. Measurement over a range of spot frequencies provides the frequency response of the Conducted Susceptibility characteristic.

A prediction of the response of this configuration can be achieved with the model of Figure 4.19. Since the assembly-under-review is exactly the same as that used for the conducted emission test, the passive components of the two models are identical. The only difference lies in the fact that the voltage source is now in the common-mode loop, whilst the output current is now in the differential-mode loop.

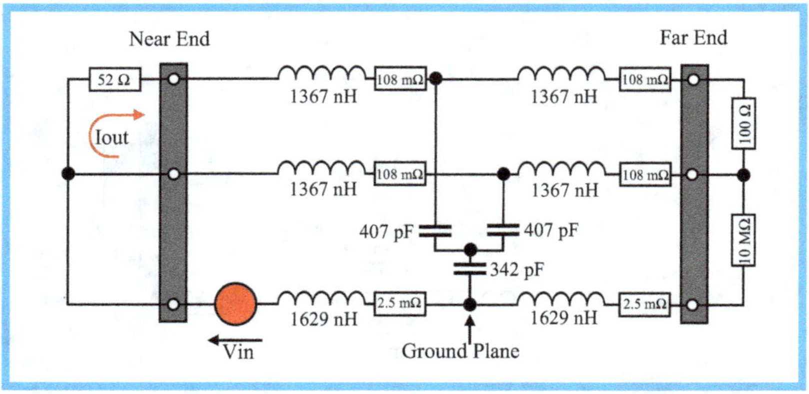

Figure 4.19 Representative circuit model of conducted susceptibility test setup

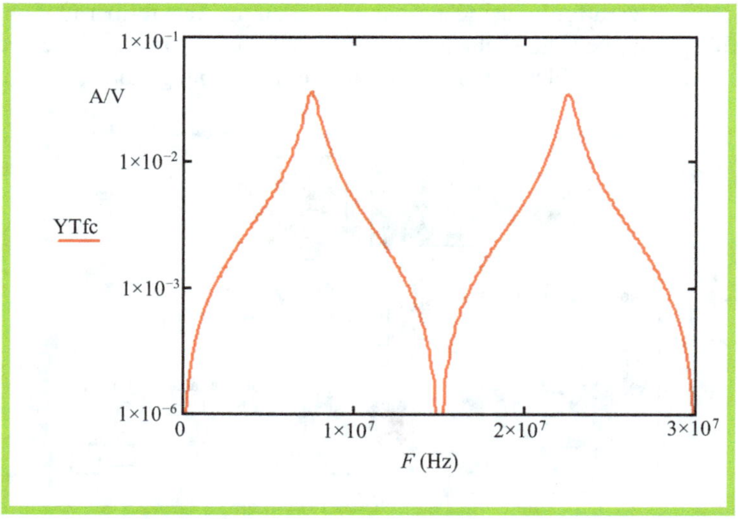

Figure 4.20 Transfer admittance of floating configuration

An analysis of the frequency response of either model can be carried out by utilising the program set out in Figures 4.10 and 4.11. It is only necessary to redefine the component values, the input voltage, the output current, and the frequency range.

Analysing the response of Figure 4.17 gives the characteristic shown in Figure 4.20. Analysing the response of Figure 4.19 gives exactly the same characteristic. This illustrates the fact that the transfer admittance for conducted susceptibility is exactly the same as for conducted emission. With the circuit models, the responses are identical. This confirms the conclusion of section 2.9.

Actual tests would show some differences, due to the effect of the test equipment and due to the radiated emission.

The response curve also illustrates that although the floating configuration gives excellent performance at low frequencies, it actually amplifies the level of interference at the quarter-wave frequency.

If the configuration-under-review is changed, to add a link between the ground plane and the return conductor at the far end, then the circuit model for the susceptibility test changes to that shown in Figure 4.21.

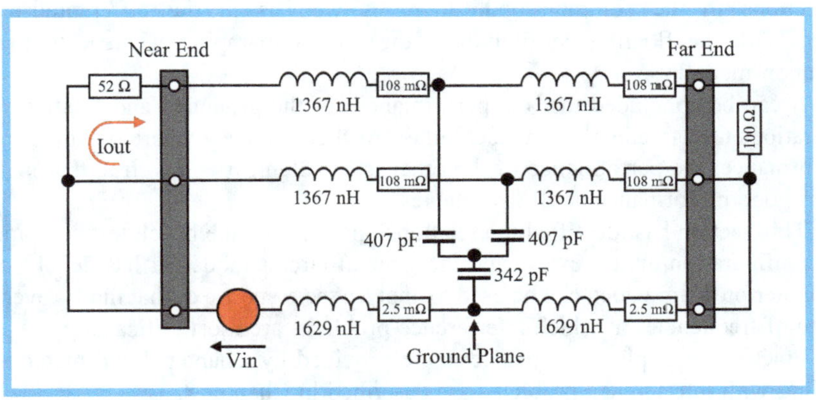

Figure 4.21 Circuit model for grounded configuration

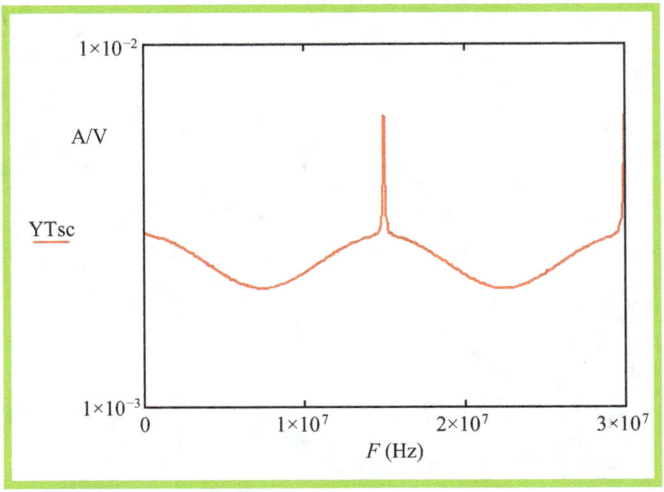

Figure 4.22 Transfer admittance of the grounded configuration

The response of this setup is given by Figure 4.22. This illustrates the fact that, for the grounded configuration, minimum interference occurs at the quarter-wave frequency, whilst the peak level occurs at the half-wave frequency.

This response can be compared with that of Figure 4.20, where peak interference with the floating configuration is experienced at quarter-wave frequency, whilst the minimum level occurs at the half-wave frequency.

There is another significant difference in the responses. The difference between the maxima and minima of the response curve of the grounded configuration is much less.

This is because the 100 Ω load resistance is damping the amplitude of the reflections in the common-mode loop, as well as in the differential-mode loop. With the floating configuration, there is no damping whatsoever in the common-mode loop.

It can be concluded that the performances of the grounded and floating configurations tend to complement each other. At the frequency where one gives poor performance, the performance of the other is excellent. It is also true that neither gives good performance at all frequencies.

This section has identified the relationships between the bench test equipment, the configuration-under-review and the general circuit model. It has developed a computer program which can be used to analyse the response of that model over the range of frequencies at which interference problems are most critical.

Practical examples of actual tests are provided by Chapter 7 and methods of dealing with interference problems are described in Chapter 8.

Chapter 5
Antenna models

EMI extends well beyond the confines of the assembly-under-review. Analysis of radiated emission and radiated susceptibility calls for a review of the properties of antennae. Fortunately, this review allows many simplifying assumptions to be invoked.

Section 5.1 provides a brief overview of the textbook analysis of the half-wave dipole and identifies a very significant parameter, when connected to the output of a radio transmitter which is generating a signal at the half-wave frequency, the dipole exhibits the same characteristics as a resistive load. This has been named the 'radiation resistance' and its value has been calculated to be 73 Ω.

Combining the radiation resistance with the primitive capacitance and primitive inductance of each monopole allows a circuit model to be created: a model which simulates the coupling between the dipole and the environment.

Developing the model to simulate antenna-mode coupling with a twin-conductor cable involves the same process that was used to derive the three-conductor model in section 2.7. This represents each conductor as a T-network and the environment as another T-network.

Since the environment must now be assigned values for inductance, capacitance, and resistance, it is fair to say that is behaves like a virtual conductor. Section 5.2 derives formulae for all the components of the circuit model.

Antenna-mode current is induced in any cable which is exposed to external radiation. Simulating the effect of this radiation calls for the introduction of a voltage source in series with the virtual conductor. Section 5.3 relates this 'threat voltage' to the electric field strength of the radiation.

If the power density of the external radiation is constant, but the frequency varies, then the graph of the amplitude of the threat voltage will exhibit a series of peaks and troughs. As far as the analysis of interference coupling is concerned, 'worst-case' conditions can be defined as the envelope curve which touches the peaks.

Section 5.4 provides an illustration of the differential-mode current which would be induced in a twin-conductor cable when the frequency of the external radiation is varied over a wide range.

If the structure is treated as a perfectly reflecting ground plane, then this will represent worst-case conditions. Since a circuit model can be created for this configuration, then such a model can be used to simulate the structure. Section 5.5

provides an example which assigns values to the components of this model. The effect of the external field can be simulated by connecting a voltage source and a 50 Ω resistor in series with the structure. Unlike a dipole receiving antenna which directs input power to a 75 Ω resistor in the receiver, the incoming energy is stored in the assembly. Since more energy is delivered, the effective value of the radiation resistance decreases. The test described in section 7.5 measures this value to be 74.49 Ω. This value is a good round figure and can be modified if experience shows it to be too pessimistic.

For radiation susceptibility assessment, it is possible to relate the power output of the test transmitter to the threat voltage applied to the equipment-under-test. Section 5.6 provides the relevant formulae.

Since mathematical software is not restricted to the analysis of circuit models, it can be used to compute the response of the equipment to the electromagnetic environment.

It is also possible to relate the antenna-mode current created by the equipment-under-test to the power received by a test receiver during a radiated emission test. Relevant formulae are provided in Section 5.7.

Since the approach used in this chapter is always to assume worst-case conditions, the simulated results are likely to err on the side of caution. The most significant benefit derived from the use of worst-case conditions is that the relevant formulae are dramatically simpler than those necessary to derive field distribution patterns.

5.1 The half-wave dipole

There is a trove of theory involved in the design of antennae, and there are many textbooks which deal with the subject. Repeating the derivations of the various formulae would be unnecessarily tedious. However, it is useful to summarise the methods used in arriving at the more important equations. Consideration is limited to the half-wave dipole [18].

5.1.1 Radiated power

Figure 5.1 illustrates a configuration where a half-wave dipole is driven from a radio-frequency transmitter. This setup will result in an oscillating current in the vertically oriented conductors. Current amplitude will be high at the junction between the dipole and the transmission line, and will be zero at the tips of the antenna. The relationship between current and time is defined as:

$$I = Io \cdot \cos(\omega \cdot t) \quad \text{(A)} \tag{5.1}$$

Current in a small element dz will create a magnetic potential A at the point P. This magnetic potential is a vector, and its direction is parallel to the axis of the conductor. A relationship can be established between A and the current in the element dz.

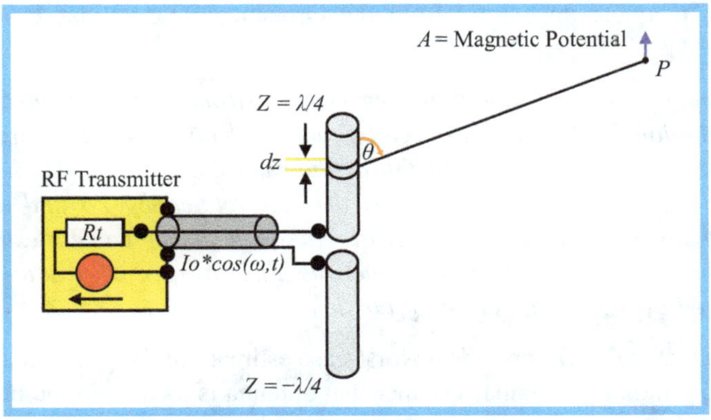

Figure 5.1 Analysis of half-wave dipole

Spherical co-ordinates, centred at the mid-point of the antenna, are then used to analyse the vectors involved in propagating the field. Figure 5.3 illustrates this. Since the vector A is aligned in the same direction as the current, there is no longitudinal component.

That is, $A_\phi = 0$. Equations for the radial and latitudinal components A_r and A_θ are derived. It is then possible to define formulae for the magnetic field vector H_ϕ. H_ϕ is a longitudinal vector and in Figure 5.1 it would be directed into the page. Then, using one of Maxwell's equations, the components of the E-field are calculated.

Since the only components of the total field which are capable of propagating power are those which are at right angles to each other, the analysis focuses on the latitudinal electric field E_θ and the longitudinal magnetic field H_ϕ. Multiplying the amplitudes of these components together gives an expression for the power density vector S_r. Integrating the value of this vector over the surface of the sphere gives a value for the total radiated power Pt:

$$Pt = \frac{1}{2} \cdot Rrad \cdot Io^2 \quad \text{(W)} \tag{5.2}$$

The parameter *Rrad* is named the radiation resistance. It is not a resistor in the conventional sense of the word, but a mathematical constant which happens to have the dimensions of resistance. Its value can be calculated, and for the half-wave dipole:

$$Rrad = 73\ \Omega \tag{5.3}$$

This value for *Rrad* is derived by assuming that the medium is lossless, meaning that the total radiated power does not decrease with distance from the antenna. Nor can it increase with distance. This being so, (5.2), with a value of 73 Ω assigned to *Rrad*, defines the maximum power which can be delivered to the environment by a dipole antenna.

It is useful to quote an extract from the Ordnance Board Pillar Proceeding at this point:

> *'At distances closer to about two wavelengths from a single linear aerial, the radiation field is accompanied by further electric and magnetic components the strengths of which fall as the square or cube of the distance from the aerial. These components are not related by Zo and do not radiate power away from the aerial. Although power can be extracted from them by very close receiving aerials, this power is always less than predicted by the extrapolation of far field theory'* [19].

This means that (5.2) will provide a worst-case estimate of the power that would be delivered to a monitor antenna, whether that antenna is located in the near-field or the far-field. For the purposes of EMC analysis, it is desirable to have available a worst-case estimate.

This leads to the circuit model of Figure 5.2, where the antenna acts as a resistive load to the co-axial transmission line. It should be emphasised that this relationship applies at only one frequency: one which is slightly higher than that of half-wave resonance. At all other non-resonant frequencies, the antenna impedance is a resistance in series with a reactance.

Equation (5.2) defines the average power which can be transmitted by the antenna when the frequency is such that this power is at a maximum value. Under all other conditions, the transmitted power is less than this maximum.

If *Irms* is the root-mean-square value of the current, then:

$$Irms = \sqrt{2} \cdot Io \quad \text{(A)} \tag{5.4}$$

And

$$Pt = Rrad \cdot Irms^2 \quad \text{(W)} \tag{5.5}$$

Since optimum transmission of power is obtained when the output impedance of the transmitter is the same as that of the load, the optimum value for *Rt* and the characteristic impedance of the cable is 73 Ω: making 75 Ω co-axial cable a good choice for signal links between radio-frequency equipment and antennae.

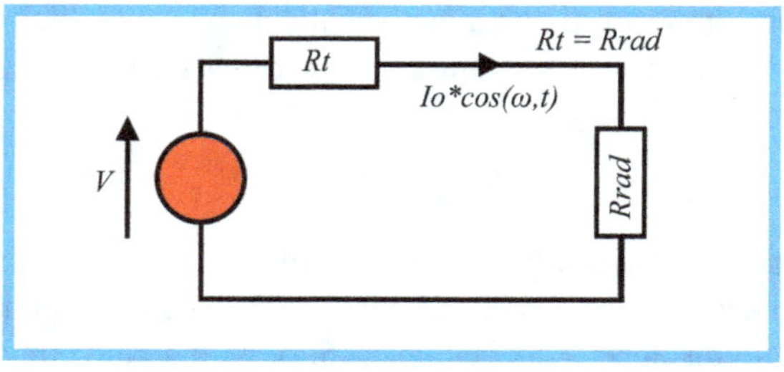

Figure 5.2 Equivalent circuit of a half-wave dipole at resonance

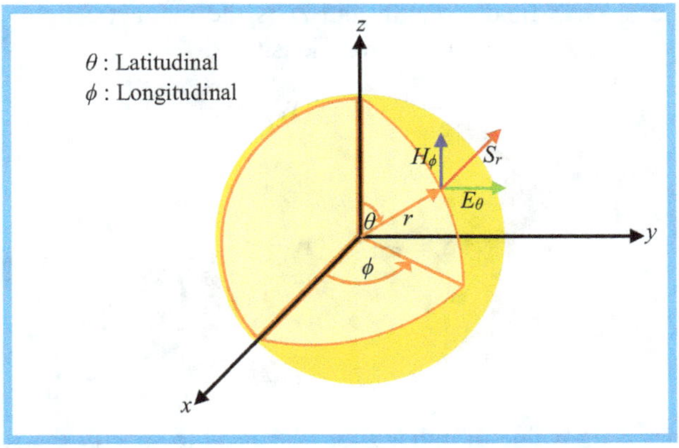

Figure 5.3 Power density on a spherical surface

5.1.2 Power density

At the surface of the sphere illustrated in Figure 5.3, the relationship between power density, *S*, and the transmitted power, *Pt*, is:

$$S = \frac{Gt \cdot Pt}{4 \cdot \pi \cdot r^2} \quad (\mathrm{W/m^2}) \tag{5.6}$$

The term *Gt* takes into account the fact that power does not radiate uniformly in all directions. It is defined as,

$$Gt = \frac{\dfrac{\text{maximum power radiated}}{\text{unit solid angle}}}{\dfrac{\text{total power delivered to antenna}}{4 \cdot \pi}} \tag{5.7}$$

It is described as the antenna gain and is a measure of the concentration of the power density of the radiated wavefront in a given direction. For a half-wave dipole,

$$Gt = 1.64 \tag{5.8}$$

5.1.3 Field strength

Field vectors *H* and *E* are tangential to the surface of the sphere, and at right angle to each other. The power density vector *S* is radial, and the flow of power is outward.

The characteristic impedance of free space is:

$$Z_o = \frac{E}{H} = \sqrt{\frac{\mu_o}{\varepsilon_o}} = 377\ \Omega \tag{5.9}$$

where E is the electric field strength and H is the magnetic field strength. The relationship between E, H, and the power density is:

$$S = E \cdot H \quad (\mathrm{W/m^2}) \tag{5.10}$$

Hence

$$S = \frac{E^2}{Z_o} \quad (\mathrm{W/m^2}) \tag{5.11}$$

And

$$S = Z_o \cdot H^2 \quad (\mathrm{W/m^2}) \tag{5.12}$$

This means that if the value of either E or H is used to specify the strength of the electromagnetic field in the far field, it is always possible to determine the value of the power density.

5.1.4 Power received

The dipole can also act as a receiving antenna. For a configuration such as Figure 5.4, the power received by the RF receiver is related to the power density Sr of the incoming wavefront by:

$$Pr = Aeff \cdot Sr \quad (\mathrm{W}) \tag{5.13}$$

Where $Aeff$ is the effective receiving area of the dipole.

A general relationship can be established between the effective receiving area, the gain Gr of a dipole antenna, and the wavelength λ.

$$Aeff = \frac{Gr \cdot \lambda^2}{4 \cdot \pi} \quad (\mathrm{m^2}) \tag{5.14}$$

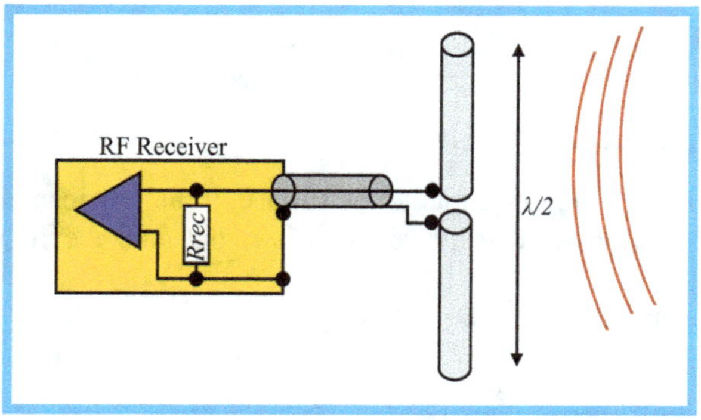

Figure 5.4 Receiver assembly

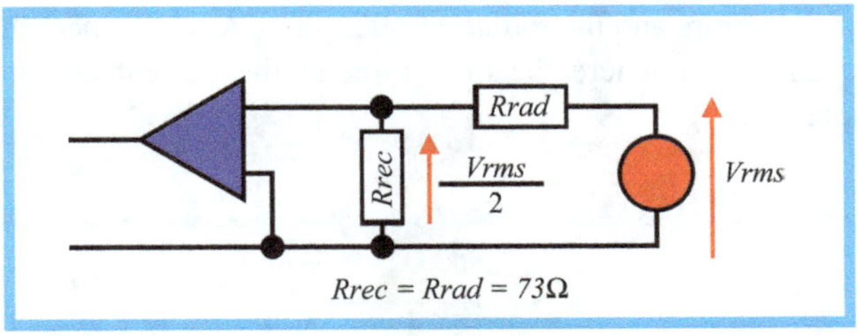

Figure 5.5 Simple circuit model of dipole receiver assembly

As with the transmitting dipole, *Gr* is equal to 1.64. Under optimum conditions, the power delivered to the receiver terminals is:

$$Pr = \frac{Gr \cdot \lambda^2 \cdot Sr}{4 \cdot \pi} \quad \text{(W)} \tag{5.15}$$

For maximum sensitivity of the receiver, it is necessary for the input resistor to match the output resistance of the antenna. This being so, the voltage along the antenna can be deduced from Figure 5.5.

Since power at the receiver input is the product of current in the resistor *Rrec* and the voltage across that resistor, then

$$Pr = \frac{1}{Rrec} \cdot \left(\frac{Vrms}{2}\right)^2 \quad \text{(W)}$$

Replacing *Rrec* with *Rrad* and re-arranging:

$$Vrms = \sqrt{4 \cdot Rrad \cdot Pr} \quad \text{(V)} \tag{5.16}$$

For a half-wave dipole receiving antenna, the radiation resistance is 73 Ω, exactly the same as for a transmitting antenna.

5.2 The virtual conductor

In the previous section, the analysis focused on the behaviour of conductors when they are designed to act as antennae. The scope of the analysis was limited to a restricted bandwidth where the frequency was at or near resonance. Exactly, the same mechanisms apply to conductors when antenna action is unintentional. However, the scope of the analysis needs to expand to cover situations at all frequencies.

At non-resonant frequencies, the reactive components of the conductors play a predominant role. Including these parameters in the model involves the use of primitive inductance and primitive capacitance.

Figure 5.6 shows how primitives can be used to create a model of the transmitting dipole. The two monopoles are effectively connected by two capacitors in series with two inductors and the radiation resistance. At resonance, the circuit acts as a series tuned circuit where the only limit to the current is provided by the resistive components.

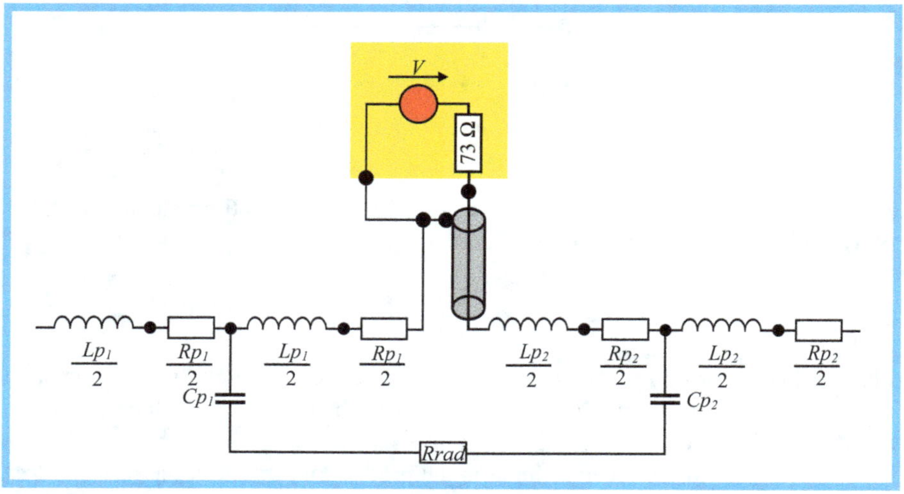

Figure 5.6 Circuit model of a half-wave dipole

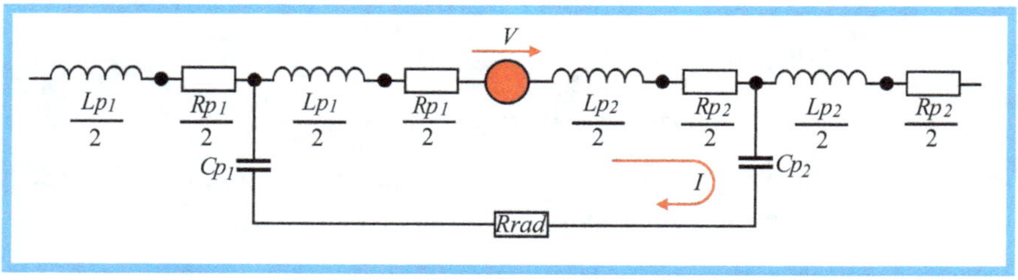

Figure 5.7 General circuit model of an isolated conductor

Figure 5.7 illustrates the situation where the co-axial link to the transmitter has been removed and the two halves of the dipole are shorted together to form a single conductor. Signals can be introduced into this conductor by threading it through a ferrite core which acts as a step-down voltage transformer. In the figure, this is illustrated as a voltage source *V*.

With such a configuration, there will be a flow of current forward and backwards along the conductor. Any current flowing into the right-hand monopole can only come from the left-hand monopole and vice versa. Since the currents and voltages in the two halves of the dipole are balanced, it is useful to redraw the model, giving the circuit of Figure 5.8.

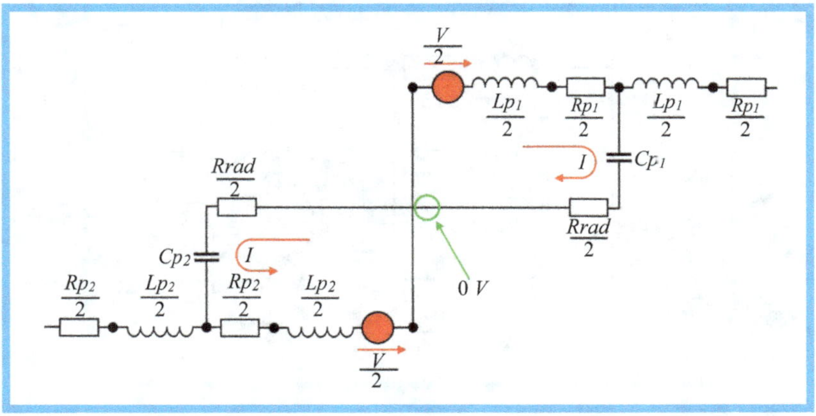

Figure 5.8 Balance between the two halves of dipole

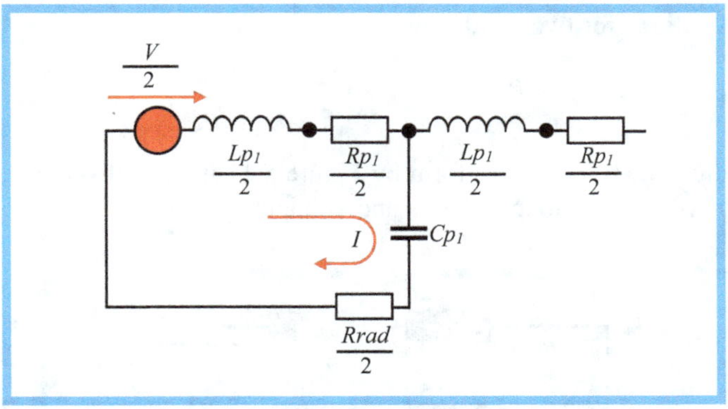

Figure 5.9 Focus on right-hand monopole

This is effectively a bridge circuit with voltages in one arm balancing voltages in the other arm. If the voltage at the centre of the conductor is defined as zero volts, then the voltage at the mid-point of *Rrad* must also be zero.

Hence, it is legitimate to join these two points together in the model. When this is done and attention is focused on the right-hand monopole, the model simplifies to that shown by Figure 5.9.

Now, if the single conductor were to be replaced by a pair of parallel conductors connected together at the mid-point and a voltage source inserted in series with both conductors, then the circuit model of the right-hand monopole would be as shown in Figure 5.10. This correlates closely with Figure 2.19, with the characteristics of the environment replacing the third conductor. The environment is acting as a virtual conductor.

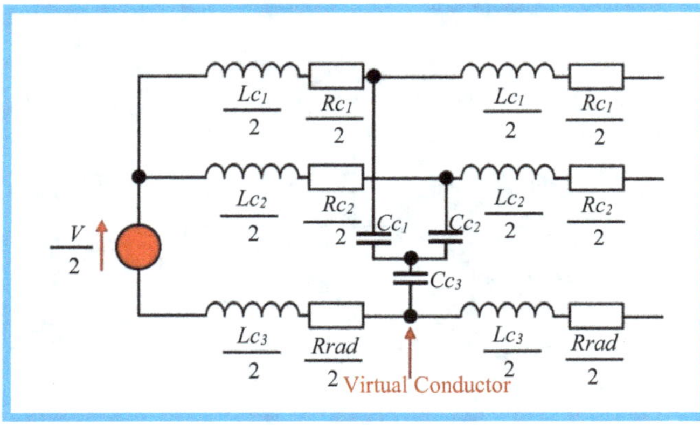

Figure 5.10 Circuit model of right-hand section of conductor pair

Deriving formulae for the components of this conductor is simply a matter of repeating the procedure introduced in Chapter 2. The starting point is to set out the primitive equations for two conductors:

$$
\begin{aligned}
Vp_1 &= Zp_{1,1} \cdot Ip_1 + Zp_{1,2} \cdot Ip_2 \\
Vp_2 &= Zp_{2,1} \cdot Ip_1 + Zp_{2,2} \cdot Ip_2
\end{aligned}
\tag{5.17}
$$

These are illustrated in diagram form by Figure 5.11, which also relates primitive currents and voltages to loop currents and voltages.

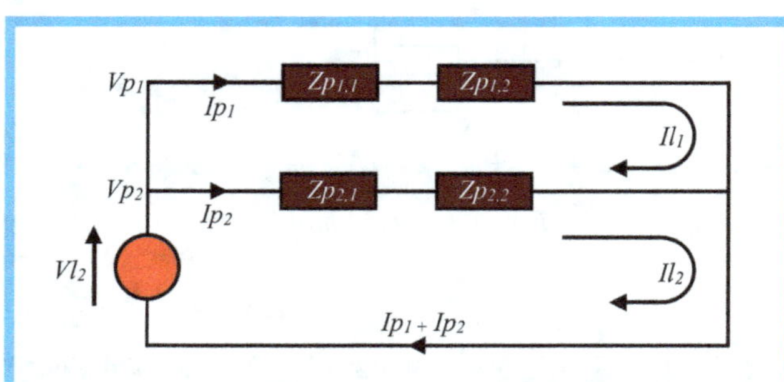

Figure 5.11 Diagrammatic representation of primitive equations

From the diagram:

$$
\begin{aligned}
Ip_1 &= Il_1 \\
Ip_2 &= Il_2 - Il_1
\end{aligned}
\tag{5.18}
$$

And

$$
\begin{aligned}
Vl_1 &= Vp_1 - Vp_2 \\
Vl_2 &= Vp_2
\end{aligned}
\tag{5.19}
$$

Using (5.18) to substitute for Ip_1 and Ip_2 in (5.17):

$$\begin{aligned} Vp_1 &= Zp_{1,1} \cdot Il_1 + Zp_{1,2} \cdot (Il_2 - Il_1) \\ Vp_2 &= Zp_{2,1} \cdot Il_1 + Zp_{2,2} \cdot (Il_2 - Il_1) \end{aligned}$$

Re-arranging:

$$\begin{aligned} Vp_1 &= \left(Zp_{1,1} - Zp_{1,2}\right) \cdot Il_1 + Zp_{1,2} \cdot Il_2 \\ Vp_2 &= \left(Zp_{2,1} - Zp_{2,2}\right) \cdot Il_1 + Zp_{2,2} \cdot Il_2 \end{aligned}$$

Using (5.19) to determine loop voltages:

$$\begin{aligned} Vl_1 &= \left(Zp_{1,1} - Zp_{1,2} - Zp_{2,1} + Zp_{2,2}\right) \cdot Il_1 + \left(Zp_{1,2} - Zp_{2,2}\right) \cdot Il_2 \\ Vl_2 &= \left(Zp_{2,1} - Zp_{2,2}\right) \cdot Il_1 + Zp_{2,2} \cdot Il_2 \end{aligned}$$

Defining the loop impedances as:

$$\begin{aligned} Zl_{1,1} &= Zp_{1,1} - Zp_{1,2} - Zp_{2,1} + Zp_{2,2} \\ Zl_{1,2} &= Zp_{1,2} - Zp_{2,2} \\ Zl_{2,1} &= Zp_{2,1} - Zp_{2,2} \\ Zl_{2,2} &= Zp_{2,2} \end{aligned} \tag{5.20}$$

Leads to the loop equations:

$$\begin{aligned} Vl_1 &= Zl_{1,1} \cdot Il_1 + Zl_{1,2} \cdot Il_2 \\ Vl_2 &= Zl_{2,1} \cdot Il_1 + Zl_{2,2} \cdot Il_2 \end{aligned} \tag{5.21}$$

If the cross-section of the conductor pair is symmetrical:

$$Zp_{1,2} = Zp_{2,1}$$

From (5.20):

$$Zl_{1,2} = Zl_{2,1} \tag{5.22}$$

Figure 5.12 shows the circuit model which creates a similar pair of equations:

$$\begin{aligned} Vc_1 &= (Zc_1 + Zc_2) \cdot Ic_1 - Zc_{1,2} \cdot Ic_2 \\ Vc_2 &= -Zc_{1,2} \cdot Ic_1 + (Zc_2 + Zc_3) \cdot Ic_2 \end{aligned} \tag{5.23}$$

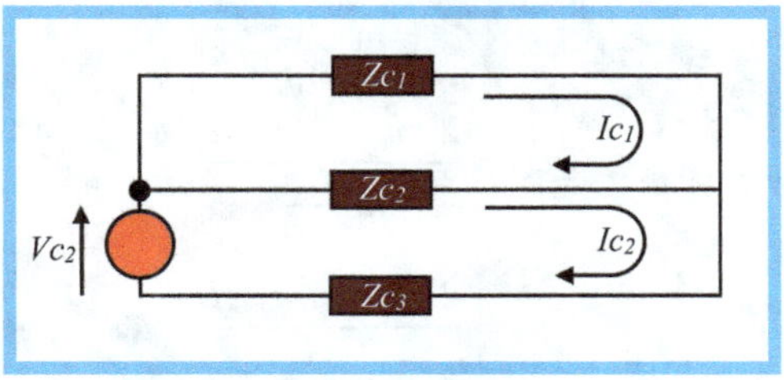

Figure 5.12 Circuit model of coupling parameters

Correlating (5.21) and (5.23) gives:

$$\begin{aligned} Zl_{1,1} &= Zc_1 + Zc_2 \\ Zl_{1,2} &= -Zc_2 \\ Zl_{2,2} &= Zc_2 + Zc_3 \end{aligned}$$

Defining circuit impedances in terms of loop impedances:

$$\begin{aligned} Zc_1 &= Zl_{1,1} + Zl_{1,2} \\ Zc_2 &= -Zl_{1,2} \\ Zc_3 &= Zl_{2,2} + Zl_{1,2} \end{aligned}$$

Using (5.20) to substitute primitive impedances for loop impedances:

$$\begin{aligned} Zc_1 &= \left(Zp_{1,1} - 2 \cdot Zp_{1,2} + Zp_{2,2}\right) + \left(Zp_{1,2} - Zp_{2,2}\right) \\ Zc_2 &= Zp_{2,2} - Zp_{1,2} \\ Zc_3 &= Zp_{2,2} + \left(Zp_{1,2} - Zp_{2,2}\right) \end{aligned}$$

This leads to:

$$\begin{aligned} Zc_1 &= Zp_{1,1} - Zp_{1,2} \\ Zc_2 &= Zp_{2,2} - Zp_{1,2} \\ Zc_3 &= Zp_{1,2} \end{aligned} \tag{5.24}$$

Letting $Zp_{i,j} = j \cdot \omega \cdot Lp_{i,j}$ and invoking (2.27) gives:

$$\begin{aligned} Lc_1 &= \frac{\mu_o \cdot \mu_r \cdot l}{2 \cdot \pi} \cdot \ln \frac{r_{1,2}}{r_{1,1}} \\ Lc_2 &= \frac{\mu_o \cdot \mu_r \cdot l}{2 \cdot \pi} \cdot \ln \frac{r_{1,2}}{r_{2,2}} \\ Lc_3 &= \frac{\mu_o \cdot \mu_r \cdot l}{2 \cdot \pi} \cdot \ln \frac{l}{r_{1,2}} \end{aligned} \tag{5.25}$$

Comparing this set of equations with that of (2.51) reveals that the inductance value associated with each conductor is the same, whether the conductor pair is viewed as a transmission line or as an antenna. The new parameter introduced by (5.25) is the circuit inductance of the virtual conductor, Lc_3. This is the primitive inductance of a conductor with the same radius as the separation between the conductors.

Letting $Zp_{i,j} = \dfrac{1}{j \cdot \omega \cdot Cp_{i,j}}$ and invoking (2.26) gives:

$$\begin{aligned} Cc_1 &= \frac{2 \cdot \pi \cdot \varepsilon_o \cdot \varepsilon_r \cdot l}{\ln \dfrac{r_{1,2}}{r_{1,1}}} \\ Cc_2 &= \frac{2 \cdot \pi \cdot \varepsilon_o \cdot \varepsilon_r \cdot l}{\ln \dfrac{r_{1,2}}{r_{2,2}}} \\ Cc_3 &= \frac{2 \cdot \pi \cdot \varepsilon_o \cdot \varepsilon_r \cdot l}{\ln \dfrac{l}{r_{1,2}}} \end{aligned} \tag{5.26}$$

Comparing this set of equations with that of (2.55) reveals that the capacitance value associated with each conductor is the same, whether the conductor pair is viewed as a transmission line or as an antenna. The new parameter introduced by (5.26) is the circuit capacitance of the virtual conductor, Cc_3.

As would be expected, there is a duality between the primitive capacitance and the primitive inductance of the cable assembly.

A necessary condition for the derivation of a value for the radiation resistance *Rrad* is that the length of the conductor is much greater than its radius.

Since the separation between conductors of a transmission line is much less than the length, the same condition is met for a transmission line as for an antenna. At this point, it is reasonable to assume that the radiation resistance of the cable is 73 Ω, the same as defined by (5.3).

Having derived formulae for all the parameters associated with the conductor pair, it becomes possible to construct a general circuit model which simulates the coupling between an isolated length of twin-conductor cable and the environment (Figure 5.13).

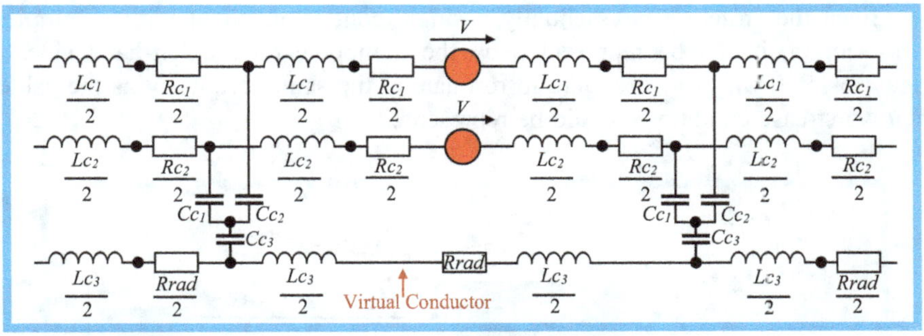

Figure 5.13 General circuit model of an isolated cable

Following on from the creation of this model, it is possible to define the virtual conductor as an imaginary conductor which enables the coupling between the cable and the environment to be simulated. It behaves as a return conductor for antenna-mode current.

It has the same properties as an actual conductor: capacitance, inductance and resistance. Numerical values for the reactive parameters can be derived from data on the physical construction of the cable. An initial assumption can be made that the resistance is 73 Ω, the same as that of the half-wave dipole. Tests, such as those described in Chapter 7, can be devised to refine the values of the three parameters.

As far as differential-mode signal transmission is concerned, the values of the reactive components of the circuit model remain exactly the same as those of a pair of conductors. As far as antenna-mode coupling is concerned, the cable can be treated as a single conductor with the same radius as the separation between the conductor pair.

In Figure 5.13, it is assumed that two identical voltage sources are used to inject a signal into the cable. Such a configuration could be implemented by clamping a toroidal transformer around the cable at its mid-point.

5.3 The threat voltage

The most severe threat to any electronic system will occur in those situations when it is completely exposed to its environment.

In the setup illustrated by Figure 5.14, the only connection between the remote transducer and the signal processing unit is a twin-conductor cable. One example could be the link between a proximity sensor and the input circuitry of the signal processing unit. One conductor would carry the signal current whilst the other carries the return current. Normal wiring practice is to connect the return conductor to the zero-volt reference in the processing unit, and for this reference conductor to be connected to the structure.

The conductors of the cable can be represented by the right-hand section of Figure 5.13.

Since the structure is essentially a single conductor, albeit with a complex cross-section, it can be represented by the monopole on the left-hand side of Figure 5.15. If the structure were to resonate at the same frequency as the cable, then worst-case conditions would be represented.

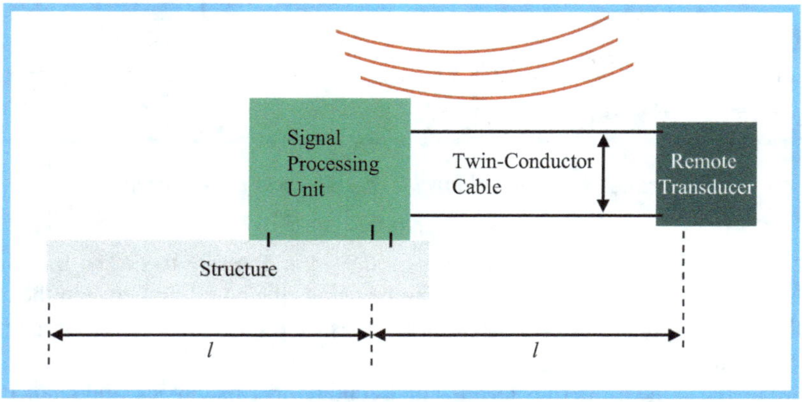

Figure 5.14 Cable exposed to external radiation

This model differs from that shown by Figure 5.13 in that the voltage sources in the conductors have been replaced by a single source located in series with the virtual conductor.

The source in Figure 5.15 now represents the effect of the external electromagnetic field. The triple-T network representing the left half of the twin-conductor cable has been replaced by a single-T network representing the structure.

Such a representation allows the circuit to be analysed from the point of view of radiated susceptibility. But before such an analysis can proceed, it is necessary to relate the amplitude of the voltage source to the strength of the field in which the system is immersed.

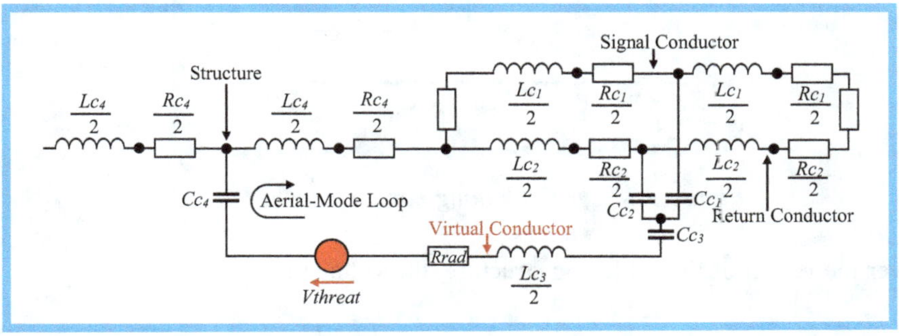

Figure 5.15 General circuit model of exposed cable and structure

Figure 5.16 depicts a length of conductor which is subjected to an external field and the associated circuit model. An incremental voltage dV is induced in series with each element dz of the conductor.

$$dV = E \cdot dz \tag{5.27}$$

Where E is the strength of the electric field at that location.

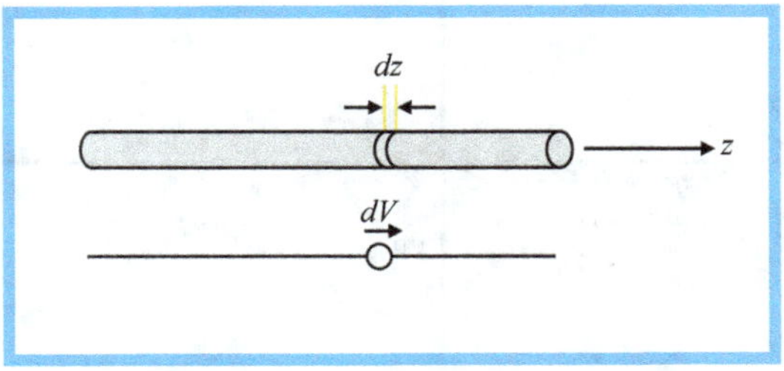

Figure 5.16 Effect of electric field

The velocity of propagation is assumed to be constant. So, if the waveform of the electric field is sinusoidal with respect to time, then the distribution in space will also be sinusoidal. Figure 5.17 illustrates the relationship between the strength of the E-field and distance along the conductor.

The relationship can be defined as:

$$E = Emax \cdot \cos\left(\frac{2 \cdot \pi}{\lambda} \cdot z\right) \tag{5.28}$$

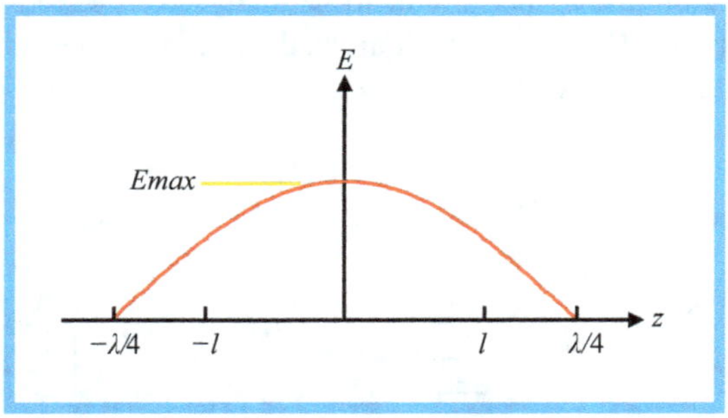

Figure 5.17 E-field along conductor when $l < \lambda/4$

Over the length of the cable and structure, the voltage is:

$$Vthreat = \int_{-l}^{l} Emax \cdot \cos\left(\frac{2 \cdot \pi}{\lambda} \cdot z\right) \cdot dz$$

This leads to the relationship between threat voltage and electric field strength:

$$Vthreat = \frac{\lambda}{\pi} \cdot Emax \cdot \sin\left(\frac{2 \cdot \pi}{\lambda} \cdot l\right) \tag{5.29}$$

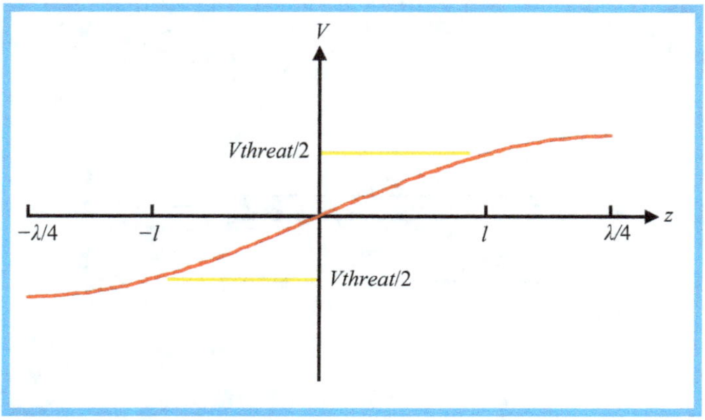

Figure 5.18 Variation of threat voltage with length of cable

Figure 5.18 illustrates the relationship when the length l is less than a quarter wavelength.

From (5.11):

$$Emax = \sqrt{S \cdot Zo} \quad \text{where } Zo = 377\ \Omega \tag{5.30}$$

The relationship between wavelength and frequency is as defined in (2.31)

$$v = \lambda \cdot f$$

If the velocity of propagation is assumed to be the speed of light in vacuum, then:

$$\lambda = \frac{c}{f} \tag{5.31}$$

Worksheet 5.3 page 1

$c := 300 \cdot 10^6\ \text{m/s}$ Zo := 377 Ω $S := 1\ \text{W/m}^2$ $l := 15\ \text{m}$

$i := 1..100$ $f_i := i \cdot 10^6\ \text{Hz}$

$$\text{Vthreat}_i := \left| \begin{array}{l} \lambda \leftarrow \dfrac{c}{f_i} \\ \text{Va} \leftarrow \sqrt{S \cdot \text{Zo}} \cdot \dfrac{\lambda}{\pi} \\ \text{Vb} \leftarrow \text{Va} \sin\left(2 \cdot \pi \cdot \dfrac{l}{\lambda}\right) \\ \text{Vc} \leftarrow |\text{Vb}| \\ \text{Vc} \leftarrow 10.1 \text{ if } \text{Vc} < 10 \end{array} \right.$$

$$\text{Venv}_i := \left| \begin{array}{l} \lambda \leftarrow \dfrac{c}{f_i} \\ \text{Va} \leftarrow \sqrt{S \cdot \text{Zo}} \cdot \dfrac{\lambda}{\pi} \\ \text{Vb} \leftarrow \text{Va} \sin\left(2 \cdot \pi \cdot \dfrac{l}{\lambda}\right) \\ \text{Vb} \leftarrow \text{Va} \text{ if } l > \dfrac{\lambda}{4} \end{array} \right.$$

Figure 5.19 Calculating the relationship between threat voltage and frequency

It is clear from (5.31) and (5.29) that the threat voltage is a function of both frequency f and length l. Equation (5.30) demonstrates that the electric field strength is proportional to the square root of the power density. Combining these equations in the way illustrated by the subroutine $Vthreat_i$ in Figure 5.19 allows the relationship between power density and threat voltage to be calculated over a wide range of frequencies. This response is illustrated by the solid curve of Figure 5.20. It is assumed that the power density S is 1 W/m^2 and that the length l is 15 m.

This curve is a series of peaks, each peak occurring at a resonant frequency. The first resonance occurs when the length l is equal to a quarter wavelength. At this frequency and at higher resonant frequencies

$$Vthreat = \frac{\lambda}{\pi} \cdot Emax \tag{5.32}$$

Joining these points together creates an envelope curve, *Venv*. Since this envelope is of threat voltages which are equal to, or higher than, all voltage points on the *Vthreat* curve, it represents worst-case conditions. This curve is defined by that of the $Venv_i$ function of Figure 5.20 and is illustrated by the dashed curve.

When the frequency is lower than that of quarter-wave resonance, both curves are co-incident.

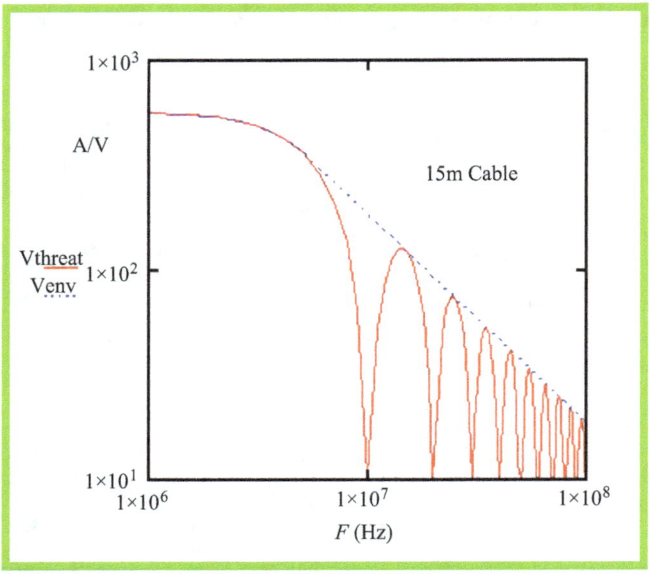

Figure 5.20 Relationship between threat voltage and frequency

Assigning this envelope voltage to the source of the circuit model of Figure 5.15 will allow the response of the assembly-under-review to be simulated.

5.4 The threat current

In the previous section, the simple assembly of Figure 5.14 was used to depict a situation in which an exposed cable can be subjected to an external electromagnetic field. A circuit model was created to simulate the configuration, and the source voltage was related to the power density of the interference. This section develops the analysis to determine the frequency response of the threat current in the signal line.

Under resonance conditions, the reactive components representing the structure act as a short-circuit and the only limit to the current is provided by the resistive components. From Figure 5.15, these are the radiation resistance and the resistance of the structure.

This means that worst-case conditions can be simulated by replacing the reactive components of the structure with a short circuit. Since the resistance of the structure is minimal compared to the radiation resistance, it is reasonable to represent this as a short-circuit as well.

A welcome consequence of representing the entire structure as a short-circuit is that the model is simplified, as Figure 5.21 illustrates.

Determining the response of this model involves the assignment of specific details to the cable and interface circuitry.

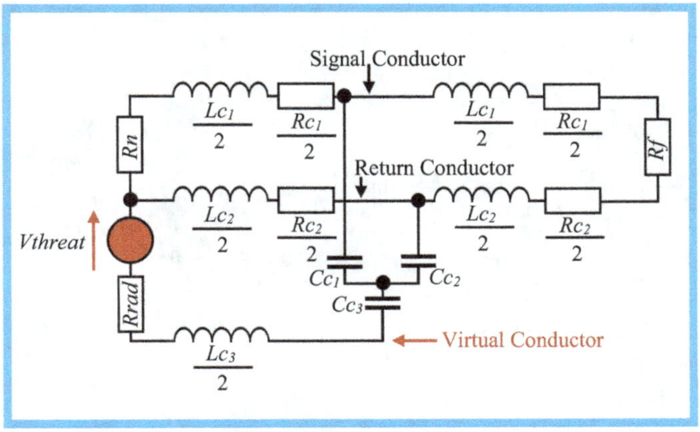

Figure 5.21 General circuit model of exposed cable – simplified

If it is assumed that the length of the cable is 15 m and that the two conductors are 0.8 mm in diameter with a spacing between centres of 1.2 mm, then all the circuit parameters can be calculated.

Figure 5.22 is a copy of the first page of the Mathcad worksheet used to analyse the response. This page determines all the component values. Inductors Lc_1, Lc_2, and Lc_3 are obtained from (5.25). The steady-state resistor values Rss_1 and Rss_2 of the two conductors are obtained from (2.57). Since the virtual conductor does not have the properties of a conventional conductor, Rss_3 is set to zero. The crossover frequency *Fx* is obtained from (2.60) and the capacitor values are derived from (2.33). Conductance values are set at zero, since it is assumed that there are no losses in the insulation.

Values for inductance, capacitance and resistance of the representative circuit model are then displayed on the sheet. To minimise reflections at the near and far ends of the line, the values of *Rn* and *Rf* are set at a value equal to the characteristic impedance. If the values of the terminating impedances are any other value, the peaks in the response curve would be higher. Terminating the line with perfectly matched impedances does not eliminate interference.

The final parameter to be calculated on this page is the value of the quarter-wave frequency *Fq*. This invokes the use of (2.34).

Assigning these values to the general circuit model of Figure 5.21 leads to the representative circuit model of Figure 5.23

Inspection of this model allows the equations of the *Zloop(f)* function to be defined, and this function is listed in Figure 5.25. It invokes the *Zbranch(f)* function, which is identical to that in Figure 4.10.

The *Vthreat(f)* function is essentially a copy of the envelope function *Venv* of Figure 5.19. It is assumed that the power density of the interference signal is 1 W/m^2.

Worksheet 5.4 page 1

$\mu_o := 4 \cdot \pi \cdot 10^{-7}$ H/m $\qquad \varepsilon_o := 8.854 \cdot 10^{-12}$ F/m $\qquad c := 2.998 \cdot 10^{8}$ m/s

$\rho := 1.7 \cdot 10^{-8}$ Ω/m $\qquad l := 15$ m

$r_{1,1} := 0.4 \cdot 10^{-3}$ m $\qquad r_{2,2} := r_{1,1} \qquad r_{1,2} := 1.2 \cdot 10^{-3}$ m

$$\mathrm{Lc}_1 := \frac{\mu_o \cdot l}{2 \cdot \pi} \cdot \ln\left(\frac{r_{1,2}}{r_{1,1}}\right) \qquad \mathrm{Lc}_2 := \frac{\mu_o \cdot l}{2 \cdot \pi} \cdot \ln\left(\frac{r_{1,2}}{r_{2,2}}\right) \qquad \mathrm{Lc}_3 := \frac{\mu_o \cdot l}{2 \cdot \pi} \cdot \ln\left(\frac{l}{r_{1,2}}\right)$$

$$\mathrm{Rss}_1 := \frac{\rho \cdot l}{\pi \cdot (r1,1)^2} \qquad \mathrm{Rss}_2 := \mathrm{Rss}_1 \qquad \mathrm{Rss}_3 := 0$$

$$\mathrm{Fx} := \frac{4 \cdot \rho}{\mu_o \cdot \pi \cdot (r_{1,1})^2} = 1.077 \times 10^{5}\ \mathrm{Hz}$$

$$\mathrm{Cc} := \left(\frac{l}{c}\right)^2 \cdot \frac{1}{\mathrm{Lc}} \qquad \mathrm{Gc} := \begin{pmatrix} 0 \\ 0 \\ 0 \end{pmatrix}$$

$$\frac{\mathrm{Lc}}{2} = \begin{pmatrix} 1.648 \times 10^{-6} \\ 1.648 \times 10^{-6} \\ 1.415 \times 10^{-5} \end{pmatrix} \qquad \mathrm{Cc} = \begin{pmatrix} 7.595 \times 10^{-10} \\ 7.595 \times 10^{-10} \\ 8.846 \times 10^{-11} \end{pmatrix} \qquad \frac{\mathrm{Rss}}{2} = \begin{pmatrix} 0.254 \\ 0.254 \\ 0 \end{pmatrix} \qquad \mathrm{Rrad} := 73$$

$$\mathrm{Zo} := \sqrt{\frac{\mathrm{Lc}}{\mathrm{Cc}}} \qquad \mathrm{Zo} = \begin{pmatrix} 65.873 \\ 65.873 \\ 565.632 \end{pmatrix} \qquad \mathrm{Rn} := \mathrm{Zo}_1 + \mathrm{Zo}_2 \quad \mathrm{Rf} := \mathrm{Rn}$$

$$\mathrm{Rn} = 131.746\ \Omega$$

$$\mathrm{Fq} := \frac{1}{4 \cdot \sqrt{\mathrm{Lc}_1 \cdot \mathrm{Cc}_1}} = 4.997 \times 10^{6}\ \mathrm{Hz}$$

Figure 5.22 Calculating values for the circuit components

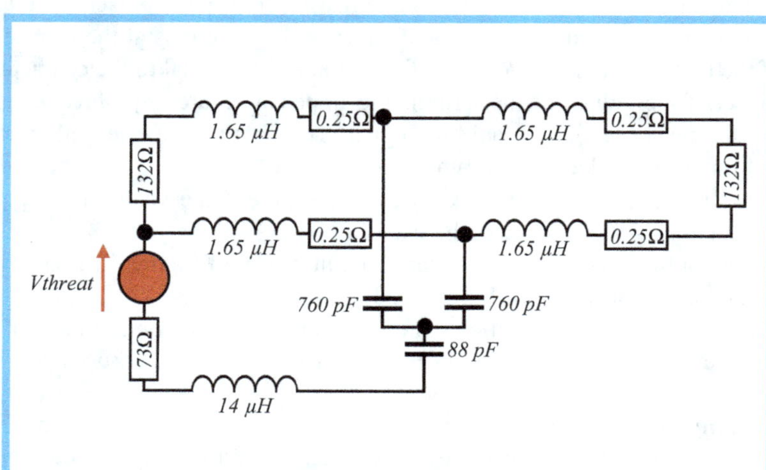

Figure 5.23 Representative circuit model of exposed cable (simplified)

These three functions are then used to calculate the frequency response of the threat current *Ithreat* in the signal loop. Figure 5.26 defines the main program and Figure 5.27 illustrates the response curve.

The curve itself is very similar to that of Figure 5.20, the main difference being that the response falls sharply as the frequency reduces below that of quarter-wave resonance.

This means that the configuration of Figure 5.14 is most susceptible to interference at about 5 MHz. At that frequency, an incoming wave with a power density of 1 W/m^2 will generate a threat current of about 1 A in the signal circuitry.

This current could trigger an electro-explosive device or could cause overheating in transducers or semiconductor devices, with consequential damage to the system.

The subsequent peaks at higher frequencies could also pose a threat; they could induce false signals into the system or prevent actual signals from reaching the intended input. System upset could occur. However, the designer is now in possession of information not previously available. Given a detailed understanding of both the effect of the interference on the signal-under-review and the bandwidth, amplitude, and function of that signal, it should be possible to design the system to be immune to radiation in any defined EMI environment.

Re-drawing Figure 5.21 in terms of distributed parameters results in Figure 5.24.

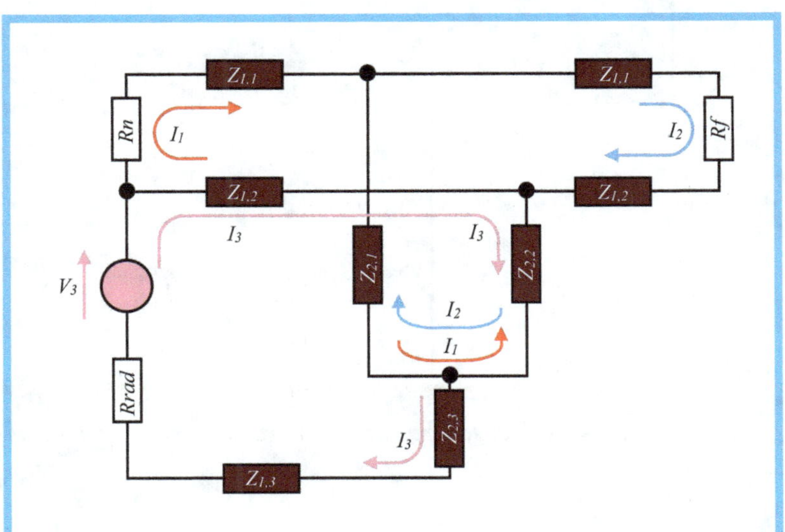

Figure 5.24 Distributed parameter model for Figure 5.23

Worksheet 5.4 page 2

$$\text{Zbranch}(f) := \begin{array}{|l} \omega \leftarrow 2\cdot\pi\cdot f \\ \text{for } i \in 1..3 \\ \quad \begin{array}{|l} \text{Rc}_i \leftarrow \text{Rss}_i\cdot\sqrt{1+\dfrac{f}{\text{Fx}}} \\ \theta \leftarrow \sqrt{(\text{Rc}_i + j\cdot\omega\cdot\text{Lc}_i)\cdot(\text{Gc}_i + j\cdot\omega\cdot\text{Cc}_i)} \\ \text{Zo} \leftarrow \sqrt{\dfrac{\text{Rc}_i + j\cdot\omega\cdot\text{Lc}_i}{\text{Gc}_i + j\cdot\omega\cdot\text{Cc}_i}} \\ Z_{1,i} \leftarrow \text{Zo}\cdot\tanh\left(\dfrac{\theta}{2}\right) \\ Z_{2,i} \leftarrow \text{Zo}\cdot\text{csch}\,(\theta) \end{array} \\ Z \end{array}$$

$$\text{Zloop}(f) := \begin{array}{|l} Z \leftarrow \text{Zbranch}(f) \\ \text{Z11} \leftarrow Z_{1,1} + Z_{2,1} + Z_{2,2} + Z_{1,2} + \text{Rn} \\ \text{Z12} \leftarrow -(Z_{2,2} + Z_{2,1}) \\ \text{Z13} \leftarrow -(Z_{1,2} + Z_{2,2}) \\ \text{Z22} \leftarrow Z_{1,2} + Z_{2,2} + Z_{2,1} + Z_{1,1} + \text{Rf} \\ \text{Z23} \leftarrow Z_{2,2} \\ \text{Z33} \leftarrow Z_{1,2} + Z_{2,2} + Z_{2,3} + Z_{1,3} + \text{Rrad} \\ \begin{pmatrix} \text{Z11} & \text{Z12} & \text{Z13} \\ \text{Z12} & \text{Z22} & \text{Z23} \\ \text{Z13} & \text{Z23} & \text{Z33} \end{pmatrix} \end{array}$$

$S := 1 \qquad \text{Zo} := 377$

$$\text{Vthreat}(f) := \begin{array}{|l} \lambda \leftarrow \dfrac{c}{f} \\ \text{Va} \leftarrow \sqrt{\text{S}\cdot\text{Zo}}\cdot\dfrac{\lambda}{\pi} \\ \text{Vb} \leftarrow \text{Va}\,\sin\left(2\cdot\pi\cdot\dfrac{l}{\lambda}\right) \\ \text{Vb} \leftarrow \text{Va} \ \text{ if } l > \dfrac{\lambda}{4} \end{array}$$

Figure 5.25 Calculating values for impedance matrix and threat voltage at each frequency

Worksheet 5.4 page 3

$n := 100 \qquad s := 1..20 \cdot n \qquad \mathrm{F}_s := s \cdot \frac{\mathrm{Fq}}{n}$

$$\mathrm{Ithreat}_s := \left| \begin{array}{l} f \leftarrow \mathrm{F}_s \\ Z \leftarrow \mathrm{Zloop}(f) \\ V \leftarrow \begin{pmatrix} 0 \\ 0 \\ \mathrm{Vthreat}(f) \end{pmatrix} \\ I \leftarrow \mathrm{Isolve}(Z, V) \\ \mathrm{Iout} \leftarrow \left| I_1 \right| \\ \mathrm{Iout} \leftarrow 10^{-2} \text{ if } \mathrm{Iout} \le 10^{-2} \end{array} \right. \qquad \max(\mathrm{Ithreat}) = 1.031$$

Figure 5.26 Calculating the frequency response of the threat current

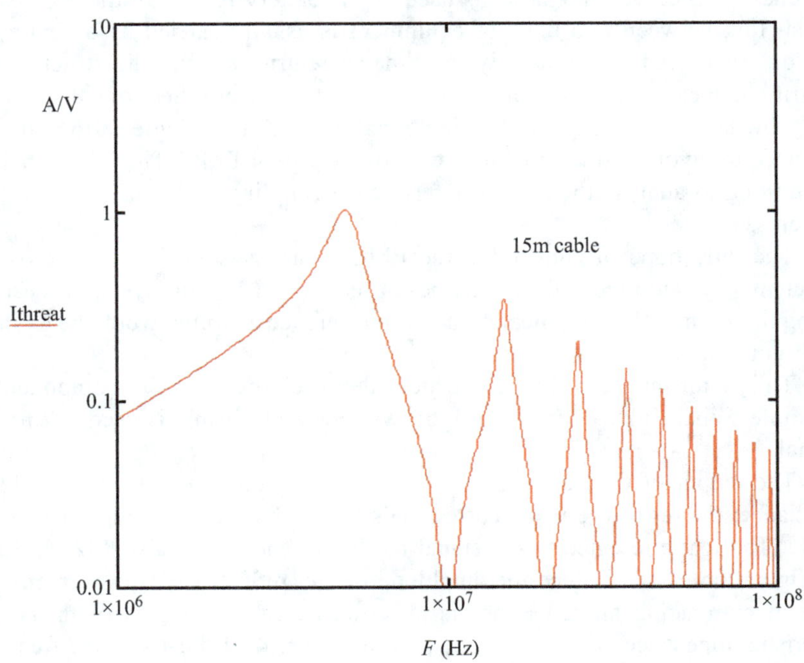

Figure 5.27 Frequency response of threat current

5.5 Coupling via the structure

In the vast majority of configurations, the signal link between two units of equipment is carried by a cable routed along the structure, the signal processing circuitry is mounted on printed circuit boards, the boards are supported by a conducting framework, and that framework is connected to a conducting structure via low-resistance joints. This is the case for most vehicles, whether they are cars, lorries, aircraft, spacecraft, ships or submersibles.

The same is true of fixed installations. Any equipment unit with a conductive outer surface which is connected to the mains supply must also be connected to the earth conductor for safety reasons. This applies to most signal generators, spectrum analysers, oscilloscopes, refrigerators, and computer towers. This earth wire provides a low-resistance path between all equipment units in the system. In industrial situations, the cables are routed along metal conduits which are connected to the earth wires, to the structure of the buildings, and to lightning conductors. The earthed conductors effectively form a mesh which encloses the equipment. Section 8.7 provides a brief description of the shielding effect this provides.

Military vehicles and many land, sea and air vehicles are not permitted to bond the zero-volt reference conductors of the printed circuit boards directly to the chassis, because the chassis is used as the battery return. In these cases, the signal link between two units of equipment is usually carried by a single conductor. Such configurations rely on either a return path via the structure or a return conductor routed to some star point in the system. Sections 8.1, 8.2 and 8.3 show how the concepts of 'equipotential ground', the 'single earth point' and the need to 'avoid earth loops' lead to extremely poor EMC. There is no point in attempting to analyse the myriad interference coupling paths that exist in these systems.

The only hope of controlling the EMC of any system in an effective and efficient way is to invoke the guidelines of section 1.8.1. If the system is designed along these lines, then a typical mode of interference coupling would be as shown by Figure 5.28.

This is an interface diagram which identifies the essential components of a simple signal link. A 15 m length of twin conductor cable is used to carry the signal.

The output of the near-end unit is a voltage source in series with a 132 Ω resistance; the resistance value corresponds to the characteristic impedance of the cable. The load presented to the signal by the far-end unit is also 132 Ω. Such a configuration should reduce the amplitude of any reflections. However, the only way of eliminating all reflections is to terminate each conductor with its characteristic impedance. The flat line of *Ycrit* in Figure 4.14 illustrates the frequency response when there are no reflections.

It is assumed that the return conductor is connected to the framework of each unit via a short length of conductor. That is, the configuration is 'grounded'.

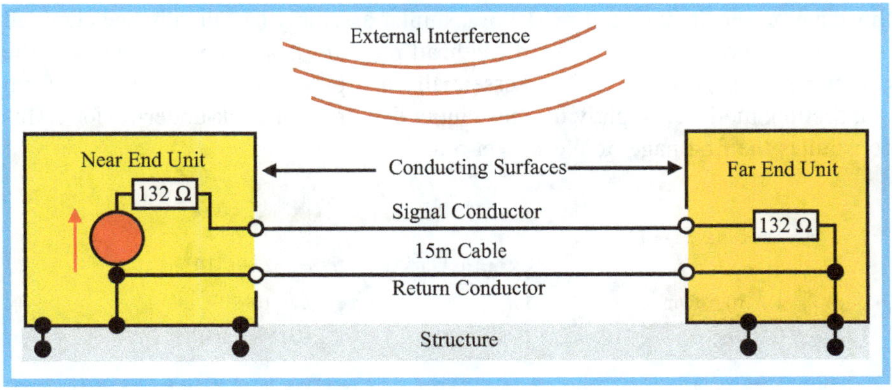

Figure 5.28 Interface diagram for the signal link under review

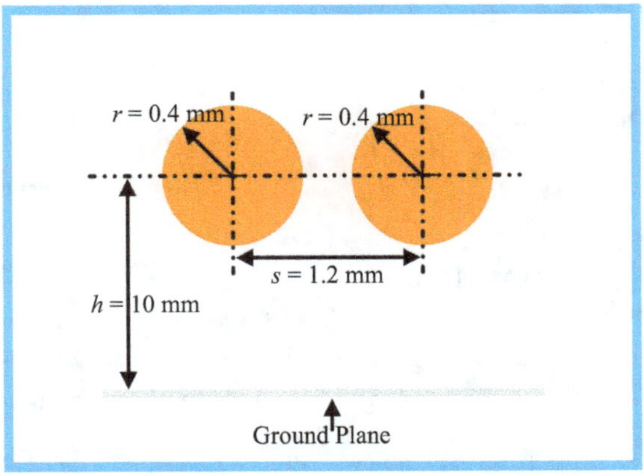

Figure 5.29 Cross-sectional view of two conductors over a ground plane

In most situations, the routing of the cable is fairly convoluted. So, the geometry of the assembly becomes complicated and the task of defining the various cross-sections becomes onerous.

One simple way of overcoming this problem is to invoke the concept of the 'ground plane' in its intended sense, as a purely reflecting surface. The ground plane does not need to be infinite. It just needs to be able to create an image of the conductors routed above it. A glass mirror does not need to be infinite in size to provide a good quality image of the object in front. An estimate can be made of the average separation between cable and structure and that value used to define the cross-section. Figure 5.29 illustrates the geometry of a section of such an assembly.

Given knowledge of the cross-section of the cable assembly and the details of the interface circuitry, enough information now exists to create a circuit model.

Since a representative circuit model of a similar assembly has already been derived in the section on cross-coupling, a Mathcad program is available to calculate the component values. Figure 5.30 is essentially a copy of the worksheet page from that section, modified slightly to utilise input data from the link-under-review. This constitutes the first page of Worksheet 5.5.

Worksheet 5.5 page 1

$\mu_o := 4\cdot\pi\cdot10^{-7}$ H/m $\varepsilon_o := 8.854\cdot10^{-12}$ F/m $c := 2.998\cdot10^{8}$ m/s

$\rho := 1.7\cdot10^{-8}$ Ω/m $l := 15$ m $h := 10\cdot10^{-3}$ m

$s := 1.2\cdot10^{-3}$ m $r := 0.4\cdot10^{-3}$ m See Figure 4.7

$$\mathrm{Rss}_1 := \frac{\rho\cdot l}{\pi\cdot r^2}\ \Omega \qquad \mathrm{Rss}_2 := \mathrm{Rss}_1 \qquad \mathrm{Rss}_3 := 0.005\ \Omega \qquad \text{Equation (2.57)}$$

$$\mathrm{Fx} := \frac{4\cdot\rho}{\mu_o\cdot\pi\cdot r^2} = 1.077\times10^{5}\ \mathrm{Hz} \qquad \text{Equation (2.60)}$$

$$\mathrm{Lc}_1 := \frac{\mu_o\cdot l}{2\cdot\pi}\cdot\ln\left(\frac{2\cdot h\cdot s}{r\cdot\sqrt{s^2+4\cdot h^2}}\right) \qquad \mathrm{Lc}_2 := \mathrm{Lc}_1$$

$$\mathrm{Lc}_3 := \frac{\mu_o\cdot l}{2\cdot\pi}\cdot\ln\left(\frac{\sqrt{s^2+4\cdot h^2}}{s}\right) \qquad \text{Equation (2.96)}$$

$$\mathrm{Cc} := \frac{1}{\mathrm{Lc}}\cdot\left(\frac{l}{c}\right)^2 \qquad \text{Equation (2.33)}$$

$$\mathrm{Fq} := \frac{1}{4\cdot\sqrt{\mathrm{Lc}_1\cdot\mathrm{Cc}_1}} = 4.997\times10^{6}\ \mathrm{Hz} \qquad \text{Equation (2.34)}$$

Component values for three-conductor assembly of Figure 5.29:

$$\frac{\mathrm{Rss}}{2} = \begin{pmatrix}0.254\\0.254\\2.5\times10^{-3}\end{pmatrix} \qquad \frac{\mathrm{Lc}}{2} = \begin{pmatrix}1.645\times10^{-6}\\1.645\times10^{-6}\\4.223\times10^{-6}\end{pmatrix} \qquad \mathrm{Cc} = \begin{pmatrix}7.608\times10^{-10}\\7.608\times10^{-10}\\2.964\times10^{-10}\end{pmatrix}$$

$$\mathrm{Zn} := \begin{pmatrix}132\\0\\0\end{pmatrix}\Omega \qquad \mathrm{Zf} := \begin{pmatrix}132\\0\\0\end{pmatrix}\Omega \qquad \mathrm{Gc} := \begin{pmatrix}0\\0\\0\end{pmatrix}S \qquad \mathrm{Rrad} := 74.49\ \Omega$$

Figure 5.30 Calculating parameter values for the circuit model

In this setup, the objective is to analyse the response of the circuitry to an external field rather than the cross-coupling between two signals. So, the program identifies signal, return, and structure as conductors 1, 2, and 3 respectively.

Data from this first page of the worksheet can be used to assign values to most of the components of the triple-T model; most, but not all. It is still necessary to relate the model to the external field.

The action of the incoming electromagnetic field is to create a current in the structure, and this current develops a voltage in the common-mode loop. If it is assumed that the shielding effectiveness of the structure is zero, then it can also be assumed that all the power in the interference field is transferred to the structure. Such a transfer of power can be represented by a voltage source *Vthreat* in series with the radiation resistance *Rrad* and with the structure.

Under worst-case conditions, the amplitude of the voltage source *Vthreat* can be defined by the envelope curve of Figure 5.19.

In section 5.2 on the virtual conductor, it is assumed that the value of the radiation resistance is 73 Ω; the same as that of a half-wave dipole.

However, an actual test on a single length of conductor revealed that, with an assumed value of 73 Ω, the response of the model was much lower than the actual response. Figure 7.27 illustrates this discrepancy. Section 7.4 describes a method for obtaining an accurate value for Rrad, with environmental considerations, providing a value of 62.075 Ω.

A subsequent test on a twin-conductor cable revealed that the measured value of the radiation resistance was slightly greater than 73 Ω. Figure 7.40 illustrates the response of the cable-under-test on the same graph as the response of the model and shows that the peak values coincide when it is calculated that the radiation resistance is 74.49 Ω. This is the value assumed for *Rrad* in Figure 5.31.

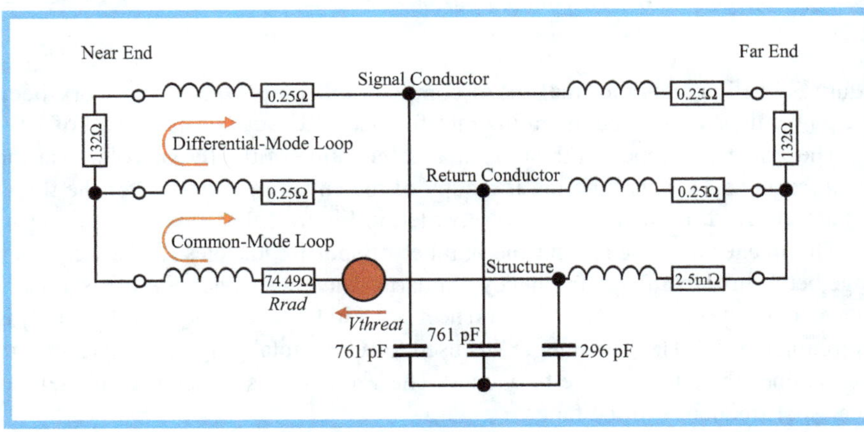

Figure 5.31 Representative circuit model of radiation coupling via the structure

Having established a rationale for a method of linking the circuit model to the external field, it becomes possible to complete the model. Figure 5.31 shows the representative circuit model for a twin-conductor cable routed along the structure.

It is a fairly simple step to convert this lumped-parameter model to one which uses distributed parameters. This is illustrated by Figure 5.32. This distributed-parameter model also defines the four circuit loops used in the subsequent mesh analysis.

There is now enough data to compile the second page of the worksheet, and this page is illustrated by Figure 5.33. The *Zbranch(f)* function calculates numerical values for the six distributed parameters of Figure 5.32. These data are provided as input to the *Zloop(f)* function which compiles a four-by-four array of loop impedance values for the circuit at the spot frequency *f*. This page of the worksheet is a slightly modified version of Page 2 of the Worksheet 4.3 (Figure 4.10).

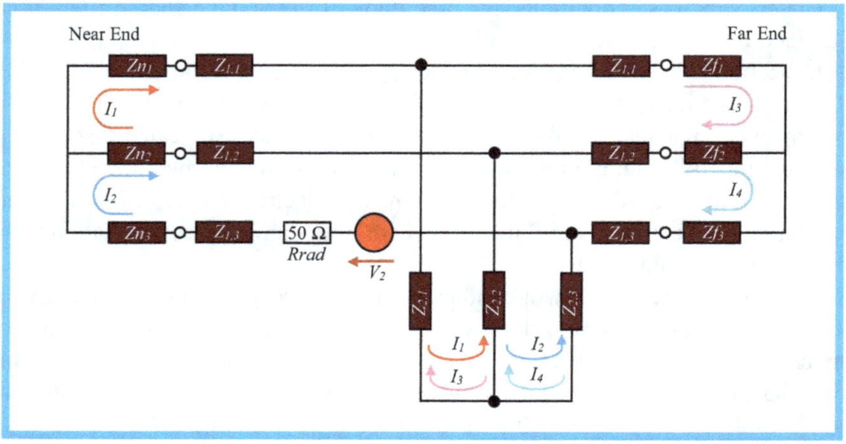

Figure 5.32 Distributed parameter model for Figure 5.31

Figure 5.34 illustrates the final set of computations on page 3 of the worksheet. The power density *S* of the incoming interference is set at a constant level of 1 W/m^2. The *Vthreat(f)* function then calculates the value of the threat voltage at the given frequency *f*. This function is a copy of the envelope function for the threat voltage, derived in Section 5.3 and recorded in Figure 5.19.

The integer *n* is used to set the number of spot frequencies in the frequency range between zero and the frequency of quarter-wave resonance *Fq*. It has already been calculated (in Page 1 of the Worksheet 5.5, Figure 5.30) that the value of *Fq* is approximately 5 MHz. The integer *s* is used to set the total range of frequencies to twenty times the quarter-wave frequency. The variable *F* is a vector which defines every spot frequency in the range.

The main program is defined by the *Iout* function. This picks out a value for the frequency *f* from row *s* of the *F* vector, assigns a four-by-four matrix of loop impedance values to the variable *Z*, sets the voltage V_2 to the value of the threat voltage at that frequency, calculates values for the four loop currents, provides the magnitude of the current I_2 as an output variable, and stores the results in the vector *Iout*.

Since I_2 is a measure of the current at the near end of the common-mode loop, the vector *Iout* is effectively a table of the common-mode current at every spot frequency. The two vectors *Iout* and *F* can then be used to display a graph which defines the frequency response of the common-mode current. This is replicated by Figure 5.35.

This response follows the expected pattern: a null at the quarter-wave frequency followed by a peak at the half-wave frequency. This is followed by a series of peaks and nulls of ever-decreasing amplitude.

Worksheet 5.5 page 2

$$\text{Zbranch}(f) := \left| \begin{array}{l} \omega \leftarrow 2\cdot\pi\cdot f \\ \text{for } i \in 1..3 \\ \quad \left| \begin{array}{l} \text{Rc}_i \leftarrow \text{Rss}_i \cdot \sqrt{1+\dfrac{f}{\text{Fx}}} \\ \theta \leftarrow \sqrt{(\text{Rc}_i + j\cdot\omega\cdot\text{Lc}_i)\cdot(\text{Gc}_i + j\cdot\omega\cdot\text{Cc}_i)} \\ \text{Z}_\text{o} \leftarrow \sqrt{\dfrac{\text{Rc}_i + j\cdot\omega\cdot\text{Lc}_i}{\text{Gc}_i + j\cdot\omega\cdot\text{Cc}_i}} \\ Z_{1,i} \leftarrow \text{Zo}\cdot\tanh\left(\dfrac{\theta}{2}\right) \\ Z_{2,i} \leftarrow \text{Zo}\cdot\text{csch}(\theta) \end{array} \right. \\ Z \end{array} \right.$$

$$\text{Zloop}(f) := \left| \begin{array}{l} Z \leftarrow \text{Zbranch}(f) \\ Z11 \leftarrow Zn_1 + Z_{1,1} + Z_{2,1} + Z_{2,2} + Z_{1,2} + Zn_2 \\ Z12 \leftarrow -(Z_{1,2} + Z_{2,2} + Zn_2) \\ Z13 \leftarrow -(Z_{2,1} + Z_{2,2}) \\ Z14 \leftarrow Z_{2,2} \\ Z22 \leftarrow Zn_2 + Z_{1,2} + Z_{2,2} + Z_{2,3} + Z_{1,3} + Zn_3 + \text{Rrad} \\ Z23 \leftarrow Z_{2,2} \\ Z24 \leftarrow -(Z_{2,2} + Z_{2,3}) \\ Z33 \leftarrow Zf_2 + Z_{1,2} + Z_{2,2} + Z_{2,1} + Z_{1,1} + Zf_1 \\ Z34 \leftarrow -(Z_{1,2} + Z_{2,2} + Zf_2) \\ Z44 \leftarrow Zf_3 + Z_{1,3} + Z_{2,3} + Z_{2,2} + Z_{1,2} + Zf_2 \\ \begin{pmatrix} Z11 & Z12 & Z13 & Z14 \\ Z12 & Z22 & Z23 & Z24 \\ Z13 & Z23 & Z33 & Z34 \\ Z14 & Z24 & Z34 & Z44 \end{pmatrix} \end{array} \right.$$

Figure 5.33 Calculating the values of the branch and loop parameters

By selecting the value of I_1 as the output variable of the main program, a graph can be created which defines the frequency response of the differential-mode current. This is illustrated by Figure 5.36. Several features of this response are worth noting:

Between 100 kHz and 1 MHz, the amplitude of the current in the differential-mode loop rises. This is because the current in the common-mode loop is relatively constant. Constant current at increasing frequency flowing in the inductance of the return conductor causes the voltage across that inductance to rise.

Worksheet 5.5 page 3

$S := 1$

$$\text{Vthreat}(f) := \left| \begin{array}{l} \lambda \leftarrow \dfrac{c}{f} \\ \text{Va} \leftarrow \sqrt{\text{S} \cdot 377} \cdot \dfrac{\lambda}{\pi} \\ \text{Vb} \leftarrow \text{Va} \sin\left(2 \cdot \pi \cdot \dfrac{l}{\lambda}\right) \\ \text{Vb} \leftarrow \text{Va} \ \text{if} \ l > \dfrac{\lambda}{4} \end{array} \right.$$

Calculating the threat voltage for a power density *S* at a frequency *f*.

See Figure 5.19.

$n := 100 \qquad s := 1..20 \cdot n \qquad \text{F}_s := s \cdot \dfrac{\text{Fq}}{n}$

Defining the frequency range

Main program

$$\text{Iout}_s := \left| \begin{array}{l} f \leftarrow \text{F}_s \\ Z \leftarrow \text{Zloop}(f) \\ V \leftarrow \begin{pmatrix} 0 \\ \text{Vthreat}(f) \\ 0 \\ 0 \end{pmatrix} \\ I \leftarrow \text{lsolve}(Z, V) \\ \left| I_2 \right| \end{array} \right.$$

Figure 5.34 Calculating the frequency response of the common-mode current

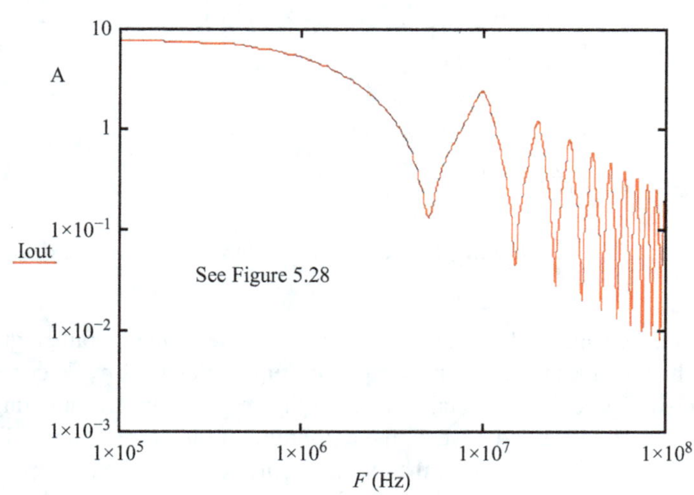

Figure 5.35 Frequency response of current in the common-mode loop

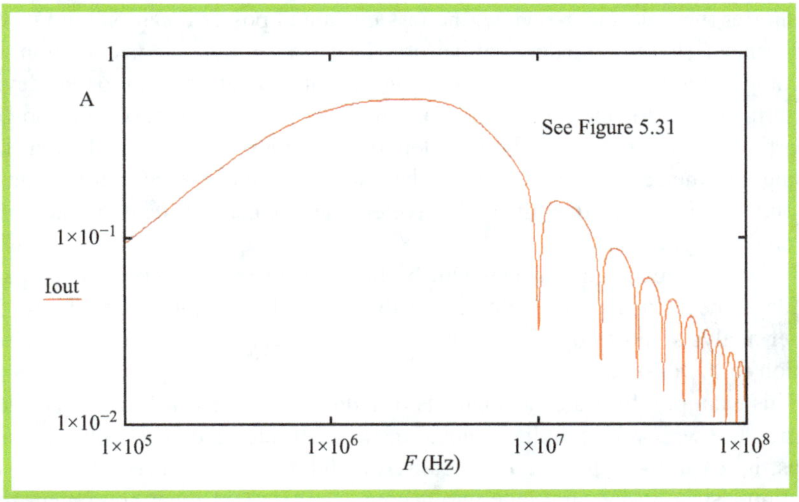

Figure 5.36 Frequency response of current in the differential-mode loop

Since this voltage is applied to the differential-mode loop, the current in that loop increases.

As the frequency rises above 2.5 MHz, the response flattens out. This is because the 132 Ω resistors act to minimise the amplitude of the resonant peaks. It is worth noting that 2.5 MHz corresponds to 1/8 of a wavelength and that the first crossover point in Figure 4.14 also occurs at 1/8 of a wavelength.

Although the common-mode current drops to a minimum at the quarter-wave frequency of 5 MHz, the amplitude of the differential-mode current remains relatively constant. This is also a feature of the use of impedance-matching resistors at the terminations. The *ycrit* curve in Figure 4.14 illustrates an ideal configuration where the use of the characteristic impedance at each interface prevents both nulls and peaks in the response.

Probably the most interesting feature is the fact that, although the common-mode current reaches a peak value at the half-wave frequency (10 MHz), the differential-mode current drops to a null. At this frequency, the common-mode loop currents at the near end and at the far end both reach a resonant peak. But they are of opposite polarity. Whilst the voltage developed along one half of the length of the return conductor is positive, the voltage developed along the other half is negative. Since the common-mode voltages across the return conductor cancel out, the net voltage along the complete length of the return conductor is minimal. Since this is the voltage which appears in the differential-mode loop, the current induced by that voltage also dips to a minimum value.

The waveform pattern repeats itself every 10 MHz, but the maximum amplitude drops at 20 dBA per decade. This identifies the fact that, as the frequency increases, the maximum power which can be delivered to the victim circuit reduces.

The shorter the link-under-review, the less unwanted power it can pick up from an interference field of defined power density. This is a significant consideration when designing printed circuit boards, or firing circuits for electro-explosive devices. This conclusion is supported by (5.15), which shows that the maximum power is proportional to the square of the wavelength. Maximum power is picked up when the length of the conductor is equal to a half-wavelength of the interfering signal. If the conductor is longer, then the half-wavelength is longer and the maximum power delivered is higher.

It can be concluded that it would be better to use a screened twisted pair to provide some measure of shielding for the signal link depicted in Figure 5.28. Further analysis and testing would be needed to check the integrity of any modification to the design.

This example illustrates an analysis of a uniform cable routed over a perfectly flat structure with very simple terminations at each interface. This could be regarded as the easiest, simplest, most idealised solution. It also represents worst-case conditions, since it does not take into account the fact that other equipment in the system is also absorbing interference energy.

If the cross-section of the cable is not uniform, it would be necessary to analyse the properties of each section individually and invoke the concept of equivalent circuits, as described in section 1.3.3. The general circuit model of Figure 5.32 can cater for impedances of any value at the terminations. If any doubt exists as to the accuracy of any particular simulation, then electrical measurements can be carried out on a test rig.

5.6 Radiation susceptibility

In the previous section, a model was created which can simulate the response of the equipment-under-review to the presence of an external electromagnetic field at the surface of the assembly. To extend the scope of the simulation, it is necessary to invoke the use of electromagnetic theory, and this can involve some very complicated mathematics.

Even so, it is possible to identify a few simple relationships which can be used to obtain a first estimate of the response.

For the analysis of the susceptibility of the equipment to external radiation, the need is to relate the threat voltage in the antenna-mode loop to the output of a remote transmitter. Since the formal EMC Requirements usually specify the setup and test levels for Radiation Susceptibility, the immediate objective is to simulate the output of the Test Transmitter.

Figure 5.37 illustrates the type of setup used for this form of testing. An antenna is located at a defined distance r from the equipment-under-test, oriented to deliver maximum field strength at the surface of the equipment, and a specified power Pr is applied to the test antenna over a specified range of frequencies. The antenna gain, Gt, is also specified.

The power density S at the surface of the equipment can be calculated using (5.6), and the intensity of the electric field E can be derived by using (5.11).

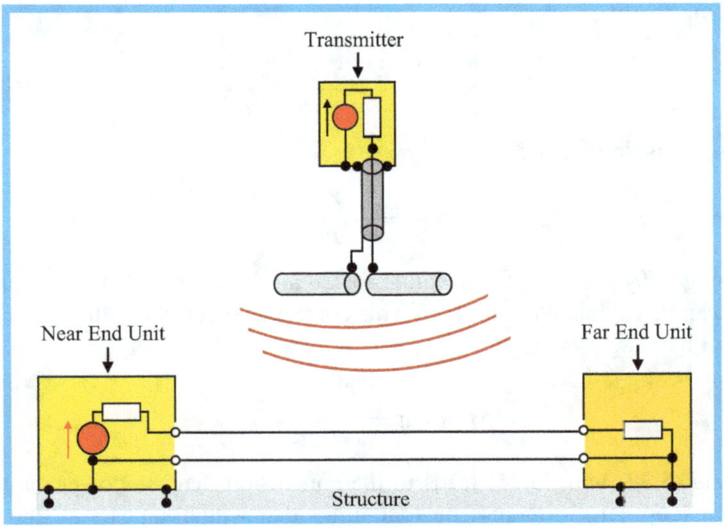

Figure 5.37 Setup for radiation susceptibility test

The threat voltage *Vthreat* is obtained from (5.32) if the length l of the cable is greater than a quarter wavelength. If the length is less than a quarter wavelength, (5.29) applies. This gives the response of the envelope curve of Figure 5.19.

Grouping these equations together allows the threat voltage to be related to the transmitter power:

$$\begin{aligned} S &= \frac{Gt \cdot Pt}{4 \cdot \pi \cdot r^2} \\ E &= \sqrt{Z_o \cdot S} \\ Vthreat &= \frac{\lambda}{\pi} \cdot E \ \text{ if } l > \frac{\lambda}{4} \\ Vthreat &= \frac{\lambda}{\pi} \cdot E \cdot \sin\left(\frac{2 \cdot \pi \cdot l}{\lambda}\right) \text{ if } l \leq \frac{\lambda}{4} \end{aligned} \tag{5.33}$$

The maximum power which can be delivered to the victim loop is limited only by the radiation resistance and the threat voltage. From Figure 5.31:

$$Pthreat = \frac{Vthreat^2}{Rrad} \tag{5.34}$$

Using (5.19) to substitute for *Vthreat*:

$$Pthreat = \left(\frac{\lambda}{\pi}\right)^2 \cdot \frac{E^2}{Rrad}$$

Substituting for E:

$$Pthreat = \frac{\lambda^2 \cdot S}{\pi} \cdot \frac{Z_o}{\pi \cdot Rrad} \tag{5.35}$$

Hence, if Ge is defined as:

$$Ge = \frac{Z_o}{\pi \cdot Rrad} \tag{5.36}$$

Then, using (5.9) and (5.3) to substitute for Zo and $Rrad$ in (5.36), the value of the gain Ge can be calculated to be 1.64. This correlates precisely with (5.8). Equation (5.35) can be re-written as:

$$Pthreat = \frac{Ge \cdot \lambda^2 \cdot S}{\pi} \tag{5.37}$$

Comparing (5.37) with (5.15) leads to the conclusion that the power delivered to the victim loop can, in theory, be four times that which appears at the receiver terminals of a matched antenna/load assembly. However, in the real world environmental factors would ameliorate the received threat voltage.

The tests described in sections 7.4 and 7.5 measure the value of the radiation resistance to be 62.075 Ω and 74.49 Ω respectively. It can be surmised that the discrepancy is due to environmental line and frequency losses along with the prevailing and fluctuating ambient environmental parameters, during the extended time-frame of experimentation.

It can also be reasoned that the test of section 7.5 represents worst-case conditions, since all the antenna-mode power is delivered to the cable. This would not be the case in a practical situation. Some of the power of the input radiation would be absorbed by the adjacent wiring in the system. Since the test on the twin-conductor cable results in a measured value of 74.49 Ω for *Rrad*, this would seem to be a reasonable starting point for any analysis of radiation susceptibility.

5.7 Radiated emission

The circuit model derived in Section 5.5 can also be used to determine the radiated emission of the equipment-under-test. Figure 5.38 illustrates the setup.

If the equipment being tested is the same as that of Figure 5.28, then the model used to predict emissions can be based on Figure 5.31. The only changes necessary are to replace *Vthreat* and *Rrad* with a short-circuit and to insert a voltage source *Vdiff*.

In Figure 5.39, the voltage source *Vdiff* is located in the differential-mode loop to simulate the output signal generated by the near-end equipment. This will create a common-mode loop current, and the amplitude of this current can be calculated.

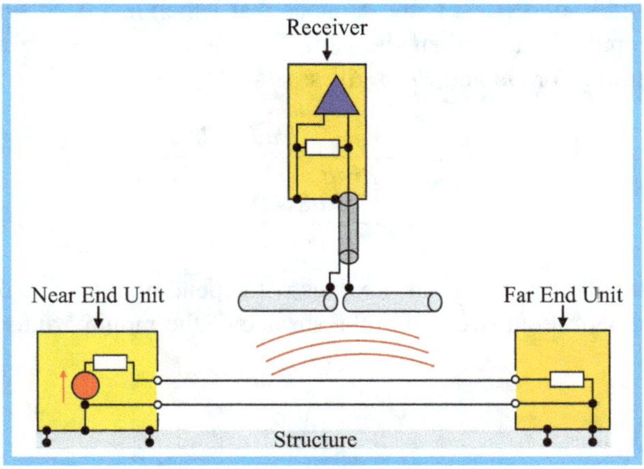

Figure 5.38 Setup for radiated emission test

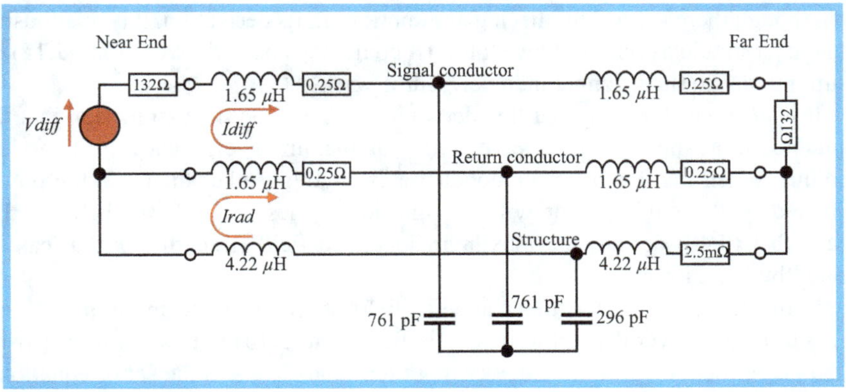

Figure 5.39 Representative circuit model for analysis of radiated emission

Under worst-case conditions, it can be assumed that the shielding effectiveness of the structure is zero. In which case, the common-mode current will be the radiation source. In Figure 5.39, it is identified as *Irad*.

The power *Prad* transmitted into the environment by the current *Irad* would be:

$$Prad = Irad^2 \cdot Rrad \tag{5.38}$$

From (5.6), the power density at a monitor antenna located a distance *r* away from the assembly would be:

$$Sm = \frac{Ge \cdot Prad}{4 \cdot \pi \cdot r^2} \tag{5.39}$$

Since the assembly-under-test is the same as that which had been subjected to the radiation susceptibility test, then the gain *Ge* would be the same as defined by (5.36). Substituting for *Ge* and *Prad* gives:

$$\begin{aligned} Sm &= \frac{Z_o}{\pi \cdot Rrad} \cdot \frac{Irad^2 \cdot Rrad}{4 \cdot \pi \cdot r^2} \\ &= Z_o \cdot \frac{Irad^2}{4 \cdot \pi^2 \cdot r^2} \end{aligned}$$

It is worth noting that this relationship does not depend on the value of *Rrad*. From (5.12), the magnetic field strength at the surface of the monitor antenna is:

$$Hm = \sqrt{\frac{Sm}{Z_o}}$$

Substituting for *Sm*:

$$Hm = \frac{Irad}{2 \cdot \pi \cdot r} \tag{5.40}$$

Even though there are no simplifying assumptions in its derivation, this relationship is about as simple as one could wish for. (It could have been derived from (2.12) by assuming an infinitely long conductor. But it wasn't).

It is also worth noting that this derivation is based on the assumption that the conductor is acting as a half-wave dipole at maximum efficiency in a lossless medium. When the length of the conductor is anything other than $\lambda/2$, the power delivered to the environment will be significantly less than *Prad*. In turn, this means that (5.40) defines the maximum magnetic field strength *Hm* that can be created by the current *Irad*.

Using (5.40) to define the magnetic field strength at the monitor antenna, (5.12) to relate power density to magnetic field, and (5.15) to relate power density to the power delivered to the receiver input terminals leads to the set of equations:

$$\begin{aligned} Hm &= \frac{Irad}{2 \cdot \pi \cdot r} \\ Sm &= Z_o \cdot Hm^2 \\ Pm &= \frac{Gm \cdot \lambda^2 \cdot Sm}{4 \cdot \pi} \end{aligned} \tag{5.41}$$

Where *Gm* is the gain of the monitor antenna used by the Test House.

This establishes a relationship between the current radiated into the environment by the equipment-under-test and the power delivered to the input of the monitor receiver.

Adding this set of equations to the subroutine which calculates the value of the common-mode current will provide a clear relationship between the signal output voltage of the near-end unit of Figure 5.38 and the level of interference measured by the test receiver.

Chapter 6
Transient analysis

Transients are an ever-present source of glitches in electronic systems. Sources such as relays, switches, motors, and power supplies can cause the sudden release of stored magnetic energy when the current maintaining that magnetisation is interrupted. Electrostatic discharges can cause a momentary creation of high-amplitude current spikes. Whatever the source, brief bursts of high-intensity electromagnetic radiation can occur almost anywhere in an electrical system.

Such transients can easily corrupt the data streams handled by microprocessors. In fact, any analogue or digital circuit link can be upset. Depending on the susceptibility of the processing circuitry, such events could be inconsequential, annoying, dangerous or catastrophic.

This section describes the use of time-step analysis to simulate transient behaviour. It derives a model which can simulate the emission produced by a step pulse propagated along a twin-conductor line. Given an insight into the mechanisms involved, it should be possible for the designer to devise circuits which minimise this emission. Such circuits should also be capable of minimising susceptibility. Chapter 8 identifies a few of the techniques deriving from this approach.

As with frequency response analysis, the effects of an incoming field can be represented by a voltage source, whilst the outgoing field can be defined in terms of the antenna-mode current. The amplitude of the voltage source is a function of the E-field of the external radiation, whilst the H-field of the emitted radiation is a function of the antenna-mode current. To a first approximation, it can be assumed that the amplitude of both fields is inversely proportional to the separation between transmitter and receiver. All the passive parameters (resistance, inductance, capacitance and conductance) remain unchanged. The electrical properties of a particular cable assembly are independent of the waveform of the signal or power carried by that assembly.

Section 6.1 provides an introduction to the basics of time-step analysis and contains two examples of how to develop programs to simulate transient waveforms. An essential feature of time-step analysis is that the input voltage is defined and the current is calculated instant-by-instant. The computations involve absolute, real values. Unlike frequency response analysis, time-step analysis never invokes the concepts of phasors, phase angles, imaginary numbers, reactance, or susceptance. A further implication is that impedances can only be resistive. So, to avoid any misunderstanding of the terms used during the course of this type of analysis, the variable *Ro* is used to define the characteristic impedance.

The equations in Section 6.2 are based on the concepts of partial currents and partial voltages used in textbook analysis of transients in transmission lines. A program is described which simulates the response of a twin conductor line when a step pulse is applied to the near end. It is possible to develop this model to include the response of interface circuitry at the terminations.

Section 6.3 examines the characteristics of the delay-line model and shows that the properties of inductance and capacitance are still present.

When a sinusoidal source at the quarter-wave frequency is switched on to a twin-conductor cable which is open-circuit at the far end, the current will gradually build up as the energy in successive half-cycles is stored in the cable. This is illustrated by Figure 6.22.

If a signal at half-wave frequency is connected, the current delivered to the cable will gradually decrease. This is because the stored energy in the cable increases to a level where energy delivered from the source to the cable is balanced by energy delivered from cable to the source. Figure 6.23 illustrates this.

These two illustrations show that an analysis of the transient response of a system to a sinusoidal signal helps us to better understand frequency response analysis.

Section 6.4 tackles the problem of analysing antenna-mode transients. An experiment is described where a step pulse was injected into a twin-conductor cable which was open-circuit at the far end. After a trial-and-error process, a circuit model was identified: one which provided a fair reproduction of the response. The existence of such a model raised several questions about the mechanisms involved, questions which led to a further development of the model.

In Section 6.5, the picture emerges of a wavefront propagating along the signal conductor in a form similar to the bow wave of a ship. Given a separation of 2 mm between the conductors, the wavefront does not arrive at the return conductor until the forward edge has progressed a further 2 mm along the signal conductor. Electric charge is deposited on the return conductor. Induced voltage causes a current to flow back towards the source. This current creates an electromagnetic field which couples with the signal conductor and that also radiates outwards. Since the current in the return conductor does not completely balance that in the signal conductor, there is an unbalanced charge which propagates along the cable, just behind the wavefront.

Since the return current is always less than the signal current, there must be a net flow of aerial-mode current out towards the far end. This outgoing current must be balanced by an inflow of current from the ground connections. Effectively, the system is behaving like a dipole, with the cable acting as one monopole and the ground conductors acting as the other.

This means that there are at least three current components involved:

- That which carries a differential-mode signal.
- That which deposits an unbalanced charge on the cable.
- That which flows out radially in the form of electromagnetic radiation.

Section 6.6 derives a general circuit model which allows the amplitude of all three of these components to be calculated. A program is described which carries out the computations. Comparing the response of the model with the waveform displayed on an oscilloscope allows the two waveforms to be aligned. The values so derived for component values allow a representative model to be derived for the cable-under-test (Figure 6.38). The tests themselves are described in Section 7.6.

6.1 Time-step analysis

6.1.1 Basic concept

The classical approach to transient analysis is to use transforms, either Laplace transforms or Heaviside transforms. However, this approach becomes extremely complex when dealing with the type of circuits commonly encountered. Hence, the approach adopted here is essentially the same as that used by Simulation Programs with Integrated Circuit Emphasis (SPICE) software – time-step analysis [20].

This form of analysis is similar to the iterative method used in mathematics to solve non-linear equations. The essential difference is that each calculation defines the condition of the system at a later time. If the time steps are sufficiently small, then the variation in the amplitude of each current can be assumed to be linear.

Two factors are involved in each set of calculations:

- The state of the system immediately prior to the time of the calculation.
- Incremental changes in the applied voltages.

Either nodal analysis or mesh analysis is possible. Since mesh analysis has been used for calculations in the frequency domain, it is logical to continue with that type of analysis in the time domain.

6.1.2 Basic equations

The basic equations are the simplest possible:

For an inductor

$$V = L \cdot \frac{dI}{dt} \tag{6.1}$$

For a resistor

$$V = R \cdot I \tag{6.2}$$

For a capacitor

$$V = \frac{Q}{C} \tag{6.3}$$

Where:

$$Q = \int_0^t I \cdot dt \tag{6.4}$$

Given that:

t = elapsed time.
dt = finite increment of time.
dI = finite increment of current.

6.1.3 Series LCR circuit

For the circuit illustrated by Figure 6.1, the total voltage at any instant is the sum of the instantaneous voltages across each component. This gives the loop equation:

$$V = L \cdot \frac{dI}{dt} + R \cdot I + \frac{Q}{C} \tag{6.5}$$

Analysis, based on the application of a Heaviside transformation, leads to an equation which defines the relationship between current and time when a step voltage is applied to the input terminals of the circuit. If the circuit is initially in a quiescent state, then:

$$I = \frac{V}{L} \cdot \frac{1}{a-b} \cdot \left(e^{-b \cdot t} - e^{-a \cdot t}\right)$$

Where $a = \alpha + \beta$, $b = \alpha - \beta$, $\alpha = \dfrac{R}{2 \cdot L}$ and $\beta = \sqrt{\dfrac{R^2}{4 \cdot L^2} - \dfrac{1}{L \cdot C}}$

A less elegant, but simpler approach would be to use time-step analysis to formulate a short computer program. The steps necessary to compile such a program are outlined below.

During the time increment dt, the current changes by the increment dI, where

$$dI = \frac{dt}{L} \cdot \left(V - R \cdot I - \frac{Q}{C}\right) \tag{6.6}$$

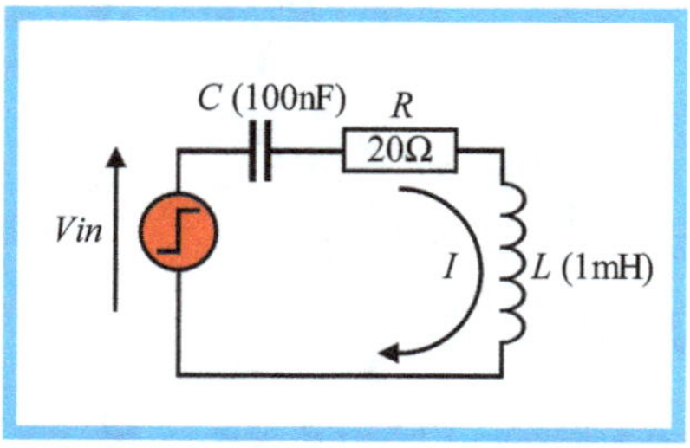

Figure 6.1 Series LCR circuit

At the end of that increment of time, the value of the current will have changed. In a Mathcad computer program, such as that illustrated in Appendix A, the new value for the current would be defined as:

$$I \leftarrow I + dI \tag{6.7}$$

During this time, the charge on the capacitor will have changed, and the program statement which updates the value of Q would be

$$Q \leftarrow Q + I \cdot dt \tag{6.8}$$

These last three equations form the basis of a simple subroutine '*next(D)*' in the Mathcad worksheet of Figure 6.2. The variable D is a two-element vector containing the value of the current and the charge in the circuit of Figure 6.1 at any particular instant.

The subroutine picks out these two values, uses (6.6)–(6.8) to update them, and then returns the new value of I and Q. The value of the time increment dt has been set at 1 µs on the top line of the worksheet, together with the other circuit parameters.

The control variable i for the main program is set to vary from 2 to N, where N has been preset at 100. The main program itself is defined immediately below the subroutine. It simply updates the data variable D at each time step, picks out the value of D_1 (the current I) and transfers this to the element $Iout_i$, where the vector $Iout$ is the record of the output current.

The variable t records the time of each calculation and allows the graph of *Iout* versus time to be displayed (Figure 6.3). Not surprisingly, this is a damped sine wave.

Worksheet 6.1.1

$L := 1 \cdot 10^{-3}\,H$ $\qquad C := 100 \cdot 10^{-9}\,F$ $\qquad R := 20\,\Omega$

$V := 1V$ $\qquad dt := 10^{-6}\,s$ $\qquad N := 100$

$$D := \begin{pmatrix} 0 \\ 0 \end{pmatrix} \qquad \text{next}(D) := \left| \begin{array}{l} I \leftarrow D_1 \\ Q \leftarrow D_2 \\ dI \leftarrow \dfrac{dt}{L} \cdot \left(V - R \cdot I - \dfrac{Q}{C} \right) \\ I \leftarrow I + dI \\ Q \leftarrow Q + I \cdot dt \\ \begin{pmatrix} I \\ Q \end{pmatrix} \end{array} \right.$$

$$i := 2..N \qquad Iout_i := \left| \begin{array}{l} D \leftarrow \text{next}(D) \\ D_1 \end{array} \right. \qquad t_i := (i - 1) \cdot dt$$

Figure 6.2 Calculating the transient response of a series LCR circuit

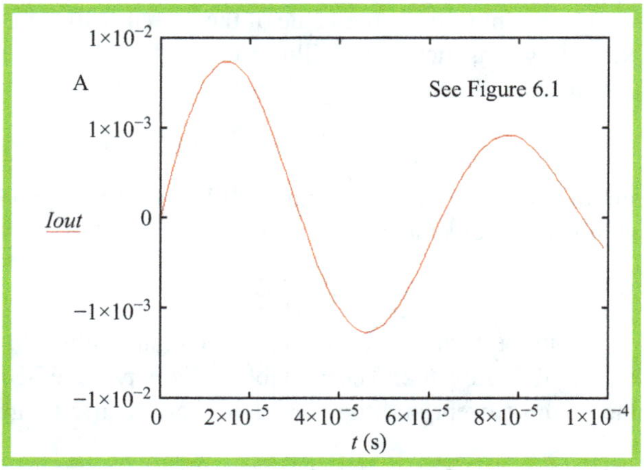

Figure 6.3 Transient response of series LCR circuit

6.1.4 Parallel LCR circuit

When the L, C and R components are in parallel, three separate loops are formed as illustrated by Figure 6.4.

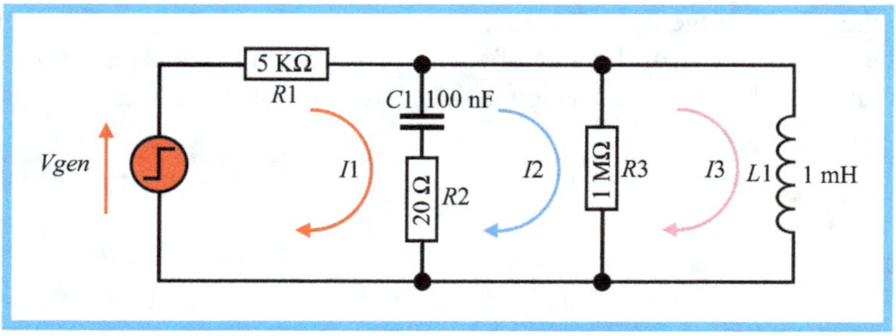

Figure 6.4 Parallel LCR circuit

The loop equations are:

$$Vgen = (R1 + R2) \cdot I_1 + \frac{Q_1}{C1} - \frac{Q_2}{C1} - R2 \cdot I_2 \tag{6.9}$$

$$0 = -R2 \cdot I_1 - \frac{Q_1}{C1} + \frac{Q_2}{C1} + (R2 + R3) \cdot I_2 - R3 \cdot I_3 \tag{6.10}$$

$$0 = -R3 \cdot I_2 + R3 \cdot I_3 + L1 \cdot \frac{dI_3}{dt} \tag{6.11}$$

In formulating the program which calculates the values of the variables at each time step, it is useful to take into consideration the characteristics of the step pulse. Theoretically, it changes from zero volts to *Vgen* in zero time. As far as the initial step is concerned, $C1$ behaves as a short circuit and $L1$ behaves as an open circuit. Most of the initial surge current I_1 flows through $C1$ and $R2$ in series. The resultant voltage across $R2$ creates a current I_2. Current I_2 flowing in $R3$ creates a voltage across $L1$.

This means that, initially, I_1 is the most significant current, I_2 is less significant, whilst the value of I_3 is inconsequential. This effectively determines the order in which parameter values should be calculated.

Since the charge on the capacitor is always due to I_1–I_2, (6.9) and (6.10) can be simplified by defining

$$Q = Q_1 - Q_2 \tag{6.12}$$

It is assumed that the initial amplitudes of I_1, I_2, and I_3 are zero.
Current I_1 can be calculated by re-arranging (6.9). This gives:

$$I_1 = \frac{1}{R1 + R2} \cdot \left(Vgen + R2 \cdot I_2 - \frac{Q}{C1} \right) \tag{6.13}$$

Since the value of I_1 is now known, it can be used to calculate a new value of I_2:

$$I_2 = \frac{1}{R2 + R3} \cdot \left(R2 \cdot I_1 + R3 \cdot I_3 + \frac{Q}{C1} \right) \tag{6.14}$$

The new value of I_2 can be used to calculate the rate of change of I_3:

$$dI_3 = \frac{dt}{L1} \cdot R3 \cdot (I_2 - I_3) \tag{6.15}$$

The new value of I_3 becomes:

$$I_3 \leftarrow I_3 + dI_3 \tag{6.16}$$

Since the values of I_2 and I_3 have also been updated:

$$Q \leftarrow Q + (I_1 - I_2) \cdot dt \tag{6.17}$$

It is then a routine matter to assemble (6.13)–(6.17) into a subroutine which updates the values of current and charge at every time step. This is illustrated in Figure 6.6.

The simplest way of checking the accuracy of the calculations is to vary the time step *dt* and re-calculate. If a reduction in the amplitude of the time-step results in no perceptible change to the output waveform, then the simulation can be assumed to be reasonably accurate.

In the worksheet of Figure 6.6, the simulation time T is set at 100 μs, and the number of iterations is defined as:

$$N = \text{ceil}\left(\frac{T}{dt}\right) \tag{6.18}$$

(In Mathcad, the function ceil(z) returns the smallest integer greater than or equal to z). No matter what value is selected for *dt*, the simulation will always run for 100 μs.

The waveform displayed in Figure 6.5 is that of the current in the inductance $L1$. This is a ringing transient, similar to that of Figure 6.3.

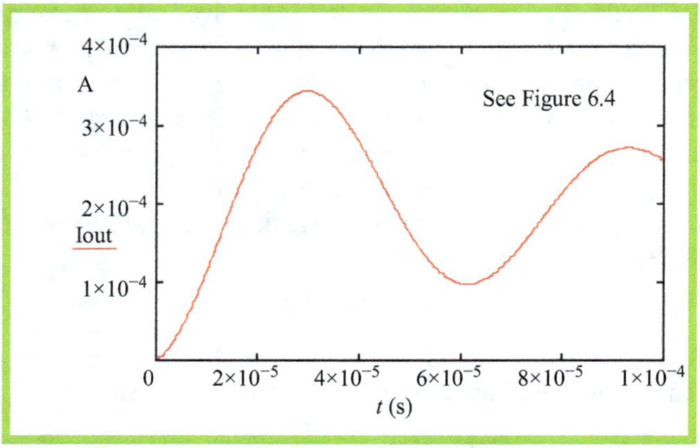

Figure 6.5 Waveform of current in the inductor of the parallel LCR circuit

Worksheet 6.1.2

$R1 := 5000\ \Omega \qquad R2 := 20\ \Omega \qquad R3 := 10^6\ \Omega$

$L1 := 1 \cdot 10^{-3}\ \text{H} \qquad C1 := 100 \cdot 10^{-9}\ \text{F} \qquad Vgen := 1\text{V} \qquad dt := 10^{-8}\ \text{s}$

$$D := \begin{pmatrix} 0 \\ 0 \\ 0 \\ 0 \end{pmatrix} \qquad \text{next}(D) := \left| \begin{array}{l} I2 \leftarrow D2 \\ I3 \leftarrow D3 \\ Q \leftarrow D4 \\ I1 \leftarrow \dfrac{1}{R1 + R2} \cdot \left(Vgen + R2 \cdot I2 - \dfrac{Q}{C1} \right) \\ I2 \leftarrow \dfrac{1}{R2 + R3} \cdot \left(R2 \cdot I1 + R3 \cdot I3 + \dfrac{Q}{C1} \right) \\ dI3 \leftarrow \dfrac{dt}{L1} \cdot R3 \cdot (I2 - I3) \\ I3 \leftarrow I3 + dI3 \\ Q \leftarrow Q + (I1 - I2) \cdot dt \\ \begin{pmatrix} I1 \\ I2 \\ I3 \\ Q \end{pmatrix} \end{array} \right.$$

$$T := 100 \cdot 10^{-6}\ \text{s} \qquad N := \text{ceil}\left(\frac{\text{T}}{\text{dt}} \right)$$

$$i := 2..N \qquad Iout_i := \left| \begin{array}{l} D \leftarrow \text{next}(D) \\ D_3 \end{array} \right. \qquad t_i := (i - 1) \cdot dt$$

Figure 6.6 Calculating the amplitude of the transient current in the parallel LCR circuit

The main difference between this and the previous response is the fact that it approaches a steady-state amplitude of 200 μA. That is, the current in a 5 kΩ resistor when a voltage of 1 V is applied.

In formulating programs similar to the two illustrated in this section, the first objective is to identify the parameter with the greatest change during the first-time increment. In the first example, this is the voltage across the inductor; in the second example, it is the current in the capacitor.

6.2 Delay line model

Transmission Line theory invokes the concept of electromagnetic waves travelling forward and backwards along a path bounded by a pair of parallel conductors. Figure 6.7 illustrates the concept.

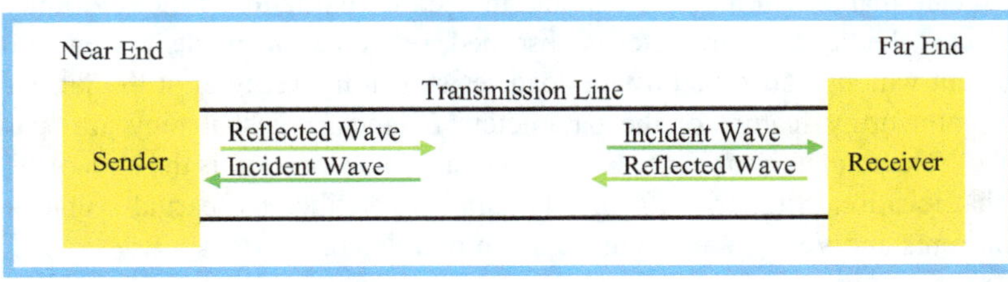

Figure 6.7 Transmission line reflections

Although the velocity of an electromagnetic wave in free space is approximately 300 m/μs, the presence of dielectric material in the cable reduces this velocity. If it is assumed that the velocity of propagation is v, the time taken for an electromagnetic wave to propagate from one end of the line to the other is the time constant T, where:

$$T = \frac{l}{v} \tag{6.19}$$

Initial assumptions are that there are no radiation losses from a transmission line and that the line does not pick up external radiation. It is worth noting that these assumptions are not identified in conventional textbooks.

If it is assumed that an electromagnetic signal has been delivered to the line by the sender at the near end, then it will arrive at the far end after an elapsed time of T seconds. Some of it will be absorbed by the load in the receiver, whilst some of it will be reflected. Figure 6.8 illustrates the components of the signal at any instant in terms of currents and voltages.

Conventional theory depicts this situation in terms of current in one conductor of a two-conductor line and assumes the properties of the line are concentrated in this one conductor. To enable the analysis to be developed to deal with three-conductor lines, it is necessary to view all currents in terms of loop currents. In the convention adopted here, a positive loop current is defined as clockwise.

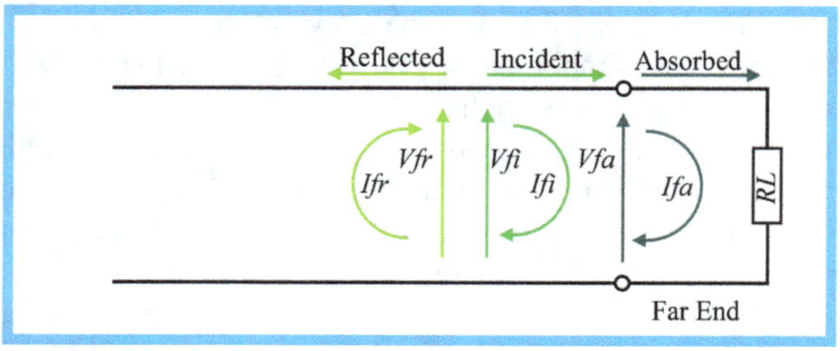

Figure 6.8 Currents and voltages at the far end of the line

It is useful to use a shorthand method of identifying each parameter to keep track of all these currents and voltages during the computation process. The method adopted here is to use the letter *f* to indicate that the parameter is at the far end, the letter *n* to indicate that it is at the near end and to use the first letter of the relevant word to indicate whether the parameter is absorbed, reflected, or incident. Hence, *Ifa* is the current which is absorbed by the load resistor in the receiver at the far end.

A noteworthy feature of the parameters *Ifi* and *Ifr* is that they are 'partial-currents'. The total current in the line at any particular location is the sum of *Ifr* and *Ifi* at that location. Similarly, *Vfr* and *Vfi* can be described as 'partial-voltages'.

Currents and voltages in Figure 6.8 can be related:

$$Vfi + Vfr = Vfa \tag{6.20}$$

$$Ifi + Ifr = Ifa \tag{6.21}$$

Voltages can be related to currents:

$$Vfi = Ro \cdot Ifi \tag{6.22}$$

$$Vfr = -Ro \cdot Ifr \tag{6.23}$$

$$Vfa = RL \cdot Ifa \tag{6.24}$$

Where *Ro* is the characteristic impedance of the line and *RL* is the load resistor. Subtracting (6.23) from (6.22) gives:

$$Vfi - Vfr = Ro \cdot (Ifi + Ifr)$$

Using (6.21)

$$Vfi - Vfr = Ro \cdot Ifa \tag{6.25}$$

Using (6.24) to substitute for *Vfa* in (6.20)

$$Vfi + Vfr = RL \cdot Ifa \tag{6.26}$$

Adding (6.25) and (6.26)

$$2 \cdot Vfi = (Ro + RL) \cdot Ifa \tag{6.27}$$

Re-arranging (6.27) and using (6.22) to substitute for *Vfi* gives:

$$Ifa = \frac{2 \cdot Ro \cdot Ifi}{Ro + RL} \tag{6.28}$$

This relates the current delivered to the receiver to the incident current arriving from the transmission line. The reflected current is determined by re-arranging (6.21).

$$Ifr = Ifa - Ifi \tag{6.29}$$

Using (6.28) to substitute for *Ifa* in (6.29) leads to the standard equation for the reflection coefficient found in many electromagnetic theory textbooks.

$$K = \frac{Ifr}{Ifi} = \frac{Ro - RL}{Ro + RL}$$

This coefficient is not utilised here because the load at the far end could be any mixture of resistors, inductors and capacitors. It is important to know the value of the current actually delivered to the receiver; *Ifa*. The magnitude of this current is a function of the state of the interface circuitry at the far end. Once *Ifa* is known, the value of *Ifr* can be obtained from (6.29).

Current reflected by the far end termination, *Ifr*, travels back along the line and manifests itself as incident current *Ini* at the terminals at the near end, as shown in Figure 6.9.

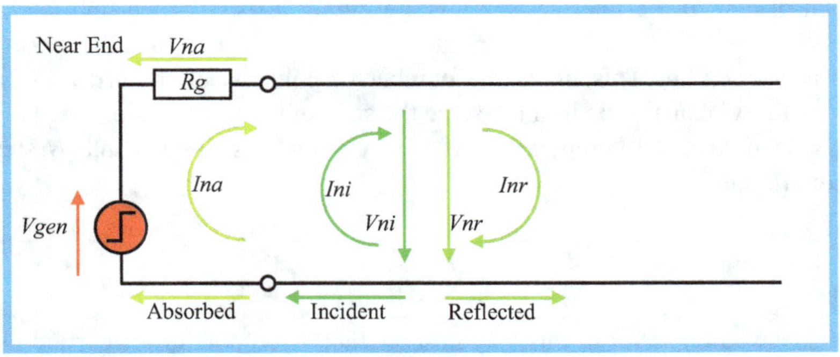

Figure 6.9 Currents and voltages at the near end of the line

Since positive loop current is defined as clockwise on all diagrams, and since power is delivered by the line to the near end terminals, the incident voltage *Vni* must necessarily be inverted with respect to the voltage *Vfr*.

At any instant, the voltages at the near end are:

$$Vni + Vnr + Vgen = Vna \tag{6.30}$$

$$Ini + Inr = Ina \tag{6.31}$$

$$Vni = Ro \cdot Ini \tag{6.32}$$

$$Vnr = -Ro \cdot Inr \tag{6.33}$$

$$Vna = Rg \cdot Ina \tag{6.34}$$

Using the same process that was used to analyse currents and voltages at the far end, the relationships at the near end become:

$$Ina = \frac{2 \cdot Ro \cdot Ini + Vgen}{Ro + Rg} \tag{6.35}$$

And

$$Inr = Ina - Ini \tag{6.36}$$

Unlike the computations involving lumped parameters, where the status of the system at time *t* is dependent on its status at time *t-dt*, the status of the equipment interface at each end of the transmission line is dependent on the status of the interface at the other end of the line at time *t-T*.

Since the current at each end of the line is time-dependent, it is necessary to store data on instantaneous values for a large number of time steps. Such a requirement calls for a table of values to be recorded, each record holding data on parameter values for a defined instant in time. An array needs to be created to hold this data.

Fortunately, it is unnecessary for this array to store a large number of variables for each instant. If the interface circuits are purely resistive, then only two parameters are required: the reflected current at the far end, *Ifr*, and the reflected current at the near end, *Inr*. This limits the number of columns of the array to 2. The number of rows can also be limited, since the state of the transmission line before *t-T* plays no part in the computation. If the duration *T* is divided into *N* steps of duration *dt*, then

$$dt = \frac{T}{N} \tag{6.37}$$

Defining the value for *dt* in this way ensures that a computation is carried out for every instant a transient arrives at a termination.

This means that the table should contain *N* records and that each record should contain two values. Since it is necessary to scan the table several times during any particular simulation, there needs to be some way of relating the time of each event to the appropriate record.

If the time interval between each sample is always *dt*, then the relationship between computation number *n* and time *t* is given by

$$t_n = n \cdot dt \tag{6.38}$$

Table 6.1 illustrates the necessary correlation between computation number, *n,* and the relevant column, *p,* of the table, given the assumption that the number of columns, *N*, is 10.

n	1	2	3	4	5	6	7	8	9	10
n	11	12	13	14	15	16	17	18	19	20
n	21	22	23	24	25	26	27	...	...	...
p	1	2	3	4	5	6	7	8	9	10
Inr										
Ifr										

Table 6.1 Relating each data sample to the computation number

The diagram of Figure 6.10 illustrates a configuration in which a signal is transmitted from one end of a twin-conductor line to the other. The time taken for a transient to propagate from one end to the other is 100 ns and the characteristic impedance is 100 Ω. The signal source at the near end is a voltage generator *Vgen* with an output impedance of 10 Ω. The resistor *RL* at the far end provides a load impedance of 1K Ω.

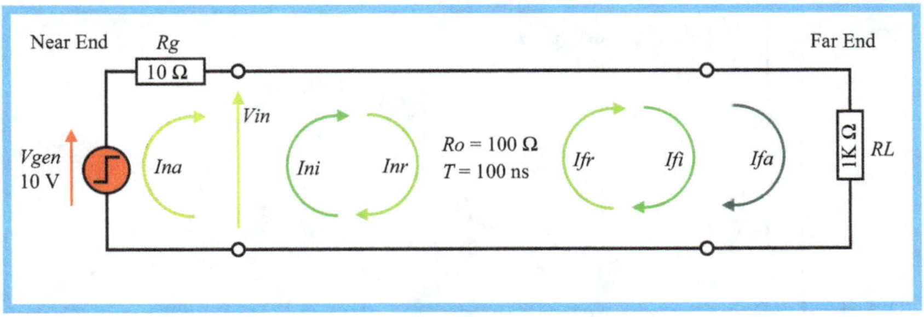

Figure 6.10 Delay line model

Determining the response of this signal link to a step voltage is simply a matter of assembling the relevant equations into an ordered sequence. The worksheet of Figure 6.11 illustrates one method.

The top two lines of the worksheet are derived directly from the characteristics of the configuration-under-review. On the third line, the number *N1* is set at 100. This defines the number of calculations to be carried out during a single traverse of a transient pulse. The value of *N1* is selected to allow the simulation of 30 such sweeps.

The function *send (INPUT, Vgen)* carries out the computations which define the currents existing at the near end of the line at any particular instant, the time *t*. It responds to two input variables.

The voltage *Vgen* is the amplitude of the voltage source. For a step pulse, its value is zero at all time before $t = 0$, and 10 V at all times thereafter. For a sine wave, it would be a sinusoidal function of time.

The variable *INPUT* is a two-element vector containing the values of *Inr* and *Ifr* at time *t-T*.

Worksheet 6.2 page 1 — See Figure 6.7

$Ro := 100\ \Omega \qquad Rg := 10\ \Omega \qquad RL := 1000\ \Omega$

$Vgen := 10\ \text{V} \qquad T := 100 \cdot 10^{-9}\ \text{s}$

$N1 := 100 \qquad dt := \dfrac{\text{T}}{\text{N1}} \qquad N2 := 30 \cdot \text{N1}$

$$\text{send}(\text{INPUT}, Vgen) := \left| \begin{array}{l} Ini \leftarrow \text{INPUT}_2 \\ Ina \leftarrow \dfrac{2 \cdot Ro \cdot Ini + Vgen}{Rg + Ro} \\ Inr \leftarrow Ina - Ini \\ \begin{pmatrix} Ina \\ Inr \end{pmatrix} \end{array} \right. \qquad \text{recv}(\text{INPUT}) := \left| \begin{array}{l} Ifi \leftarrow \text{INPUT}_1 \\ Ifa \leftarrow \dfrac{2 \cdot Ro \cdot Ifi}{RL + Ro} \\ Ifr \leftarrow Ifa - Ifi \\ \begin{pmatrix} Ifa \\ Ifr \end{pmatrix} \end{array} \right.$$

$$\text{point}(\text{n}) := \left| \begin{array}{l} m \leftarrow \text{mod}(n, N1) \\ m \leftarrow N1 \ \text{ if } \ m = 0 \end{array} \right.$$

$$Ina := \left| \begin{array}{l} \text{data}_{2,N1} \leftarrow 0 \\ \text{for } i \in 1..N2 \\ \quad \left| \begin{array}{l} p \leftarrow \text{point}(\text{i}) \\ \text{INPUT} \leftarrow \text{data}^{\langle \text{p} \rangle} \\ \begin{pmatrix} Ina \\ Inr \end{pmatrix} \leftarrow \text{send}(\text{INPUT}, Vgen) \\ \begin{pmatrix} Ifa \\ Ifr \end{pmatrix} \leftarrow \text{recv}(\text{INPUT}) \\ \text{OUTPUT} \leftarrow \begin{pmatrix} Inr \\ Ifr \end{pmatrix} \\ \text{data}^{\langle \text{p} \rangle} \leftarrow \text{OUTPUT} \\ I_i \leftarrow Ina \end{array} \right. \\ I \end{array} \right.$$

$n := 1..N2 \qquad t_n := (n-1) \cdot dt$

Figure 6.11 Calculating the waveform of the current at the near end of the line

The first line of this subroutine picks out the value of *Ifr*. Since it is assumed that there are no losses in the line, then this value is the amplitude of the current *Ini* arriving at the near-end terminals at time *t*.

Given that the values of *Vgen* and *Ini* are now available to the subroutine, the values of *Ina* and *Ini* can be calculated using (6.35) and (6.36). The output of the subroutine is a two-element vector containing the values of these two parameters at time *t*.

The subroutine *recv(INPUT)* performs a similar set of calculations for currents at the far end of the line and returns a two-element vector with the values of *Ifa* and *Ifr*.

The *point(n)* function takes as input the number *n* of the sequence of computations and uses it to calculate a value for *p*, a pointer to the appropriate record in Table 6.1.

In Mathcad, the function *mod(n,N)* returns the remainder of *n* when divided by *N*.

The main program is used to calculate a set of values for *Ina*, the current delivered to the input terminals of the transmission line.

The first action of the main program is to define the array *data* as having two rows and *N1* columns. Initially, all the values are set to zero.

A control variable *i* is set to run from 1 to *N2*, the total number of computations involved.

For each computation, the value of the integer *p* is calculated. This points to the appropriate column in the data array, and the record stored in that column is defined as the vector *INPUT*. This is a two-element vector holding the values of *Inr* and *Ifr* at time $t - T$.

These values are treated as *Ifi* and *Ini* at time *t* by the two subroutines *send (INPUT, Vgen)* and *recv (INPUT)*, and values for *Inr* and *Ifr* are calculated. These are placed in a two-element vector, *OUTPUT*, and the contents of this vector are used to overwrite the record in column *p* of the *data* array.

The final action of each computation is to select the local variable *Ina* and place the value in element *i* of the vector *I*.

The output of the main program is a vector containing all the values computed for *Ina*. This parameter can be regarded as the current delivered to the input terminals of the transmission line.

The waveform of this current is illustrated by Figure 6.12. The amplitude of the initial step current is due to a 10 V supply loaded by 10 Ω and 50 Ω in series.

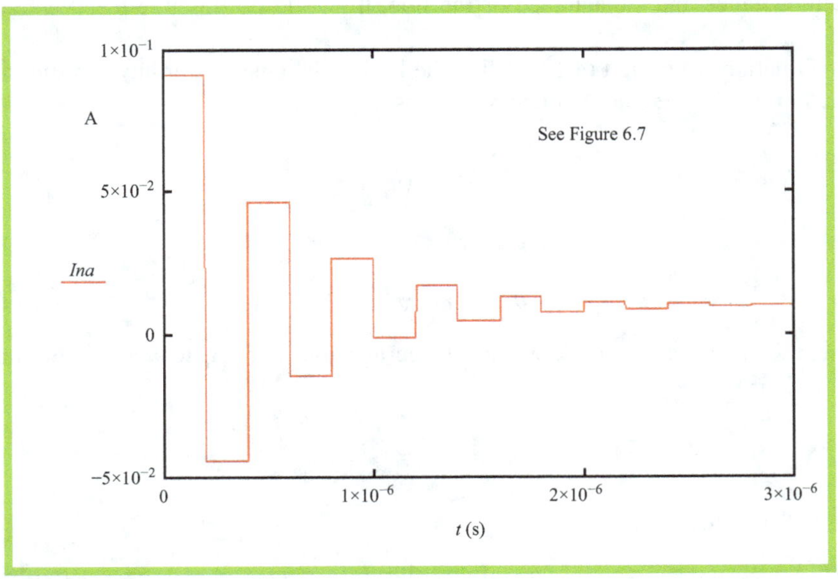

Figure 6.12 Waveform of current delivered to input of transmission line

It is only after several reflections have occurred that the current settles down to its steady state value – the current due to a 10 V supply loaded by 1000 Ω and 10 Ω in series. During the settling-in period, there is a burst of high-frequency oscillation in the system.

The configuration depicted in Figure 6.10 is indicative of many signal links in the average system. In many configurations, the output impedance of the sender is less than 10 Ω and the input impedance of the load is greater than 1000 Ω. This means that there is a brief burst of oscillation on the line every time a voltage step occurs.

Step changes in voltage occur every time a logic signal changes state and every time a device is switched on or off. If the interface circuitry is not matched to the line, then it will also carry a great number of high-frequency transient currents. Moreover, the frequency will be close to that of quarter-wave resonance, the ideal frequency for creating maximum emission.

6.3 Line characteristics

It would seem that the model for the delay line has lost two of the essential features of any circuit model; there are no inductors and no capacitors. However, these parameters have not really disappeared.

The definitive parameters for a loss-free line are the characteristic impedance *Ro* and the transition time *T*. (In transient analysis, the concept of reactance does not exist, since the parameters *L* and *C* are treated separately. Impedance parameters can only be resistive.) Since these have been derived from the values for the inductance and capacitance of the line, it should be possible to reverse the process.

Equations (4.6) and (4.18) define the basic relationships for any transmission line. For a loss-free line, frequency response analysis gives:

$$Ro = \sqrt{\frac{La}{Ca}} \tag{6.39}$$

And

$$\theta = j \cdot \omega \cdot \sqrt{La \cdot Ca} \tag{6.40}$$

Where *La* and *Ca* are the loop inductance and the loop capacitance of the line. From (2.33):

$$\sqrt{La \cdot Ca} = \frac{l}{v} = T \tag{6.41}$$

It follows that:

$$\theta = j \cdot \omega \cdot T \tag{6.42}$$

From (6.39)

$$Ca = \frac{La}{Ro^2} \tag{6.43}$$

Substituting for *Ca* in (6.41) gives:

$$T = \frac{La}{Ro}$$

This removes the *j* operator and the *ω* parameter from the equations and allows inductance to be defined in terms of *T* and *Ro*:

$$La = T \cdot Ro \tag{6.44}$$

Substituting for *La* in (6.43):

$$Ca = \frac{T}{Ro} \tag{6.45}$$

Hence, (6.44) and (6.45) allow the values of *La* and *Ca* to be derived from *Ro* and *T*.

The inductance of a transmission line can be determined by shorting the far-end terminations and applying a voltage to the near end via a resistance of known value. Figure 6.13 illustrates the configuration and Figure 6.15 shows the corresponding model when lumped parameters are used.

Analysing the response of Figure 6.13 is simply a matter of copying the Mathcad worksheet of Figure 6.11, changing the value of *RL* to zero, and re-running the program. This constitutes page 1 of a new worksheet.

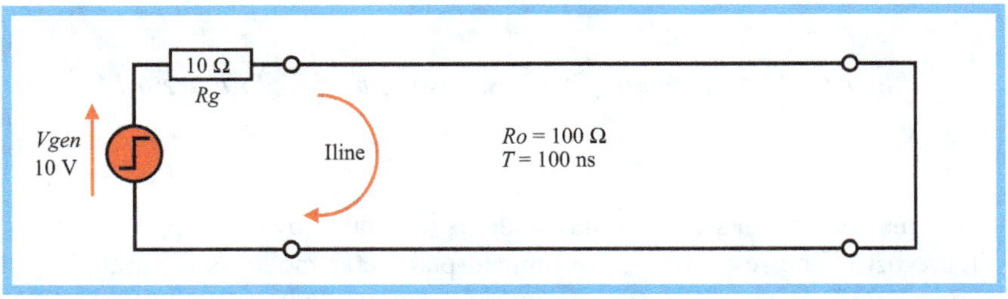

Figure 6.13 Transmission line configured as an inductance

Analysing the response of Figure 6.14 is a matter of following the process described in Section 6.1.1. The loop equation is:

$$Vgen = Rg \cdot I + L \cdot \frac{dI}{dt} \tag{6.46}$$

Re-arranging this equation gives

$$dI = \frac{L}{dt} \cdot (Vgen - Rg \cdot I) \tag{6.47}$$

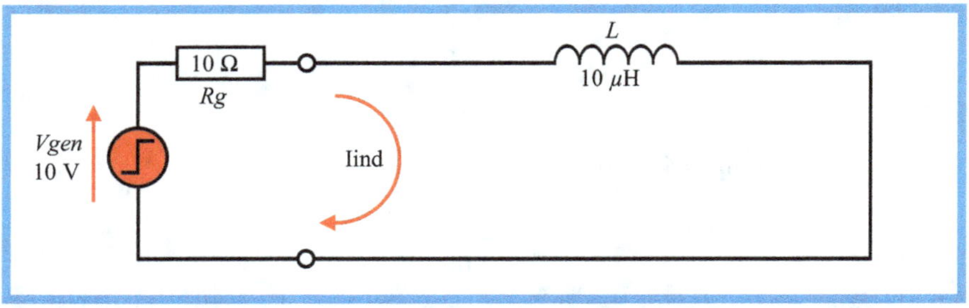

Figure 6.14 Lumped parameter model of short-circuited line

Equation (6.47) forms the core calculation of the program illustrated by Figure 6.15. This is Page 2 of the new worksheet.

Worksheet 6.3.1 page 2

(Page 1 of worksheet is a copy of Figure 6.11 with the value of RL set at zero)

$Rg = 10\ \Omega$ $\qquad RL = 0\ \Omega$ $\qquad Vgen = 10\ \text{V}$

$dt = 1 \times 10^{-9}\ \text{s}$ $\qquad L := T \cdot Ro = 1 \times 10^{-5}\ \text{H}$ $\qquad I := 0\ \text{A}$

$$next\,(I) := \left| \begin{array}{l} \mathrm{dI} \leftarrow \dfrac{\mathrm{dt}}{\mathrm{L}} \cdot (Vgen - \mathrm{Rg} \cdot \mathrm{I}) \\ I \leftarrow Id + I \end{array} \right.$$

$$Iind_n := \left| \begin{array}{l} I \leftarrow next\,(I) \\ I \end{array} \right.$$

Figure 6.15 Calculating the response of model of Figure 6.14

The response of the transmission line model is illustrated by the stepped curve *Iline* of Figure 6.16. The response of the lumped-parameter model is illustrated by the solid curve *Iind*. As expected, the current rises exponentially towards 1 A, which is the steady-state current in 10 Ω due to a constant voltage of 10 V.

Correlation between the responses of the two models is quite evident. It illustrates the fact that the transmission line model indeed has inductive properties, and that (6.44) defines the value of that inductance.

The capacitance of the line can be revealed by open circuiting the terminations at the far end of the line and by increasing the value of *Rg*.

In Figure 6.17, the value of *RL* is changed to 10 MΩ which is an open circuit compared to *Ro*. The value of *Rg* is changed to 1000 Ω, ten times the value of *Ro*.

For this configuration, the lumped-parameter model would be as shown in Figure 6.18, and the value of *C* can be calculated using (6.45).

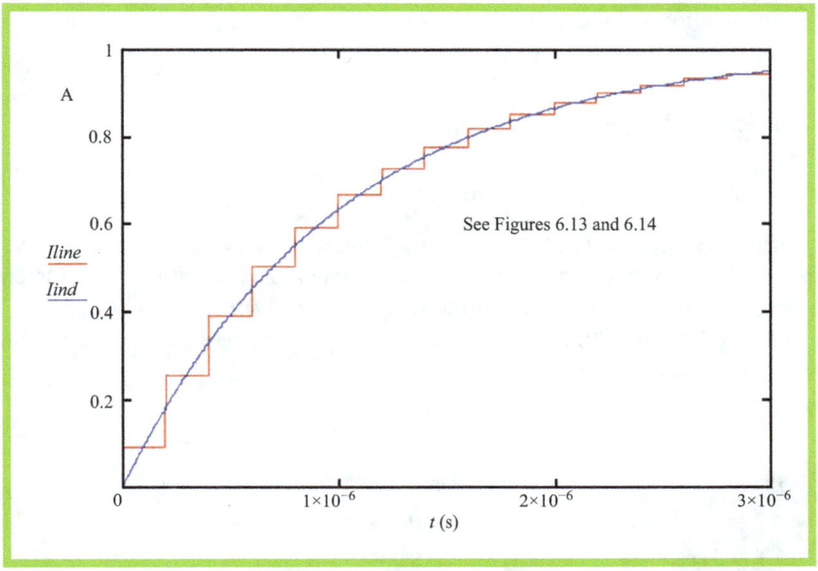

Figure 6.16 Response of short-circuited line and of 10 μH inductor

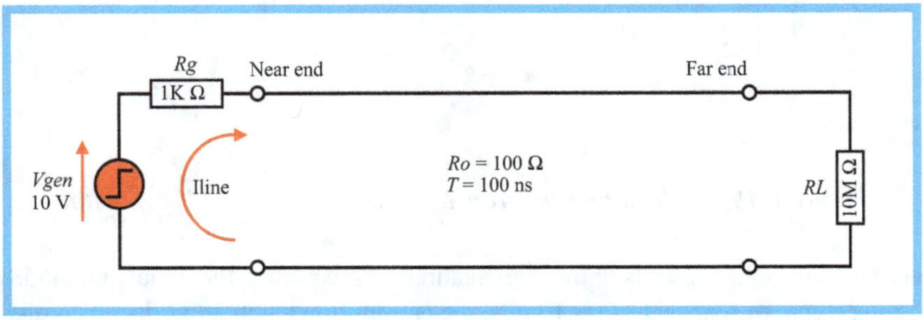

Figure 6.17 Transmission line configured as a capacitor

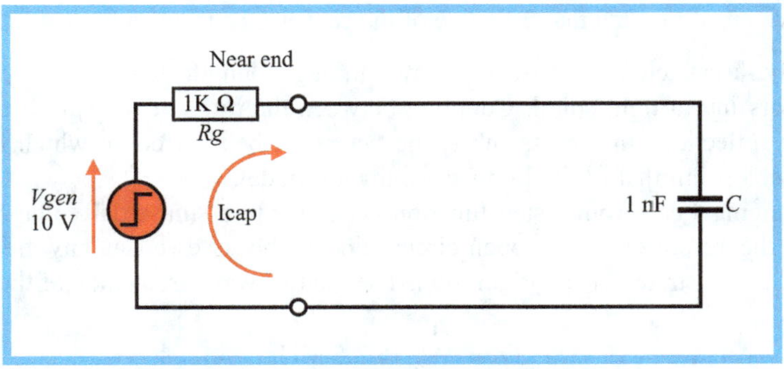

Figure 6.18 Lumped-parameter model of open-circuit line

The loop equation for this model is:

$$Vgen = Rg \cdot I + \frac{Q}{C} \tag{6.48}$$

And this leads to:

$$I = \frac{1}{Rg} \cdot \left(Vgen - \frac{Q}{C} \right) \tag{6.49}$$

Comparing the responses of the open circuit line with that of the R–C circuit can be carried out by invoking a similar procedure to that employed with the assessment of inductance. Figure 6.19 is a copy of the second page of the relevant worksheet. The new values of *Rg* and *RL* are recorded at the top of the page, and the core calculation of the program is derived from (6.49).

Worksheet 6.3.2 Page 2

(Page 1 of this worksheet is a copy of Figure 6.11 with the value of RL set to 10 MΩ)

$Rg = 1 \times 10^{3}\ \Omega \quad RL = 1 \times 10^{7}\ \Omega \quad T = 1 \times 10^{-7}\ \text{s} \quad dt = 1 \times 10^{-9}\ \text{s} \quad C := \frac{T}{Ro} = 1 \times 10^{-9}\ \text{F}$

$Icap :=$
- $Q \leftarrow 0$
- for $i \in 1..N2$
  - $Ic \leftarrow \frac{1}{Rg} \cdot \left(Vgen - \frac{Q}{C} \right)$
  - $Q \leftarrow Q + Ic \cdot dt$
  - $I_i \leftarrow Ic$
- I

Figure 6.19 Calculating the transient current in the 1 nF capacitor

Figure 6.20 shows the results of the computations. As expected, the initial amplitude of the current into the capacitor *C* is 10 mA, the current flowing in 1 kΩ due to a voltage of 10 V. The stepped curve, which tracks this response, is due to the current delivered to the open-circuit transmission line of Figure 6.17. The correlation between the two curves could not be closer. This demonstrates that the transmission line possesses capacitive properties and that the value of the capacitance is given by (6.45).

Normally, a capacitor is constructed by winding a length of two closely spaced conductors into a tight spiral. Coupling between the turns of the spiral results in multiple reflections in the assembly, the net result being a curve which is indistinguishable from that of the lumped parameter model.

Changing *Vgen* from a step-function generator to a sinusoidal voltage source enables the response of the open-circuit line to be assessed at any frequency. Figure 6.21 illustrates the configuration. The quarter-wave frequency of the line is:

$$fq = \frac{1}{4 \cdot T} = 2.5\ \text{MHz} \tag{6.50}$$

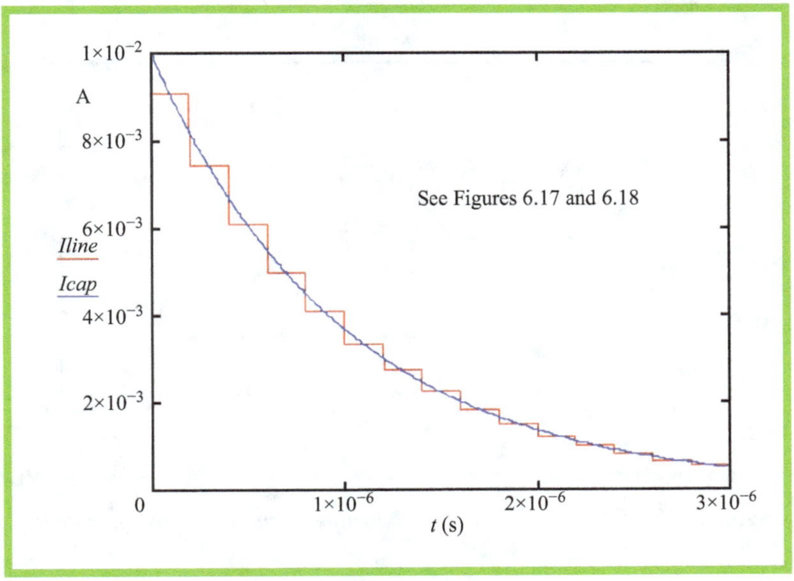

Figure 6.20 Response of open circuited line compared to that of capacitor

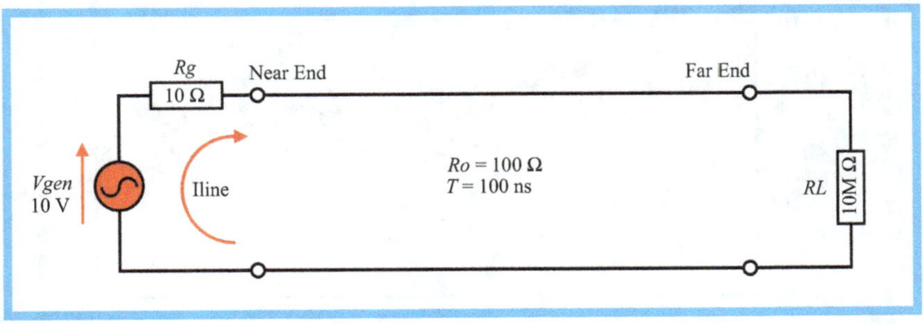

Figure 6.21 Transmission line acting as a tuned circuit

If this is the frequency of the source *Vgen*, then the current delivered would be as shown in Figure 6.22. This illustrates the fact that the amplitude of the current gradually builds up. At every half-cycle, the amplitude is increased by a small increment. A limit is reached when the peak current is 1A, the current in *Rg* when a peak voltage of 10 V is applied.

Hence, when an alternating current at the quarter-wave frequency is applied to an open circuit line, the input terminals appear to be short-circuited. Such a conclusion is exactly the same as that obtained with AC analysis. However, there is a significant difference between the two types of analysis; AC analysis does not predict the gradual build-up of current amplitude.

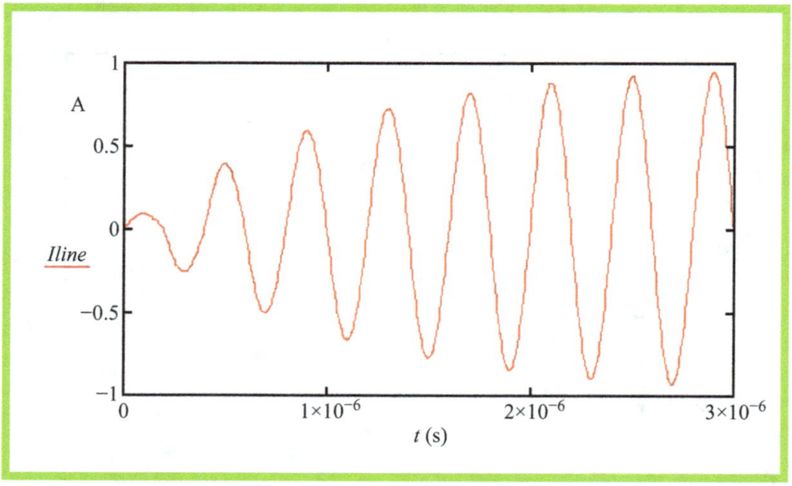

Figure 6.22 Response of open-circuit line at quarter-wave frequency

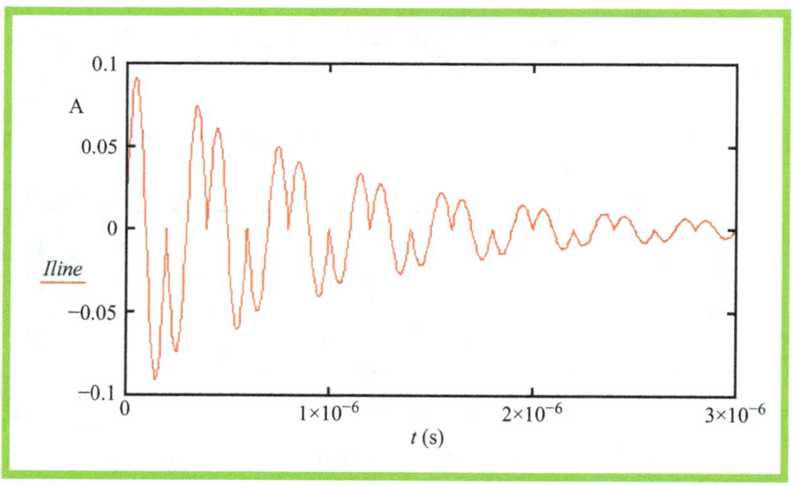

Figure 6.23 Response of open-circuit transmission line at half-wave frequency

If the frequency of the source generator were to be changed to 5 MHz, the half-wave frequency, then the current waveform at the near end of the line would be as illustrated by Figure 6.23.

For the first cycle of the waveform, the current is due to a 10 V sinusoidal signal in a resistance of 110 Ω. That is, the resistance value is equal to $Rg + Ro$. Subsequent cycles of the waveform are progressively reduced. After about 100 cycles, the current delivered to the line is effectively zero. Such a response is due to the fact that the incident current at the near-end of the line generates a voltage across *Rg* which precisely balances the voltage delivered by the source generator *Vgen*.

This means that, as far as the source generator is concerned, a transmission line which is open-circuit at the far end will appear as an open-circuit at the near end. Again, this conclusion is the same as that predicted by AC analysis. However, AC analysis does not predict the existence of a brief burst of current into the line, and that a standing wave exists along the line all the time an input signal is applied.

6.4 Antenna-mode current

Having developed a delay-line model to simulate the action of a twin-conductor cable, the next obvious step was to set up an experiment and observe how an actual line responds to a step input. Comparing the actual waveform with that of the simulation should provide some indication of the accuracy of the model. So, this was done.

A 15 m length of two-core mains cable was purchased, and a signal generator was used to inject a square wave of about 8 μs duration into one end of the line. Figure 6.24 is an illustration of the setup.

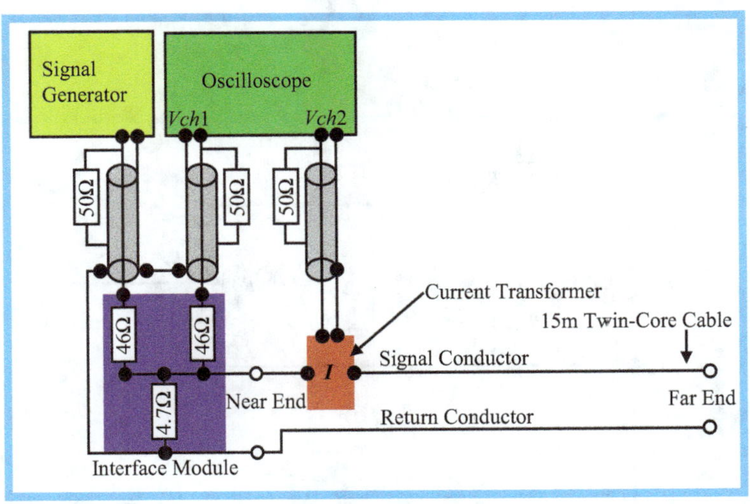

Figure 6.24 Setup for radiated transient test

The interface circuitry at the near end was designed to provide low source impedance, whilst the far end of the cable was open circuit. This provided a configuration in which several reflections could be observed for each step of the input waveform.

The input voltage was monitored by channel 1 of the oscilloscope via a simple potentiometer network, whilst the output current was monitored on channel 2 via a current transformer. Waveforms were recorded as accurately as possible.

The first circuit model created to simulate this waveform was found to be wildly inaccurate. A modified model was found to be equally unrepresentative. This was followed by a trial-and-error process which eventually resulted in the model of Figure 6.25. This produced a waveform which was a fair representation of the trace displayed on channel 2 of the oscilloscope.

What follows is a description of the circuit model and an attempt to relate its operation to the behaviour of the system-under-review. The fact that this approach also seems to be very 'trial and error' is because that's just what it was. However, the end result is the identification of features of electromagnetic coupling which could not be predicted by frequency response analysis.

In this model, *Rg* represents the source impedance of the step-function generator *Vgen,* whilst *Rcable* is a resistor which allows copper losses and dielectric losses to be catered for, and *Ro* represents the characteristic impedance of the cable.

Equations for the near end of the cable are:

$$Vni + Vnr + Vgen = Vna \tag{6.51}$$

$$Ini + Inr = Ina \tag{6.52}$$

$$Vni = Ro \cdot Ini \tag{6.53}$$

$$Vnr = -Ro \cdot Inr \tag{6.54}$$

$$Vna = (Rg + Rcable) \cdot Ina \tag{6.55}$$

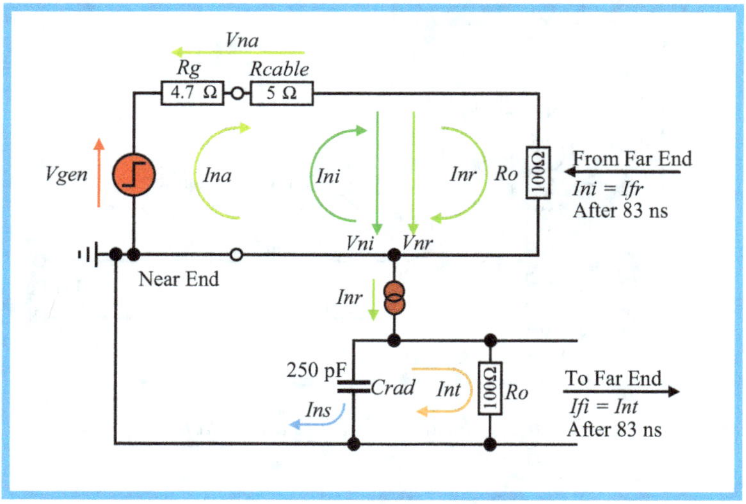

Figure 6.25 Delay line model of 15 m 2-core cable

These are essentially the same as those defined in Section 6.2, with the exception that the parameter *Rcable* is included in (6.55). Using the process described in Section 6.2, it can be deduced that:

$$Ina = \frac{2 \cdot Ro \cdot Ini + Vgen}{Ro + Rg + Rcable} \tag{6.56}$$

Re-arranging (6.52)

$$Inr = Ina - Ini \tag{6.57}$$

This current transient travels down the line at a velocity approaching that of light. During this transit, most of the current flows from the signal conductor, through the characteristic impedance *Ro*, and back via the return conductor. (Most of the current delivered to the signal conductor is derived from current flow via the conductors of the structure and the source resistance *Rg*. Section 6.5 deals with this aspect of the phenomenon in more detail.) However, not all of the differential-mode current is delivered to the terminations at the far end. The current which does not arrive is represented by *Ins* flowing in the radiation capacitor *Crad*. The energy apparently lost to the environment due to current flow in the resistor *Ro* is effectively stored as a voltage across the capacitor.

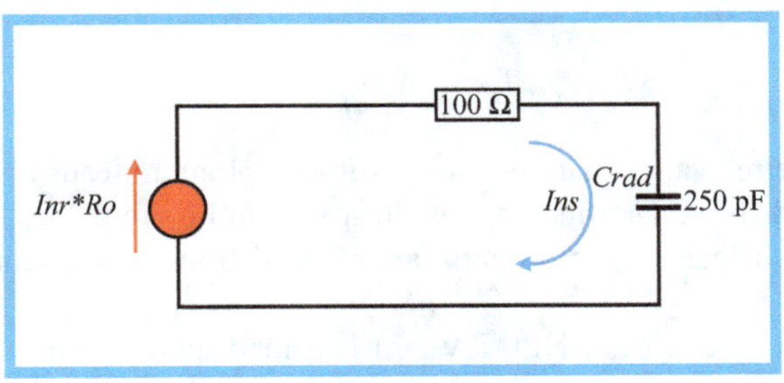

Figure 6.26 Calculating the value of the stored current

The amplitude of *Ins* can be calculated using Figure 6.26, since a current source in parallel with a resistor can be represented by a voltage source in series with that resistor.

The loop equation for this circuit is:

$$Inr \cdot Ro = Ins \cdot Ro + \frac{Qns}{Crad} \tag{6.58}$$

Hence

$$Ins = Inr - \frac{Qns}{Ro \cdot Crad} \tag{6.59}$$

After each time step, the value of *Qns* changes. Using Mathcad terminology:

$$Qns \leftarrow Qns + Ins \cdot dt \tag{6.60}$$

The current actually transmitted along the cable must be:

$$Int = Inr - Ins \tag{6.61}$$

After a delay of T seconds, this transmitted current arrives at the far end. Again, using Mathcad terminology:

$$Ifi \leftarrow Int \tag{6.62}$$

In the model, it is assumed that the open circuit is represented by a high-value resistor RL, and a value of 10 MΩ is assigned to this parameter.

The current absorbed in RL is:

$$Ifa = \frac{2 \cdot Ro \cdot Ifi}{Ro + RL} \tag{6.63}$$

And the current reflected from the far end is:

$$Ifr = Ifa - Ifi \tag{6.64}$$

After a further time delay of T seconds, the incident current arriving at the near end is given by:

$$Ini \leftarrow Ifr \tag{6.65}$$

Assembling the relevant equations into a software program leads to the creation of a Mathcad worksheet. The first page is illustrated in Figure 6.27.

The top two lines of the program are derived from component values of the circuit model.

The function send(near,INPUT,Vgen) is a modified version of send(INPUT, Vgen) of Figure 6.11. The modifications are essentially the inclusion of (6.59)–(6.61).

Since the states of the currents and voltages at the near end of the line are now a function of Ins and Qns, the values of these parameters need to be available as an input to the subroutine. This is done through the use of the vector 'near'. Since this vector is also used to provide updated values of Ina and Int to the main program, these parameters are also included as input variables (even though they're not actually used by the subroutine).

The function recv(INPUT) is essentially the same as that defined in the worksheet of Figure 6.11.

Figure 6.28 illustrates the main program. It is a development of the main program of Figure 6.11. To correlate the simulated waveform with the actual waveform displayed on channel 2 of the oscilloscope, it was necessary to define the time of the leading edge T1 and the sweep time T2. Since all the time steps in the program are equal in value, it is a simple matter to identify the step-counts N1 and N2 at which they occur.

Since the vector 'near' is used in the near-end calculations, this vector needs to be declared at the start of the main program. It contains four variables. So, the number of rows is four.

Worksheet 6.4 Page 1

$Ro := 100\,\Omega \quad Rg := 4.7\,\Omega \quad Rcable := 5\,\Omega \quad RL := 10^{7}\,\Omega \quad \mathrm{T} := 83\cdot 10^{-9}\,\mathrm{s} \quad Vg := 1\mathrm{V}$

$Crad := 250\cdot 10^{-12}\,\mathrm{F} \quad \mathrm{N} := 100 \quad \mathrm{dt} := \dfrac{\mathrm{T}}{\mathrm{N}}$

$$\text{send (near, INPUT, } Vgen) := \left| \begin{array}{l} \begin{pmatrix} Ifi \\ Ini \end{pmatrix} \leftarrow \text{INPUT} \\ \begin{pmatrix} Ina \\ Int \\ Ins \\ Qns \end{pmatrix} \leftarrow \text{near} \\ Ina \leftarrow \dfrac{2\cdot Ro\cdot Ini + Vgen}{Rg + Ro + Rcable} \\ Inr \leftarrow Ina - Ini \\ Ins \leftarrow Inr - \dfrac{Qns}{Ro\cdot Crad} \\ Int \leftarrow Inr - Ins \\ Qns \leftarrow Qns + Ins\cdot dt \\ \begin{pmatrix} Ina \\ Int \\ Ins \\ Qns \end{pmatrix} \end{array} \right.$$

$$\text{recv (INPUT)} := \left| \begin{array}{l} \begin{pmatrix} Ifi \\ Ini \end{pmatrix} \leftarrow \text{INPUT} \\ Ifa \leftarrow \dfrac{2\cdot Ro\cdot Ifi}{RL + Ro} \\ Ifr \leftarrow Ifa - Ifi \\ \begin{pmatrix} Ifa \\ Ifr \end{pmatrix} \end{array} \right.$$

Figure 6.27 Subroutines for near-end and far-end calculations

The first step of the iterative subroutine of the main program is to set the initial value of *V*gen to zero. At time *T*1, it is switched to the value *V*g. The value of *V*g is defined on the first page of the worksheet. The most significant modification to the subroutine of Figure 6.11 is due to the need to update the values of the parameters *Ins* and *Qns*. So, the number of rows in the 'near' vector changes from two to four.

After all the values have been updated, the final action of the iterative subroutine is to place the value of Ina in the appropriate row of the vector Out. When the iterations have been completed, the contents of this vector are transferred to the vector 'Idiff'.

Worksheet 6.4 page 2

$$\text{point}(n) := \begin{vmatrix} m \leftarrow mod(n, N) \\ m \leftarrow N \text{ if } m = 0 \end{vmatrix}$$ See Table 6.1

$T1 := 100 \cdot 10^{-9}$ s start time $N1 := \dfrac{T1}{dt}$ $T2 := 2 \cdot 10^{-6}$ s end time $N2 := \dfrac{T2}{dt}$

```
Idiff := | data[2,N] ← 0
         | near[4] ← 0
         | for i ∈ 1..N2
         |   | Vgen ← Vg if i > N1
         |   | p ← point(i)
         |   | INPUT ← data<p>
         |   | near ← send(near, INPUT, Vgen)
         |   | (Ina, Int, Ins, Qns)ᵀ ← near
         |   | (Ifa, Ifr)ᵀ ← recv(INPUT)
         |   | OUTPUT ← (Int, Ifr)ᵀ
         |   | data<p> ← OUTPUT
         |   | Out[i] ← Ina
         | Out
```

$n := 1..N2$ $t_n := (n-1) \cdot dt$

Figure 6.28 Calculating the waveform of the current at the near end of the line

Figure 6.29 shows the waveform of the current *Ina* of the circuit model of Figure 6.25. This simulates the current that is being monitored by the current transformer in the setup of Figure 6.24. This waveform is extremely informative.

The first rising edge is exactly as would be predicted by the simple delay-line model of Section 6.2. After this first step, the source delivers a constant current to the line. The amplitude is due to the application of the voltage *Vgen* to *Rg, Rcable*, and *Ro* in series.

This constant current is maintained throughout the time taken by the transient edge to travel to the far end and for the reflected pulse to travel back to the near end. During this period, current delivered to the capacitance *Crad* is totally invisible to the source generator at the near end. It does not appear in the waveform monitored by channel 2 of the oscilloscope.

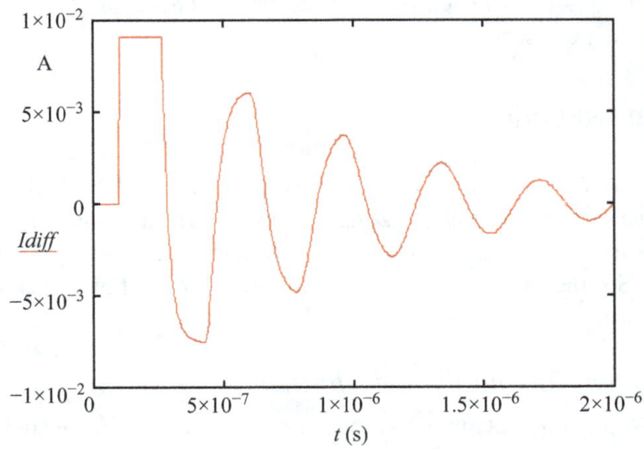

Figure 6.29 Waveform of current at near end of transmission line

When the first leading edge reaches the far end, all the current is reflected straight back into the line. Its amplitude is unchanged, but its sign is reversed. Positive current leaving the near end appears as a negative current on its return.

Since the load at the near end is much less than *Ro*, the amplitude of the reflected current is almost doubled.

If the simple delay-line model had been used, then this current would have caused a step change in the line current from positive to negative. This does not happen in practice. The trailing edge is an exponential curve, providing a clear indication that some of the current did not reach the far end. The lost energy is effectively stored as a voltage across *Crad*.

When the voltage step reaches the far end, current *Inr* of Figure 6.26 ceases to flow, and the amplitude of the voltage source *Inr·Ro* drops to zero. Antenna mode current *Ins* now flows back from the cable into the terminals at the near end.

Since the edge of the waveform is now distributed in time, and since subsequent traverses of the transients are subjected to the same physical phenomena, the result is a gradual transition from a square waveform to a sinusoidal waveform. After only a few cycles, the mismatched line is resonating at its natural frequency.

The amplitude of the waveform gradually decays and after about 3 μs is virtually undetectable.

It should be emphasised at this point that the component values of the circuit model were adjusted iteratively until the simulated waveform bore a close resemblance to the waveform monitored on the oscilloscope. In fact, the development of the model and the explanation of how it simulated the physical phenomena went hand-in-hand.

Although the explanation so far provided is reasonably plausible, there are still some puzzling questions raised by the test results. These questions are posed and answered in the next section.

6.5 Radiated emission

Although the analysis of Section 6.4 created a circuit model which replicated the observed phenomena, several puzzling features of the test results remain. Chief among these is the fact that the antenna-mode current is not visible to the current transformer. So, the question: 'why does the transformer not detect this current?' needs an answer.

6.5.1 Current linking the transformer

There can be no doubt that antenna-mode current is flowing in the signal conductor, since that is the only conductor being energised. The return conductor is held at ground potential at the near end. The fact that the transformer of Figure 6.24 does not detect this component of the current means that there must be antenna-mode current in the return conductor as well.

Earlier tests on the same cable, using a sinusoidal source, had demonstrated quite clearly that the coupling between the conductors is close enough to guarantee that most of the current emanating from the signal conductor is picked up by the return conductor and flows in the opposite direction (see Section 7.5).

Also, the test of a single conductor, described in Section 7.4, illustrates the fact that all the current transmitted along one monopole is provided by the current delivered from the other monopole. With the setup shown in Figure 7.50, the signal conductor can be regarded as one monopole, whilst the other monopole exists in the form of the assembly of earthed conductors of the test equipment.

It is not plausible to assume that the send and return components of the differential-mode current propagate simultaneously along the cable. It is true that if the signal source in Figure 7.50 had been completely isolated from structural conductors, then the voltages applied to the input terminals of the cable would have been equal in magnitude and opposite in sign. But it is not isolated.

Since the return terminal is connected to both the return conductor and to ground, it follows that the voltages applied to the signal and return conductors are unbalanced. Since the primitive capacitance of the ground conductors is much greater than that of the return conductor, that unbalance must be significant. The amplitude of the current flowing from the return conductor into the ground conductors has to be much less than that delivered into the signal conductor.

However, the fact that the antenna-mode current must emanate from the ground conductor does not explain why it is flowing in the same direction as the signal current. A process of elimination means that the energy causing this outward flow can only come from the voltage source, via the signal conductor.

The picture emerges of a current transient propagating along the surface of the signal conductor and creating the wavefront of an electromagnetic field. This spreads out in the same way as the bow wave of a ship. Since the conductors of the transmission line are 2 mm apart, the wavefront does not reach the return conductor until the current pulse has propagated at least 2 mm along the signal conductor.

When the wavefront does reach the return conductor, it induces a voltage on the surface. This voltage creates a current which flows back towards the near-end. Current in the return conductor creates an electromagnetic field of its own, a field which spreads out to enclose the signal conductor. This field tends to neutralise the field emanating from the signal conductor, but not all of it. The coupling cannot be 100 per cent efficient. So, the amplitude of the induced current is less.

By the time the field from the return conductor reaches the signal conductor, the advancing wavefront has progressed a further 2 mm. Since this happens for every increment of length, it must happen along the whole length.

Behind the wavefront, there is an increasing length of cable in which current is flowing in one direction in the signal cable, and in the opposite direction in the return conductor. Since the current in the return conductor is slightly less, there must be an unbalanced current flow in the cable. Since both conductors are now acting as transmitting antennae, the net result is that there is a residual current flow from the near end. Mutual coupling ensures that this antenna-mode current is shared equally between the two conductors.

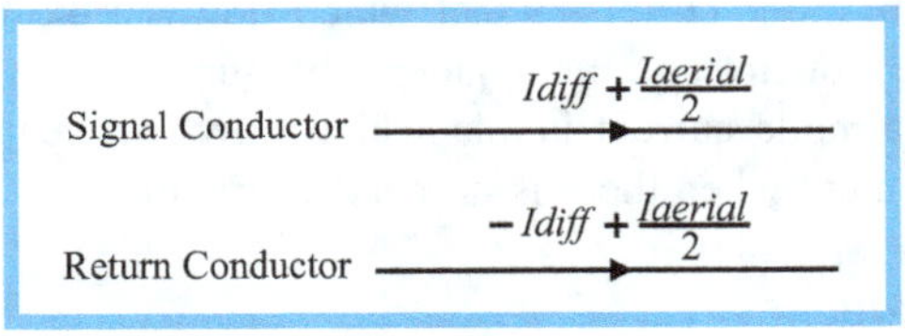

Figure 6.30 Unidirectional current in cable

Hence, there are two current components flowing in each conductor: the differential-mode current and the antenna-mode current. Figure 6.30 illustrates the current flow, viewed from the point of view of a unidirectional current. This shows that there is a net flow outward of *Iaerial*. At the wavefront, antenna-mode current is flowing in the forward direction along both conductors.

Figure 6.31 views the same two conductors, this time from the point of view of loop current. Magnetic material in the transformer responds only to the current which encloses the core. Since the antenna-mode current in one conductor balances the antenna-mode current in the other, this part of the current flow is 'invisible' to the transformer.

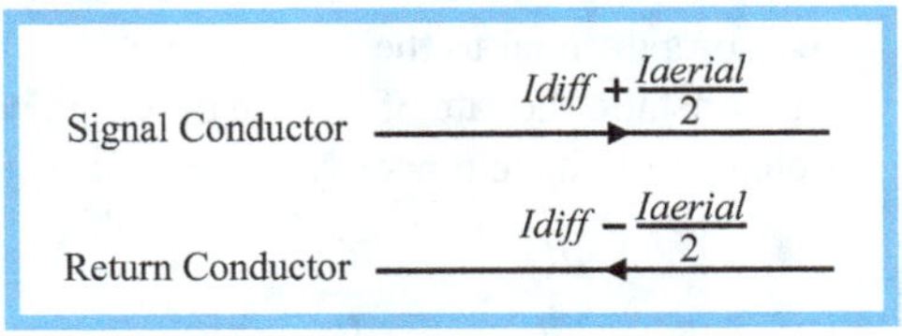

Figure 6.31 Loop current in cable

In classical transmission-line theory, the concept of partial currents was introduced to explain the behaviour of reflections at the terminations. The above reasoning indicates that there is at least one more partial current to add to the incident current and reflected current; antenna-mode current.

To summarise: Electric charge is deposited on the return conductor. Induced voltage causes a current to flow back along the return conductor towards the source. This current creates an electromagnetic field which couples with the signal conductor and that also radiates outward. Since the current in the return conductor does not completely balance that in the signal conductor, there is an unbalanced charge which propagates along the cable, just behind the wavefront. In addition, there is antenna-mode current which propagates outwards along both conductors and is converted into an electromagnetic wave. This means that there are at least three current components involved:

- That which carries a differential-mode signal.
- That which deposits an unbalanced charge on the cable.
- That which flows out radially in the form of electromagnetic radiation.

It is useful at this point to make a distinction between common-mode current and antenna-mode current:

- Common-mode current is that which flows in the loop formed by the return conductor and the adjacent structure.
- Antenna-mode current is that which flows between the cable and the environment when there is no nearby structure.

6.5.2 Line voltage

The reasoning also leads to an explanation for the use of lumped parameters in the delay-line model of Figure 6.25. Conventional delay-line theory caters for the fact that, as the wavefront progresses, the voltage between the conductors undergoes a step change. It does not cater for a situation where the step change on one conductor is less than the step change on the other.

If such a situation is taken into account, it means that there is also an unbalanced voltage component which progresses along the cable.

Conductive material behind the step is charged; conductive material in front of the step is uncharged. Effectively, the charged portion of the system capacitance is increasing, instant by instant. For an elemental length of cable:

$$dQ = V \cdot dC \tag{6.66}$$

where dC is the capacitance of that element. Since the cross-section of the cable is constant, the capacitance is proportional to the length which has been charged. Since the propagation velocity is constant, the rate of change of length with time is constant. So, the apparent rate of change of capacitance with time is constant. From (6.66):

$$\frac{dQ}{dt} = V \cdot \frac{dC}{dt} \tag{6.67}$$

Equation (6.67) is applicable to a situation where a constant voltage is being applied to a capacitance which varies with time.

Conventional circuit theory makes the assumption that the capacitance is constant and that the parameter which varies with time is the voltage. That is

$$\frac{dQ}{dt} = C \cdot \frac{dV}{dt} \tag{6.68}$$

It follows from (6.67) and (6.68) that

$$C \cdot \frac{dV}{dt} = V \cdot \frac{dC}{dt} \tag{6.69}$$

This relationship justifies the existence of lumped parameters in the delay-line model of Figure 6.25.

6.5.3 Source current and voltage

From the point of view of the source generator, another part of the picture emerges. Since *Vgen* is the only source of power in the system, it must supply both the differential-mode current and the antenna-mode current. Figure 6.32 shows that the current in the resistor *Rg* must be the sum of *Idiff* and *Iaerial*.

As far as the generator is concerned, the source of *Iaerial* must be the ground conductor. Such a deduction is entirely plausible, since this conductor contains a large surface area of conducting material. This being so, it must have a high capacitance, a low inductance, and hence, a low characteristic impedance.

This means that, at the output terminals of the supply, all the antenna-mode current is flowing in the signal conductor. The signal conductor is behaving as a transmitting antenna and the return conductor as a receiving antenna. Differential-mode current is delivered to the return conductor whilst the rest is converted into electromagnetic radiation.

From Figure 6.32, the voltage applied to the input terminals of the cable is *Vin*, where:

$$Vin = Vgen - Rg \cdot (Idiff + Iaerial)$$

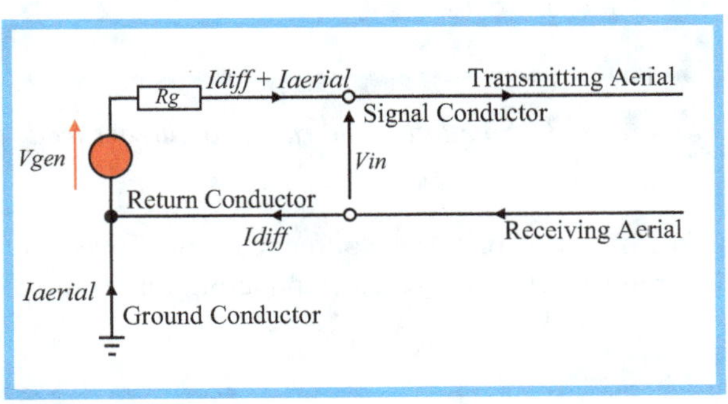

Figure 6.32 Current sharing at source generator

Comparing Figure 6.32 with Figure 6.25 allows the parameters of the two models to be correlated. That is, the current *Ina* used in the calculations corresponds to *Idiff*, and the antenna-mode current *Ins* corresponds to *Iaerial*. This gives:

$$Vin = Vgen - Rg \cdot (Ina + Ins) \tag{6.70}$$

6.5.4 Radiated current

Current used to deliver an unbalanced charge to the transmission line is only one component of antenna-mode current.

Developing the model to simulate the current which is actually being radiated involves one further step in the reasoning. In conventional textbook analysis of transmission line transients, it is necessary for the inductive and capacitive parameters to undergo a metamorphosis and reappear as a resistance *Ro* and a time delay *T*. The relationships are set out in Section 6.3.

For a two-conductor cable, normal practice is to represent the characteristic impedance as a single resistor.

However, the modelling process for alternating signals has revealed that this single resistor provides an inadequate representation of the observed phenomenon. Three conductors are necessary: the signal conductor Ro_1, the return conductor Ro_2, and the virtual conductor Ro_3. Figure 6.33 provides a more revealing picture of the impedance presented to the output terminals of the signal source.

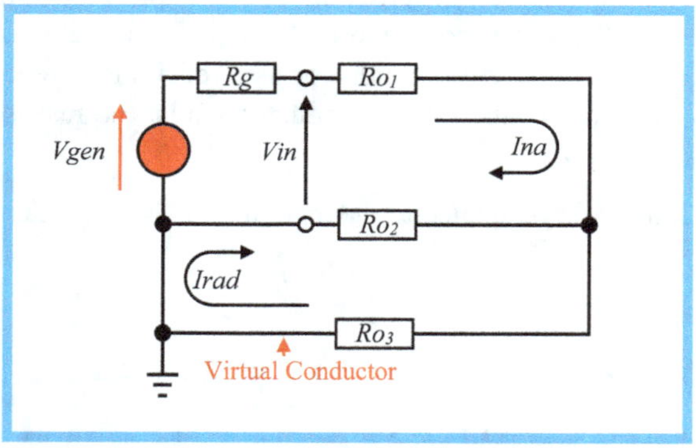

Figure 6.33 Simulating the radiated current Irad

This figure makes it clear that the source current flows along the signal conductor. Most returns via the return conductor, but some flows into the environment.

Calculating the value for these three components is relatively simple, using the equation:

$$Ro_i = \sqrt{\frac{Lc_i}{Cc_i}} \tag{6.71}$$

Where i is an integer identifying the conductor, and Lc_i and Cc_i are defined by (5.25) and (5.26).

The resistance *Ro* as 'seen' by the output terminals of the source at the near-end is:

$$Ro = Ro_1 + \frac{Ro_2 \cdot Ro_3}{Ro_2 + Ro_3} \tag{6.72}$$

The relationship between *Irad* and *Ina* is:

$$Irad = \frac{Ro_2}{Ro_2 + Ro_3} \cdot Ina \tag{6.73}$$

Since the far end is isolated from the ground, there is no third conductor available to deliver extra charge to the cable. So, the primitive voltages existing at the far-end terminals are perfectly balanced with respect to the environment. Even so, both conductors act as transmitting antennae, as far as returning current from the far end is concerned.

6.5.5 Cable losses

Any transmission line provides a highly efficient means of carrying an electromagnetic signal from one location to another. Even so, there are losses:

- Differential-mode energy is stored in the cable. This is released into the environment when the line is disconnected.
- Some energy is used to heat up the conductors, due to copper losses. This can be simulated by inserting a resistor in series with the source resistance.

  Since transient currents are involved, skin effect will cause the value of this resistor to be much higher than the steady-state resistance of the conductors.
- There could be losses in the dielectric material of the insulation. These could be simulated by placing a high-value resistor between the two conductors at the generator end of the line.
- Some of the transient differential-mode current radiates away. When the polarity of the radiation from one conductor is positive, the polarity of the radiation from the other is negative. If the cable is twisted, this differential-mode radiation tends to cancel itself out at a short distance from the line.
- There is also a transient antenna-mode current *Irad* which emanates from the cable in the form of electromagnetic radiation. Figure 6.33 shows how this can be simulated. Current flows into the cable from the ground conductor to replace this loss. This is the same phenomenon that was analysed in Section 5.2, using sinusoidal waveforms.

Section 6.4 has shown that, as well as propagating down the line, the current transient leaves a residual charge on the line. The amplitude of the stored charge can be determined by the model of Figure 6.25. If a step voltage is applied to the input terminals, current *Ins* flows into *Crad* for $2 \cdot T$ seconds; that is, the time it takes for the step to be reflected back to the input terminals. Since this charge is eventually converted back into a current and the energy dissipated in the resistors *Rg* and *Rcable*, this stored energy also represents a loss.

6.5.6 Line parameter measurements

Section 6.3 has shown that when the delay-line model is used to simulate the response of a lossless line, the inductive and capacitive parameters disappear from the equations. They are replaced by the characteristic impedance and the propagation time. That is, *L* and *C* are replaced by *Ro* and *T*.

If a step pulse is applied to a line with a mismatched termination at the far end, then a steady current *I* will flow into the line for $2 \cdot T$seconds. Figure 7.47 shows how this time period and current can be measured for an open-circuit line.

From (6.45):

$$C = \frac{T}{Ro}$$

Substituting $\frac{V}{I}$ for Ro in this equation gives:

$$C = \frac{I \cdot T}{V} \tag{6.74}$$

Similarly, from (6.44):

$$L = T \cdot Ro$$

Substituting $\frac{V}{I}$ for Ro:

$$L = \frac{V \cdot T}{I} \tag{6.75}$$

Since *V*, *I*, and *T* are known, both *L* and *C* can be calculated.

6.6 Transient emission model

The reasoning of the previous section points the way to a further development of the model used to simulate transient radiation; a development which allows all the current and voltage components to be simulated.

In Figure 6.34, the resistor representing the characteristic impedance of the transmission line is replaced by a potentiometer network. Resistors *Ro*1 and *Ro*2 represent the characteristic impedances of the signal and return conductors, respectively, whilst *Ro*3 represents the characteristic impedance of the virtual conductor.

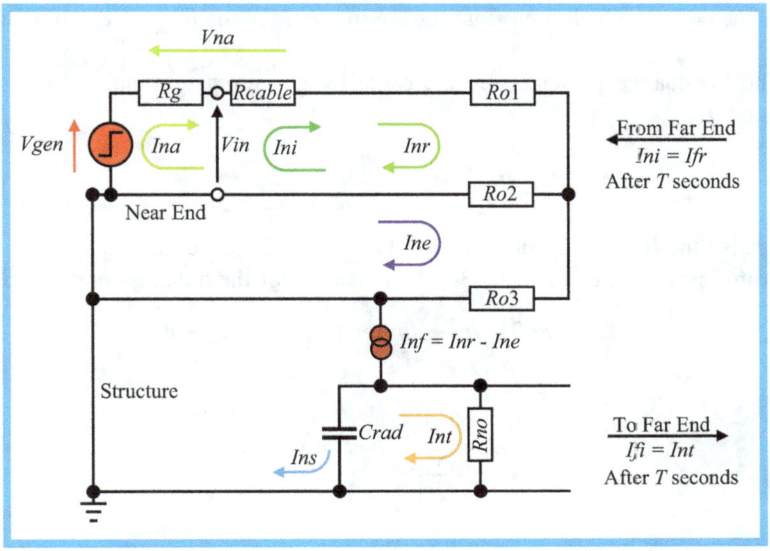

Figure 6.34 General circuit model for transient emission

As in the earlier model, *Ina*, *Ini* and *Inr* represent the absorbed, incident, and reflected currents at the near end, whilst *Ifa*, *Ifi* and *Ifr* represent currents at the far end. The current *Ine* is that which departs into the environment in the form of electromagnetic radiation. As far as the transmission line is concerned, this represents current which has disappeared from the system. Consequently, there is less current available to lay down antenna-mode charge on the conductors. This current is identified in the diagram as *Inf*. This means that there are three components to the current reflected from the near end:

- That which is converted into electromagnetic radiation; *Ine*.
- That which is stored on the outer surface of the cable; *Ins*.
- That which is actually transmitted to the far end; *Int*.

Of course, there are other losses due to

- Voltage-drop due to current in the resistance of the two conductors.
- Voltage-drop due to current in the dielectric material of the insulation.

These latter two effects are due to the conversion of electrical energy into heat energy.

Losses due to differential-mode radiation are indistinguishable from those due to losses in the dielectric material, and could be simulated by a resistor connected between the two cable conductors. In the model of Figure 6.25, all the losses are simulated by a single resistor, *Rcable*.

A comparison between Figures 6.25 and 6.34 reveals that the significant modification is the inclusion of a new circuit loop carrying the current *Ine*.

This is the current which flows into the environment in the form of electromagnetic radiation.

The impedance presented by the cable to the output terminals of the source generator is:

$$Ro = Ro1 + \frac{Ro2 \cdot Ro3}{Ro2 + Ro3} \tag{6.76}$$

This is also the impedance presented to the terminals at the far end.

From inspection of Figure 6.34, the equation for the antenna-mode loop is:

$$0 = -Ro2 \cdot (Ini + Inr) + (Ro2 + Ro3) \cdot Ine$$

Hence:

$$Ine = \frac{Ro2}{Ro2 + Ro3} \cdot (Ini + Inr)$$

So:

$$Ine = Loss \cdot Ina \tag{6.77}$$

Where:

$$Loss = \frac{Ro2}{Ro2 + Ro3} \tag{6.78}$$

The current flowing from the near-end towards the far-end is now:

$$Inf = Inr - Ine \tag{6.79}$$

The current which stores antenna-mode charge on the cable is similar to (6.59):

$$Ins = Inf - \frac{Qns}{Rno \cdot Crad} \tag{6.80}$$

And the current which is actually transmitted to the far-end terminals is:

$$Int = Inf - Ins \tag{6.81}$$

In section 6.5, it was reasoned that the current *Ins* used to create antenna-mode charge on the cable must flow through the source resistance *Rg*. This means that (6.55) needs to be modified. The modified relationship becomes:

$$Vna = (Rg + Rcable) \cdot Ina + Rg \cdot Ins \tag{6.82}$$

Since the current used to deliver stored charge varies along the length of the cable, it is assumed that no voltage drop is incurred along the length of the cable by *Ins*. Taking this new relationship into account, (6.56) changes to:

$$Ina = \frac{2 \cdot Ro \cdot Ini + Vgen - Rg \cdot Ins}{Ro + Rg + Rcable} \tag{6.83}$$

There is now sufficient information available to modify the worksheet of Section 6.4 to analyse the response of the model of Figure 6.34. However, several more relationships are needed before the response of the model can be correlated with that of the setup of Figure 6.24.

The differential-mode current delivered by the source generator to the cable is:

$$Idiff = Inr + Ini - Ine$$

That is:

$$Idiff = Ina - Ine \tag{6.84}$$

The antenna-mode current is, by definition, *Ine*.

Since the only source of energy is the generator *Vgen*, then the current *Ins* must flow through the source resistor *Rg*. Hence, the voltage actually applied to the terminals at the near end must be:

$$Vin = Vgen - Rg \cdot (Ina + Ins) \tag{6.85}$$

It can be assumed that the relationship between the voltage monitored by channel 1 of the oscilloscope and the voltage *Vin* is:

$$K = \frac{Vch1}{Vin} \tag{6.86}$$

Similarly, it can be assumed that the relationship between the current monitored by the current transformer, *Imon*, and the voltage displayed by channel 2 of the scope is:

$$ZT = \frac{Vch2}{Imon} \tag{6.87}$$

Having established all the necessary relationships, the next step is to assemble them into a program which simulates the transient response of the configuration. The first page of the worksheet is devoted to the definition of the input variables associated with the model of Figure 6.34 and with the test setup illustrated by Figure 6.24. This page is reproduced in Figure 6.35.

As with the frequency-response models, this time-domain-response model can also be used to correlate the theoretical response with that of the actual hardware. Relevant variables are identified by the phrase ‘adjust to suit’.

The second page of the worksheet (Figure 6.36) defines the two subroutines used to calculate the responses at the near and far ends of the line after each time step. It is a modified version of the subroutines illustrated in Figure 6.27.

The third page of the worksheet, illustrated by Figure 6.37, defines the main program used to calculate the response of the selected variable over the defined time.

The output variable can be:

- *Vch1*: The input voltage waveform, as observed on channel 1 of the oscilloscope. The leading edge is defined by *T1*, the trailing edge by *T2*.
- *Vdiff*: The differential-mode current waveform, as observed on channel 2 of the oscilloscope.
- *Vrad*: The antenna-mode current waveform, as observed on channel 2 of the oscilloscope.

Worksheet 6.6 page 1

$Rg := 4.7\ \Omega \quad RL := 10^7\ \Omega$ Impedances of source and load

$Vmeas := -0.41\ \text{V}$ Measured voltage between horizontal sections of square wave

$Crad := 220 \cdot 10^{-12}\ \text{F}$ Value assigned to radiation capacitor **Adjust to suit**

$Rcable := 1\ \Omega$ Resistor which simulates cable losses **Adjust to suit**

$Ro1 := 50\ \Omega \quad Ro2 := Ro1 \quad Ro3 := 600\ \Omega$ See Figure 6.34 **Adjust to suit**

$Loss := \dfrac{Ro2}{Ro2 + Ro3}$ Equation (6.78)

$Ro := Ro1 + \dfrac{Ro2 \cdot Ro3}{Ro2 + Ro3} = 96.154$ Equation (6.76)

$ZT := 2.27\ \Omega$ Equation (7.7)

$K := \dfrac{50}{96.2}$ Equation (7.9)

$Vg := \dfrac{Vmeas}{K} \quad Vg = -0.789$ Equation (6.86)

$T := 83 \cdot 10^{-9}\ \text{s}$ Measured transit time.

$N := 100$ Number of time-steps per transit

$dt := \dfrac{T}{N}$ Time of each step

$T1 := 150 \cdot 10^{-9}\ \text{s}$ Time of leading edge of square wave, as displayed on oscilloscope

$T2 := 4.1 \cdot 10^{-6}\ \text{s}$ Time of trailing edge of square wave, as displayed on oscilloscope

$T3 := 5 \cdot 10^{-6}\ \text{s}$ Sweep time of oscilloscope

$N1 := \dfrac{T1}{dt} \quad N2 := \dfrac{T2}{dt} \quad N3 := \dfrac{T3}{dt}$ Number of time steps at which each event occurs

$n := 1 .. N3 \quad t_n := (n - 1) \cdot dt$ Definition of horizontal axis of display

Figure 6.35 Definition of input variables

In the particular case of the routine illustrated by Figure 6.37, the simulation would be of *Vch1*.

The input variables have been adjusted to simulate the responses of the three tests described in section 7.6. Several iterations were carried out before a fair correlation was achieved between the test results and the model. When the final iteration had occurred, no further alteration was made to the input variables.

Worksheet 6.6 page 2

$$\text{send}(\text{near}, \text{INPUT}, \text{Vgen}) := \left| \begin{array}{ll} Ini \leftarrow \text{INPUT}_2 & \\ \begin{pmatrix} Ina \\ Int \\ Ine \\ Ins \\ Qns \end{pmatrix} \leftarrow \text{near} & \\ Ina \leftarrow \dfrac{2 \cdot Ro \cdot Ini + Vgen - Rg \cdot Ins}{Ro + Rg + Rcable} & \text{Equation (6.83)} \\ Inr \leftarrow Ina - Ini & \text{Equation (6.57)} \\ Ine \leftarrow Loss \cdot Ina & \text{Equation (6.77)} \\ Inf \leftarrow Inr - Ine & \text{Equation (6.79)} \\ Ins \leftarrow Inf - \dfrac{Qns}{Ro \cdot Crad} & \text{Equation (6.80)} \\ Qns \leftarrow Qns + Ins \cdot dt & \text{Equation (6.60)} \\ Int \leftarrow Inf - Ins & \text{Equation (6.81)} \\ \begin{pmatrix} Ina \\ Int \\ Ine \\ Ins \\ Qns \end{pmatrix} & \end{array} \right.$$

$$\text{recv}(\text{INPUT}) := \left| \begin{array}{l} Ifi \leftarrow \text{INPUT}_1 \\ Ifa \leftarrow \dfrac{2 \cdot Ro \cdot Ifi}{RL + Ro} \\ Ifr \leftarrow Ifa - Ifi \\ \begin{pmatrix} Ifa \\ Ifr \end{pmatrix} \end{array} \right. \qquad \text{Copy of function defined in Figure 6.27}$$

Figure 6.36 Subroutines for near-end and far-end calculations

The program was run three more times, each time with a different output variable selected.

Then the input variables were assigned to the general circuit model of Figure 6.34 to create the representative circuit model of the assembly-under-test (Figure 6.38).

Section 7.6 demonstrates the fact that the same circuit model can simulate three different responses of the cable-under-test. This indicates that it is a useful predictor of transient radiation.

Figure 7.53 demonstrates that there is a clear correlation between the frequency response model of Figure 7.49 and the transient response model of Figure 6.38.

```
Worksheet 6.6 page 3
point (n) := | m ← mod (n, N)
             | m ← N if m = 0

Vch1 := | data_{2,N} ← 0
        | near_5 ← 0
        | for i ∈ 1..N3
        |   | Vgen ← Vg if i > N1
        |   | Vgen ← 0 if i > N2
        |   | p ← point (i)
        |   | INPUT ← data^⟨p⟩
        |   | near ← send (near, INPUT, Vgen)
        |   | (Ina, Int, Ine, Ins, Qns)^T ← near
        |   | (Ifa, Ifr)^T ← recv (INPUT)
        |   | OUTPUT ← (Int, Ifr)^T
        |   | data^⟨p⟩ ← OUTPUT
        |   | Idiff ← Ina − Ine                 Equation (6.84)
        |   | Vin ← Vgen − Rg·(Ina + Ins)       Equation (6.85)
        |   | Vch1 ← K·Vin                      Equation (6.86)
        |   | Vdiff ← ZT·Idiff                  Equation (6.87)
        |   | Vrad ← ZT·Ine                     Equation (6.87)
        |   | V_i ← Vch1                  Channel 1 selected as output variable.
        | V
```

Figure 6.37 Main routine used to analyse transient response

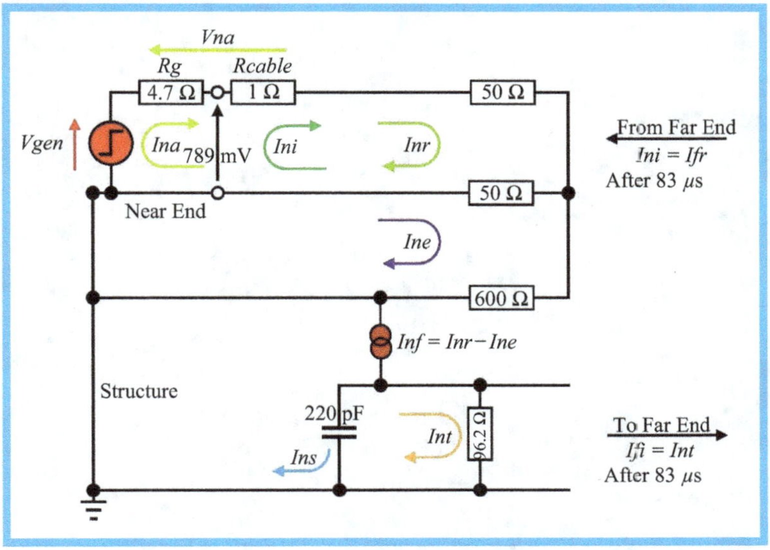

Figure 6.38 Representative circuit model of transient emission from cable

Both of these representative models have been related to electromagnetic phenomena, and both replicate the actual response of the hardware. It should be noted that Vmeas must be left as a Peak-to-Peak Voltage and not converted to an RMS value, as the capacitive and inductive elements have been 'removed' from the calculations to represent a 'loss-less' cable. This also allows for the model to produce a 1:1 representation of what is observed on an oscilloscope! So, it is fair to claim that the general circuit models on which they are based can be used with a high degree of confidence. Further testing and modelling should confirm their reliability. Sufficient information has been provided for this to be done. Section 7.6 illustrates how the model can be used to obtain first estimations of component values from the model.

Chapter 7
Bench testing

It is normal practice to carry out bench tests of the functional behaviour of prototype circuitry during the development of any new product. This identifies problems which had not been predicted during the feasibility study and provides an opportunity to rectify those faults. It also provides an early opportunity to check that the design requirements are being met.

Since EMC is also a functional requirement, it is logical to expect that this aspect of design should be checked at the prototype stage.

Two items of essential equipment are the voltage transformer and the current transformer.

Several manufacturers produce such items, but they are highly expensive and not really suitable for general-purpose use. So, the transducers described in this section were assembled from components obtained from a supplier of electronic components.

Section 7.1 describes the construction of a low-cost voltage transformer. This consists of ten turns of wire wound on a split-core ferrite assembly. A low-value resistor in parallel with the primary winding ensures that the transducer is a wide-band device. The secondary winding is the loop-under-test. A monitor winding enables the amplitude of the injected voltage to be measured.

The frequency response of the transformer was checked by applying a known voltage to the primary winding and measuring the output of the secondary. A circuit model was subjected to analysis using a Mathcad program. Only a few iterations of this program were needed to establish a one-to-one correlation between the test results and the response of the model. By this means, a model was created which could be used to simulate the response of the voltage transformer. This model can be used for analysis in the frequency domain or the time domain.

A current transformer was assembled in much the same way. This time, the primary was the loop-under-test, and the secondary provided an output voltage proportional to the input current. A circuit model was created to simulate the device characteristics. This circuit model provides the calibration curve for the device. The process is described in Section 7.2.

Section 7.3 describes the construction of a tri-axial cable which can be used to minimise interference between cables of the test equipment.

Section 7.4 describes a test on an isolated conductor, where a known voltage is injected into the centre of the conductor using the voltage transformer and the current is monitored with the current transformer. Data on the length and diameter of the conductor were used to assign value to the components of the dipole model of Section 5.2. A close match was obtained between the response of the conductor and the model. However, the peak of the emitted current was significantly higher than that of a dipole antenna. This is probably because there was no resistance in the centre of the cable to damp down the oscillations (as there would be in a dipole antenna).

Section 7.5 describes a similar test on a twin-conductor cable, where the terminations are open-circuit. A circuit model was developed which simulates both the differential-mode current and the antenna-mode current.

Section 7.6 provides details of the transient test on the 15 m cable described in Sections 6.4 and 6.5.

Photographs of the actual waveforms are compared with graphs of the simulated waveforms, providing persuasive evidence of the validity of the model. It is not claimed that this model is correct in every detail.

However, the one single model simulates the waveforms of three separate signals reasonably accurately. Every parameter is shown to be related to electromagnetic phenomena. It reveals aspects of transmission-line behaviour that are not identified in textbooks on electromagnetic theory. Sufficient information is provided to allow any electronic designer to replicate the tests and perform the analyses. Clear correlation can be established between the results obtained from the frequency response tests of Section 7.5 and those obtained from the transient tests of Section 7.6.

Although the test equipment and tests described in Sections 7.1–7.5 are limited to a bandwidth of 20 kHz to 20 MHz, the technique and approach are applicable to a much wider range of frequencies. Each designer will have test equipment which is suitable for use with the equipment under development. It is simply a matter of adapting the available equipment to carry out the type of testing described in this chapter.

The technique of circuit modelling can also be applied to the high-frequency characterisation of components such as capacitors, inductors, and filters. Section 7.7 provides an example of the characterisation of a capacitor over the range 200 kHz to 1 GHz.

7.1 Voltage transformer

This section defines the requirement of a voltage transformer intended for use with bench test equipment, describes a particular design and records details of a set of characterisation tests.

Basic requirements for a voltage transformer for use with EMC testing are:

- It should enable a known voltage to be induced in series with the loop-under-test.
- It should have a wide bandwidth.
- It should interface with 50 Ω co-axial cable.

If it is used as an item of General Purpose Test Equipment and is available for bench testing of equipment under development, then there are two more requirements:

- It should be a split-core transformer, capable of clamping around the cable-under-test.
- It should be a low-cost item.

Figure 7.1 illustrates the basic design of such a device and indicates how it can be connected to other items of bench test equipment and to the loop-under-test. In order to meet the requirement for a wide bandwidth, the impedance of the primary winding should be higher than *R*1 at the lowest operating frequency.

The requirement to measure the voltage induced in the locp-under-test is met by the monitor turn. If the loop-under-test is a short-circuit, then a high current will be induced, and the applied voltage will be reduced due to the output impedance of the transformer. The voltage detected by the monitor turn will also drop. This means that the monitor turn will measure the actual voltage applied to the loop-under-test.

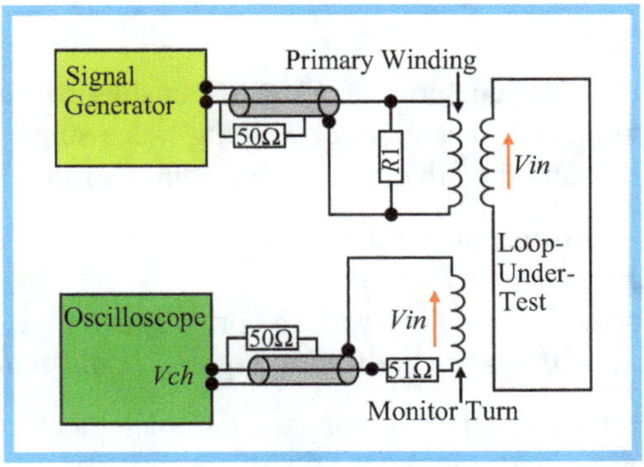

Figure 7.1 Use of voltage transformer

From the circuitry of Figure 7.1, the relationship between the voltage *Vin* applied to the loop-under-test and the voltage monitored at the appropriate channel of the oscilloscope is:

$$Vin = \frac{50 + 51}{50} \cdot Vch \tag{7.1}$$

The 50 Ω input impedance at the oscilloscope is provided by a BNC adaptor and a BNC terminator.

Such a transformer was constructed from parts purchased from suppliers of electronic components, and is illustrated by Figure 7.2. The core itself is a ‘cable suppression core assembly’ part number 04 31 173 551, supplied by Fair Rite Products Corp.

Figure 7.2 Voltage transformer

The primary winding was ten turns of 22 SWG enamelled copper wire wound around one of the split cores. The monitor winding was a single turn of enamelled copper wire wound around the other core. There were two reasons for this:

- It ensures that the magnetic field being coupled was the same as that which was coupling the loop-under-test.
- It minimises capacitive coupling with the primary winding, since it is wound on the other half of the split-core ferrite. Figure 7.2 illustrates this separation.

The windings were terminated in a standard terminal block, which provided a mount for the other components. A 68 Ω 2 W resistor was connected in parallel with a 240 Ω, 0.6W resistor and the primary winding. This provided a 53 Ω resistor in parallel with the primary.

Ideally, the co-axial cable should have been terminated by a 50 Ω resistor to match its characteristic impedance. However, it is inevitable that reflected impedance of the loop-under-test will act as a load in parallel with the termination resistor. To compensate for this loading effect, a value of 53 Ω was fitted.

Two BNC connectors were connected to the terminal block to provide an interface with 50 Ω co-axial cable. A section of a plastic box was used to provide more rigidity to the assembly.

Although this assembly does not meet the quality standard of a professionally engineered item of equipment, it does the job intended and has proved to be reliable over a period of more than 5 years.

Testing was carried out using the set-up of Figure 7.3. A splitter box was used to provide an input to channel 1 of the oscilloscope. This allowed the input to the voltage transformer to be monitored. The output of the monitor turn was monitored by channel 2. Each 16.5 Ω resistor is a parallel combination of two 33 Ω resistors.

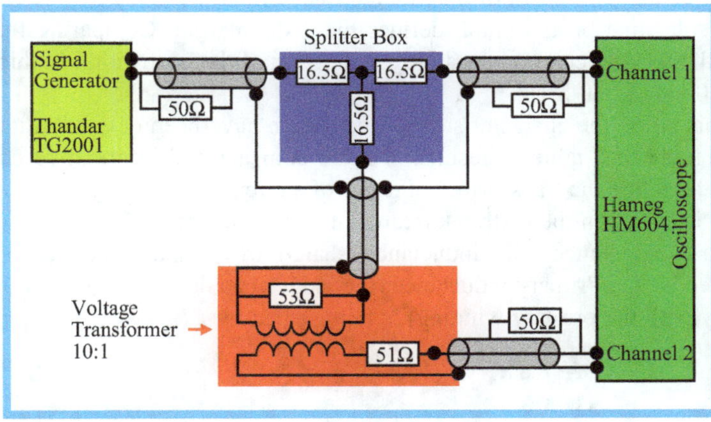

Figure 7.3 Characterisation of voltage transformer

The signal generator was set to provide a sinusoidal output at a particular frequency, and the amplitudes of the peak-to-peak signals were measured on the oscilloscope. This process was carried out over a number of defined frequencies and the results were tabulated. Column 1 of the table of Figure 7.4 gives the frequency in MHz, column 2 gives the amplitude of channel 1 in volts, and column 3 gives the amplitude of the signal at channel 2, in mV.

Worksheet 7.1 Page 1

Column 1: Frequency, MHz
Column 2: Peak-to-peak Voltage on channel 1, V
Column 3: Peak-to-peak Voltage on channel 2, mV

$$\text{data} := \begin{pmatrix} 0.01 & 2.2 & 50 \\ 0.02 & 2.6 & 90 \\ 0.05 & 3.4 & 146 \\ 0.1 & 3.8 & 170 \\ 0.2 & 3.95 & 177 \\ 0.5 & 3.95 & 180 \\ 1 & 4 & 182 \\ 2 & 4 & 180 \\ 5 & 3.9 & 172 \\ 10 & 3.8 & 158 \\ 15 & 3.6 & 140 \\ 19 & 3.45 & 124 \end{pmatrix}$$

$$f_s := \text{data}_{s,1} \cdot 10^6 \qquad s := 1 .. \text{rows}(\text{data})$$

$$Vch1_s := \text{data}_{s,2} \qquad Vch2_s := \text{data}_{s,3} \cdot 10^{-3}$$

$$TFt_s := \frac{\text{Vch2}_s}{\text{Vch1}_s} \qquad Ref_s := \frac{\max(\text{TFt})}{\sqrt{2}}$$

Figure 7.4 Using test results to calculate the transfer function (TFt)

For each spot frequency, the ratio of output voltage *Vch2* to input voltage *Vch1* was calculated, and this ratio gave the value of the Transfer Function at that frequency. Plotting this parameter against the frequency, as shown in Figure 7.5, gives an indication of the bandwidth of the voltage transformer.

The solid line on the graph defines the 3 dB margin. Comparing the dotted curve with the solid curve gives a clear indication of the bandwidth. In this case, it is from 20 kHz to 20 MHz.

At this point, the characteristics of the device have been defined in terms of a specified test and a table of results. A more compact and informative characterisation is to define the transformer in terms of a circuit model. Figure 7.6 is such a model. It is based on the textbook treatment of transformers.

Mutual inductance (the inductance shared by primary and secondary) is represented by *L2*. Primary inductance (the inductance due to magnetic flux which links only with the primary winding) is represented by *L1*.

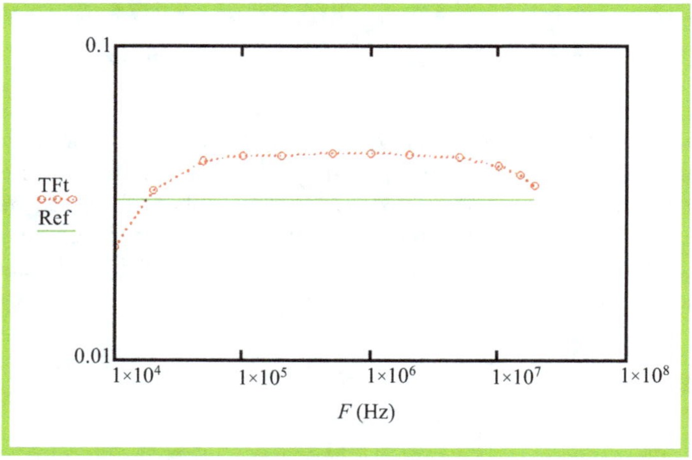

Figure 7.5 Transfer function of voltage transformer

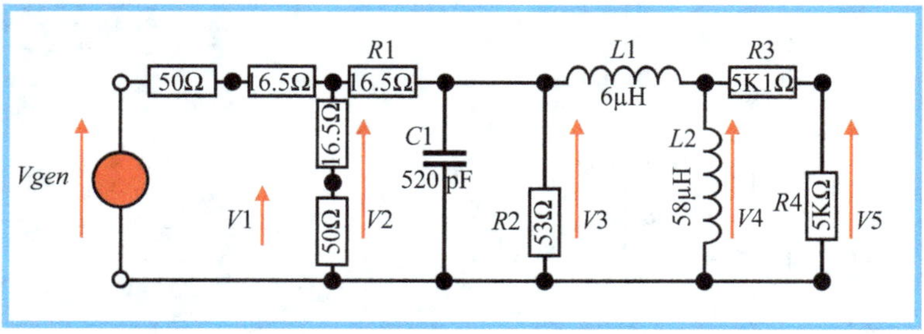

Figure 7.6 Representative circuit model of voltage transformer

Since the turns ratio is 10 to 1, the impedance ratio of the secondary reflected in the primary is 100 to 1. Resistance *R3* is the reflected value of the series resistance of the secondary circuit, and resistance *R4* is the reflected value of the load resistance at channel 2 input terminals.

Since the simulation is only concerned with amplitudes of the signals, there is no need to know the phase relationship between currents and voltages. So, as far as this model of the setup is concerned, the co-axial cable is transparent.

Resistance *R2* is the parallel combination of the 68 Ω and 240 Ω resistors. The 16.5 Ω resistors represent the splitter-box components, whilst the 50 Ω resistors represent the output impedance of the signal generator and the input impedance of channel 2 of the oscilloscope. The capacitance *C*1 represents the capacitance of the primary winding.

The voltage *V*1 represents the voltage as monitored by channel 1 of the scope, *V*2 is the voltage at the junction of the 16.5 Ω resistors, *V*3 is the voltage applied to the input terminals of the transformer, *V*4 is the voltage across the mutual inductance and *V*5 is ten times the voltage monitored by channel 2.

Simulating the frequency response of this model over the range 10 kHz to 20 MHz requires that the frequency steps are spaced logarithmically. So, a vector needs to be created which defines the set of frequencies to be used. A set of 100 frequencies provides enough data points to ensure a smooth curve on the graph. Figure 7.7 shows the section of the Mathcad worksheet which performs this function.

$$y1 := \log\left(10 \cdot 10^3\right) \qquad y2 := \log\left(20 \cdot 10^6\right) \qquad m := \frac{y2 - y1}{100}$$

$$i := 1..101 \qquad F_i := \left| \begin{array}{l} y \leftarrow m \cdot (i-1) + y1 \\ 10^y \end{array} \right.$$

Figure 7.7 Calculating a set of equally spaced frequencies over a logarithmic scale

Calculating the transfer function of the circuit model *TFm* is performed by the section of the worksheet illustrated in Figure 7.8. The objective is to determine the ratio of *V*5 to *V*1, since the former simulates the voltage at channel 2, whilst the latter simulates the voltage at channel 1. So, the amplitude of *V*1 is set at unity. This means that the transfer function of the model at any frequency is the amplitude of *V*5 divided by the turns ratio.

Equations used in the subroutine illustrated in Figure 7.8 are derived from inspection of the circuit model, using the technique of 'equivalent circuits'.

It is now possible to compare the Transfer Function derived from test results, *TFt*, with that derived from the circuit model, *TFm*, and this is done in the graph of Figure 7.9.

Worksheet 7.1 Page 3

$R1 := 16.5\ \Omega \qquad R2 := \dfrac{68 \cdot 240}{68 + 240} = 52.987\ \Omega \qquad R3 := 5100\ \Omega \qquad R4 := 5000\ \Omega$

$Lp := 64 \cdot 10^{-6}\ \mathrm{H} \qquad L1 := 6.0 \cdot 10^{-6}\ \mathrm{H} \qquad L2 := Lp - L1 = 5.8 \times 10^{-5}\ \mathrm{H}$

$V1 := 1\mathrm{V} \qquad V2 := \dfrac{50 + R1}{50} \cdot V1 = 1.33 \qquad C1 := 520 \cdot 10^{-12}\ \mathrm{F} \qquad Turns := 10$

$$
\begin{aligned}
TFm_i := \;& \omega \leftarrow 2 \cdot \pi \cdot F_i \\
& Y2 \leftarrow \frac{1}{j \cdot \omega \cdot L2} + \frac{1}{R3 + R4} \\
& Z2 \leftarrow \frac{1}{Y2} \\
& Y1 \leftarrow \frac{1}{R2} + \frac{1}{Z2 + j \cdot \omega \cdot L2} + j \cdot \omega \cdot C1 \\
& Z1 \leftarrow \frac{1}{Y1} \\
& V3 \leftarrow \frac{Z1}{Z1 + R1} \cdot V2 \\
& V4 \leftarrow \frac{Z2}{Z2 + j \cdot \omega \cdot L1} \cdot V3 \\
& V5 \leftarrow \frac{R4}{R3 + R4} \cdot V4 \\
& \frac{|V5|}{Turns}
\end{aligned}
$$

Figure 7.8 Calculating the transfer function of the model (TFm)

This graph illustrates the fact that the curve produced by the circuit model intersects all the data points derived from testing the assembly. Since the component parts of the model simulating the voltage transformer can be identified, it means that those parts can be used to define its characteristics.

It is useful to describe the process by which the two curves of Figure 7.9 were brought into co-incidence. It was basically an iterative process.

Initially, guess values were assigned to Lp, $L1$, and $C1$. Values of other circuit components were fixed. Then, the value of one of the parameters at the top of the worksheet was altered slightly, and the page was scrolled up to display the final graph. The curve on the graph did not change until the mouse was clicked. This allowed the change in the position of the curve to be noted. If the curves moved closer together, then the same incremental change was made to the parameter.

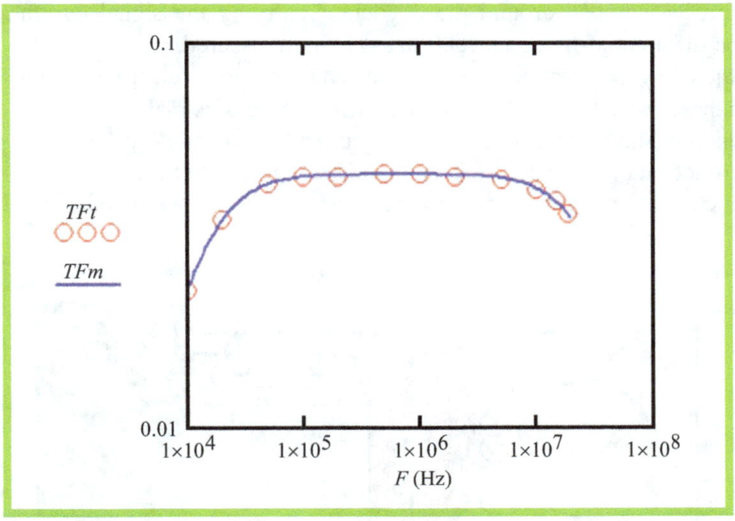

Figure 7.9 Transfer functions derived from testing and modelling

Initially, it was assumed that capacitor $C1$ was very small, and attention was focused on the low-frequency response.

The value of Lp (see third line of Figure 7.8) was adjusted to 'line up' the two curves at the low end of the frequency range, and $L1$ was adjusted to 'line up' the mid-frequency response.

It is useful to note that $L2$ is a dependent variable. This means that the total inductance of the primary winding of the transformer model can be altered with a single parameter change.

Surprisingly little iteration was needed to achieve coincidence over the lower and mid-frequency ranges. Finally, the value of $C1$ was adjusted to 'line up' the high-frequency roll-off.

The useable range of this transformer is from 20 kHz to 20 MHz. It is reasonable to expect that transformers can be constructed to match the operating frequency range of any particular system-under-review; for example, by using a smaller transformer to extend the frequency range upwards, or by adding more turns on the secondary to extend the range downwards.

7.2 Current transformer

The purpose of the current transformer is to enable measurements to be made of the current in system cables. In configurations where the loop is completed by conductive components of the structure, the measurement would be of the common-mode current. Where the far end of the cable is isolated (e.g., with loudspeaker or microphone cables), the measurement would be of antenna-mode current.

If the core is clamped around the conductor carrying the signal current, then it would be differential-mode current that is being measured.

In the analytical treatment of the preceding chapters, a clear distinction is made between these current modes. For example, with a three-core mains conductor, the live/neutral loop would carry differential-mode current, the neutral/earth loop would carry common-mode current, and the antenna-mode current would flow in a loop formed by the earth conductor and the virtual conductor representing the environment.

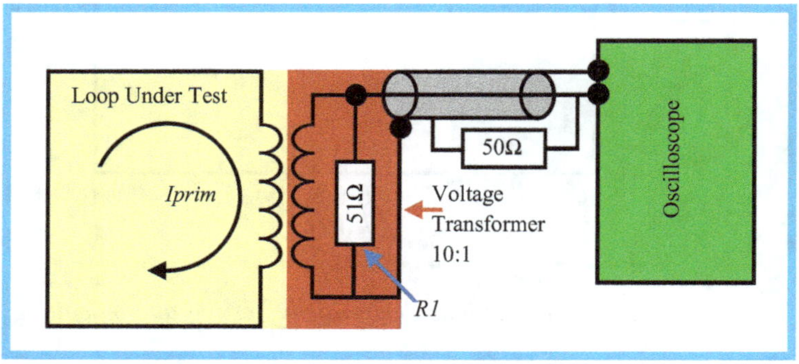

Figure 7.10 Use of current transformer

Figure 7.10 illustrates the method of coupling the test equipment to the loop-under-test. With such a configuration, the load reflected into the primary circuit (the loop-under-test) is the load presented to the secondary divided by the square of the number of turns.

Hence, the greater the number of turns, the less the value of the reflected load, and the less will be the effect of the test equipment on the system-under-review.

The load 'seen' by the secondary winding is the 51 Ω resistor *R1* in parallel with the 50 Ω impedance of the screened cable. Figure 7.10 illustrates the relationship.

As far as this load is concerned, it 'sees' a current source *Isec* (Figure 7.11), where:

$$Isec = \frac{Iprim}{Turns} \tag{7.2}$$

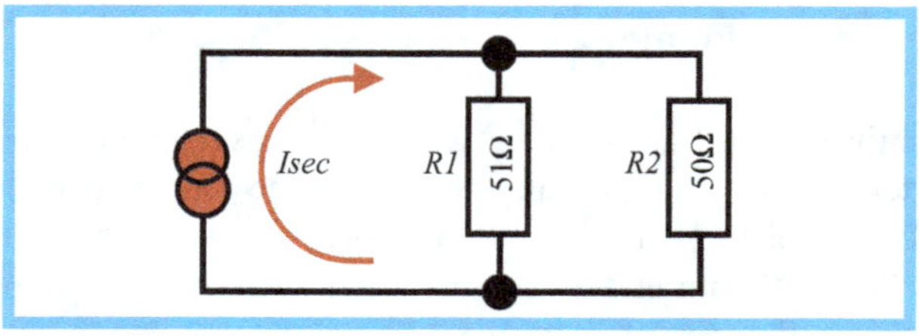

Figure 7.11 Transformer as a current source

And *Iprim* is the current in the loop-under-test. The variable *Turns* is the number of turns on the primary winding.

From the viewpoint of the input terminals of the oscilloscope, the signal is generated by a voltage source, as illustrated by Figure 7.12. In all the tests described in this chapter, a 50 Ω resistor is inserted in parallel with each scope input.

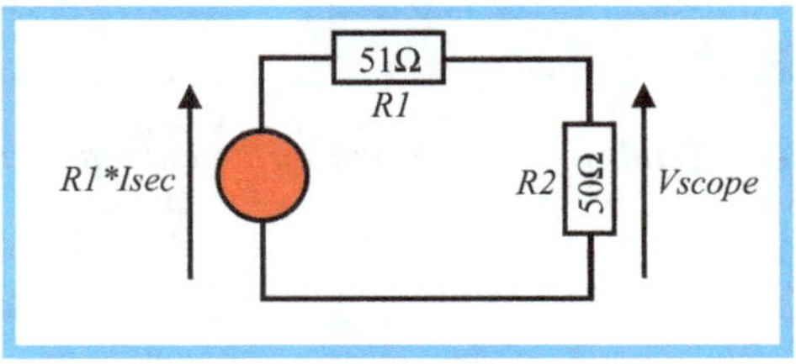

Figure 7.12 Transformer as a voltage source

In the particular transformer described here, the core is exactly the same as that used for the voltage transformer, and the secondary winding comprises ten turns of 22 SWG enamelled copper wire. Figure 7.13 illustrates the assembly. The tie wrap is removable and is used to ensure that the two halves of the core are tightly clamped together.

Figure 7.13 Current transformer assembly

To characterise this transformer, a simple coupling jig was assembled. This delivered a primary current of known amplitude and known frequency and was designed to ensure that the loop-under-test was tightly coupled to the transformer.

The test set-up is shown in Figure 7.14. Channel 1 of the oscilloscope was used to measure the current delivered to the transformer; that is, the current in the 50 Ω resistor placed at the scope input connector. Channel 2 was used to monitor the output of the transformer assembly.

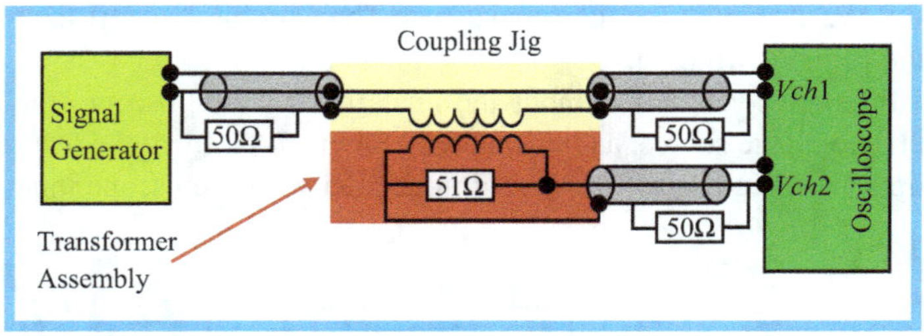

Figure 7.14 Setup for calibration of the current transformer

Worksheet 7.2 Page 1

$$\text{data} := \begin{pmatrix} 0.005 & 8 & 98 \\ 0.01 & 8 & 175 \\ 0.02 & 8 & 270 \\ 0.05 & 8 & 340 \\ 0.1 & 8 & 360 \\ 0.2 & 8 & 365 \\ 0.5 & 8 & 375 \\ 1 & 8 & 370 \\ 2 & 8 & 370 \\ 5 & 7.9 & 365 \\ 10 & 7.8 & 360 \\ 15 & 7.4 & 335 \\ 19 & 6.9 & 330 \end{pmatrix}$$

Column 1: Frequency, MHz
Column 2: Channel 1 Voltage, V
Column 3: Channel 2 Voltage, mV

$s := 1..\text{rows}(\text{data})$ $\qquad f_s := \text{data}_{s,1} \cdot 10^6$

$$ZTt_s := \left| \begin{array}{l} Vch1 \leftarrow \text{data}_{s,2} \\ Vch2 \leftarrow \text{data}_{s,3} \cdot 10^{-3} \\ Iprim \leftarrow \dfrac{Vch1}{50} \\ \dfrac{Vch2}{Iprim} \end{array} \right.$$

Figure 7.15 Using the test results to calculate the transfer impedance

As with the test on the voltage transformer, measurements were taken of the peak-to-peak amplitude of the sine waves displayed on the screen of the scope.

Figure 7.15 tabulates the results and illustrates how the transfer impedance *ZTt* is calculated, by dividing the voltage at channel 2 input by the current in the primary winding.

It is worth noting that the parameter resulting from these calculations is derived from the ratio of two measurements. It can be assumed that the amplifiers for channels 1 and 2 in the oscilloscope are identical. This being so, most of the errors in the absolute measurements are cancelled out.

Mathcad software was then used to display the results. As with the response of the voltage transformer, this was a set of points which could be joined up to create a frequency response characteristic.

Again, the use of circuit theory enables the creation of a circuit model of the link between the loop-under-test and the oscilloscope input.

Figure 7.16 is a development of the simple model of Figure 7.11. *R*2 and *R*3 represent the 51 Ω resistor in the transformer assembly and the resistor at channel-2 input connector, respectively.

*L*1 represents the inductance of the transformer winding, whilst *R*1 represents transformer losses. These losses could be due to the magnetic field from the loop-under-test which does not link with the transformer core. Another cause of losses is the eddy current in the core.

Capacitance *C*1 and resistor *R*4 were added to the model to simulate additional losses at frequencies over 2 MHz.

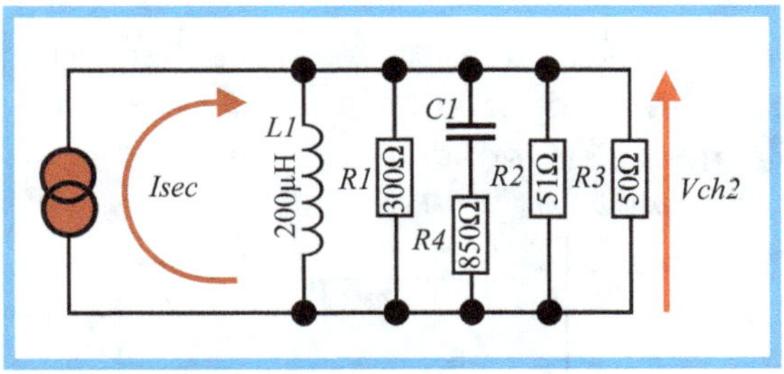

Figure 7.16 Representative circuit model of a current transformer

This model is amenable to the use of Simulation Programs with Integrated Circuit Emphasis (SPICE) software.

It would have been possible to use such software to produce a frequency response curve similar to that of *ZTt*.

Achieving close correlation between theoretical results and test results would have involved downloading the results of the analysis to a computer file, picking up that file with Mathcad software, and then using Mathcad to compare the two curves.

Such a process was avoided by carrying out the frequency response analysis in the same worksheet that was used to display the test results. Figure 7.17 illustrates the program involved.

This first calculates a set of 101 frequencies, on a logarithmic scale, between 5 kHz and 20 MHz, and stores them in the vector *F*. It then defines the component values of the circuit model as well as the number of turns on the secondary.

The equations used in the function *ZTm* are derived from an inspection of the circuit model. The impedance *Z*2 is the impedance as 'seen' by the current generator. It defines the ratio of *Vch*2 to *Isec*. Dividing *Z*2 by the number of turns gives the ratio of *Vch*2 to *Iprim*. That is, the transfer impedance of the circuit model.

Figure 7.18 illustrates the correlation between the test results and the response of the model. Although there was an initial discrepancy between the two curves, a few adjustments of *L*1 and *R*1 led to a curve which intersected the data points at the low frequencies. Varying the values of $C1$ and $R4$ led to a curve which intersected the data point above 2 MHz as well.

The existence of this model makes it possible to deduce the amplitude of the current in the loop-under-test by noting the amplitude and frequency of the signal observed on the oscilloscope. That is, the response of the model of Figure 7.18 provides the calibration curve for the current transformer.

Worksheet 7.2 page 2

$y1 := \log(5 \cdot 10^3) \qquad y2 := \log(20 \cdot 10^6) \qquad m := \dfrac{y2 - y1}{100}$

$$i := 1..101 \qquad F_i := \left| \begin{array}{l} y \leftarrow m \cdot (i-1) + y1 \\ 10^y \end{array} \right.$$

Calculating a set of 100 frequencies, equally spaced between 5 kHz and 20 MHz

$R1 := 300\ \Omega \qquad R2 := 51\ \Omega \qquad R3 := 50\ \Omega \qquad R4 := 850\ \Omega \qquad \text{Turns} := 10$

$L1 := 200 \cdot 10^{-6}\ \text{H} \qquad C1 := 60 \cdot 10^{-12}\ \text{F}$

$$ZTm_i := \left| \begin{array}{l} \omega \leftarrow 2 \cdot \pi \cdot F_i \\ Z1 \leftarrow R4 + \dfrac{1}{j \cdot \omega \cdot C1} \\ Y2 \leftarrow \dfrac{1}{R1} + \dfrac{1}{R2} + \dfrac{1}{R3} + \dfrac{1}{j \cdot \omega \cdot L1} + \dfrac{1}{Z1} \\ Z2 \leftarrow \dfrac{1}{Y2} \\ ZT \leftarrow \dfrac{|Z2|}{\text{Turns}} \end{array} \right.$$

Figure 7.17 Calculating the transfer impedance of the circuit model

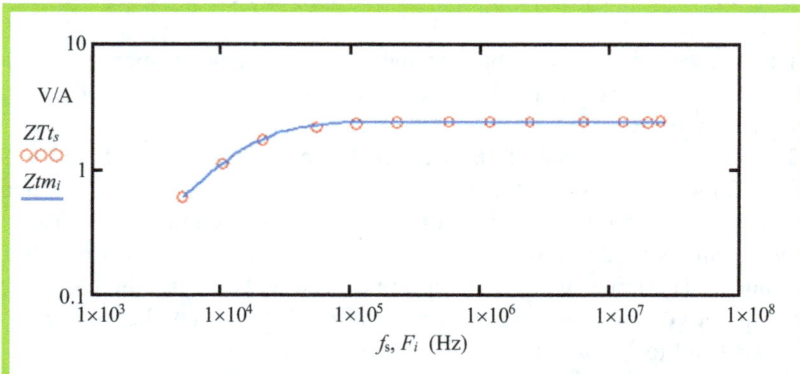

Figure 7.18 Transfer impedance from test results, ZTt, and from circuit model, ZTm

The fact that the frequency response of the transformer assembly is flat over a wide bandwidth means that it can also be used to monitor the amplitude and waveform of transient currents. There is one important proviso: the bandwidth of the waveform being monitored must lie within the bandwidth of the device.

A first approximation to the transient response characteristic can be made by assuming that the inductor *L*1 is open-circuit, whilst the capacitor *C1* is replaced by a short-circuit. The admittance of the circuit becomes *Ysec*, where:

$$Ysec = \frac{1}{R1} + \frac{1}{R2} + \frac{1}{R3} + \frac{1}{R4} \tag{7.3}$$

And

$$Rsec = \frac{1}{Ysec} \tag{7.4}$$

This gives:

$$Vch = Rsec \cdot Isec \tag{7.5}$$

Voltage at the oscilloscope input can be related to the current in the primary loop by invoking (7.2):

$$Vch = \frac{Rsec}{Turns} \cdot Isec \tag{7.6}$$

For this particular device, the transfer impedance to use with transient test analysis is:

$$RT = \frac{Rsec}{Turns} = 2.27 \tag{7.7}$$

7.3 Tri-axial cable

In any test setup, there will be cables which connect the Test Equipment to the Equipment-Under-Test (EUT). It is normal practice for these cables to be co-axial, because such cables are the most practical and efficient means of transmitting signals from one location to another, as well as being the least likely to cause interference problems.

Even so, stray coupling does exist between co-axial cables. Practical experience has shown that this coupling can create unwanted signals which completely obscure the signals from the EUT.

One way of reducing this coupling is to use tri-axial cables as illustrated by Figure 7.19. It was not particularly difficult to manufacture such an assembly. The protective boots were removed from the BNC connectors at each end of an RG58 cable assembly and one end of an 18 Ω resistor connected to each shell. A roll of wire braid was cut to length and fitted over the assembly to form an outer sheath.

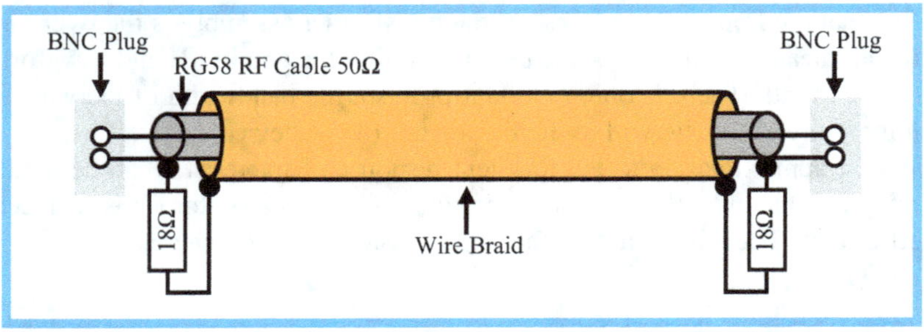

Figure 7.19 Tri-axial cable

The 18 Ω resistors were then connected to the outer-sheath, care being taken to ensure the outer braid did not touch the shells of the connectors. Finally, the assembly was sheathed in an insulating braid.

The inner and outer braids act as a transmission line and the characteristic impedance can be calculated using:

$$Ro = \frac{1}{2 \cdot \pi} \cdot \sqrt{\frac{\mu_o \cdot \mu_r}{\varepsilon_o \cdot \varepsilon_r}} \cdot \ln\left(\frac{r_{3,3}}{r_{2,2}}\right) \tag{7.8}$$

Where $r_{2,2}$ and $r_{3,3}$ are the radii of the inner and outer braids, respectively.

For RG58 cable, the diameter of the screen is 3.3 mm and that of the outer-sheath is 5 mm. Allowing for a loose-fitting braid, it can be assumed that the braid diameter is 6 mm. If it is also assumed that the relative permittivity is 4, then invocation of (7.8) gives the value of *Ro* as 18 Ω.

In the presence of an external field, antenna-mode current will flow in the outer skin of the outer braid. Voltage developed along the length of the braid will cause common-mode current to flow in the loop formed by the inner and outer braid. Since these two conductors act as a transmission line, any signal arriving at either end is absorbed in an 18 Ω resistor. Since these resistors are the same as the characteristic impedance of the line, there will be little or no reflection. This means that most of the energy delivered to the cable in the form of electromagnetic radiation will be dissipated in a resistive load. To achieve good performance at very high frequencies, several resistors would need to be assembled in an annular ring at each termination to minimise the inductance of the resistive elements.

As far as radiation from the differential-mode signal is concerned, any voltage developed along the length of the inner braid will cause a circulating current in the outer transmission line. Again, the energy of the unwanted signal will be absorbed by the two end resistors.

The net result is a significant improvement in the shielding effectiveness of the RG58 cable.

7.4 The isolated conductor

The voltage transformer provides a means of injecting a defined voltage into any conducting loop without making physical contact between the conductors of the EUT and the test equipment. The current transformer provides the means of measuring the amplitude of the induced current. This being so, it is possible to carry out tests on the simplest configuration which can be envisaged (the isolated conductor). So, this is the first test recorded here.

Since the behaviour of an isolated conductor was expected to approximate that of a half-wave dipole, and since the estimated value of the half-wave frequency of a 15 m length is 10 MHz, the selection of such a length brings the first resonant frequency within the operating range of the test equipment.

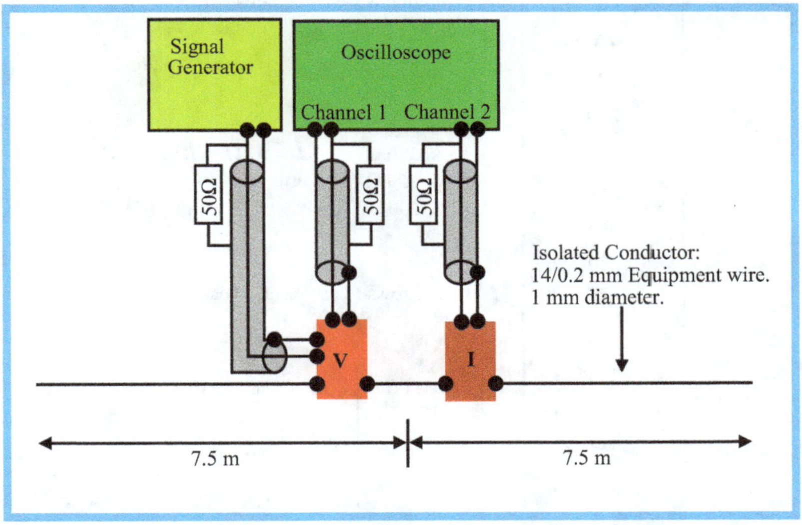

Figure 7.20 Testing the response of an isolated conductor

Figure 7.20 illustrates the setup. As with the calibration process for the current transformer, the output of the signal generator was varied over a range of spot frequencies. Note was taken of the peak-to-peak amplitude of the signal displayed on both channels of the oscilloscope, and the results were tabulated.

Although such a procedure appears more cumbersome than one which uses a programmable signal generator and spectrum analyser, the process has five distinct advantages:

- Using the same item of equipment to compare input and output signals eliminates many problems of calibration.
- Monitoring the waveforms provides a visible indication of any distortion which might be present.
- The user is not inundated with vast quantities of data to be processed and interpreted.

- Oscilloscopes and signal generators are ubiquitous items of equipment in any electronics laboratory.
- All the items of test equipment can be provided by designers with a limited budget.

Figure 7.21 displays the recorded data in the form of a three-column array and defines a function Ytest(s) which can be used to calculate the admittance characteristic. This constitutes the first page of the Mathcad worksheet which analyses the results.

Worksheet 7.4 Page 1

$$data := \begin{pmatrix} 1 & 400 & 0.4 \\ 2 & 400 & 0.4 \\ 3 & 400 & 1 \\ 4 & 395 & 1.9 \\ 5 & 390 & 3.4 \\ 6 & 390 & 5.9 \\ 6.5 & 385 & 8.4 \\ 7 & 385 & 14.5 \\ 7.5 & 380 & 26 \\ 7.83 & 360 & 36 \\ 8 & 350 & 31.5 \\ 8.5 & 345 & 16.5 \\ 9 & 345 & 10.5 \\ 10 & 345 & 6.2 \\ 11 & 345 & 6 \\ 12 & 345 & 4.4 \\ 13 & 340 & 3.4 \\ 14 & 340 & 3 \\ 15 & 330 & 3 \\ 16 & 330 & 2.4 \\ 17 & 320 & 1.4 \\ 18 & 315 & 2.2 \\ 19 & 310 & 3.4 \\ 19.6 & 305 & 4.2 \end{pmatrix}$$

Column 1: Frequency, MHz
Column 2: Channel 1 Voltage, mV
Column 3: Channel 2 Voltage, mV

Yt is the admittance derived from test results.

From Figure 7.16:

$R1 := 300\ \Omega \qquad R2 := 51\ \Omega \qquad R3 := 50\ \Omega$

$R4 := 850\ \Omega \qquad Turns := 10$

$L1 := 200 \cdot 10^{-6}\ \mathrm{H} \qquad C1 := 60 \cdot 10^{-12}\ \mathrm{F}$

$s := 1 .. \mathrm{rows}(data) \qquad f_s := data_{s,1} \cdot 10^6$

$$Ytest(s) := \left| \begin{array}{l} Vch1 \leftarrow data_{s,2} \cdot 10^{-3} \\ Vin \leftarrow \dfrac{51+50}{50} \cdot Vch1 \\ Vch2 \leftarrow data_{s,3} \cdot 10^{-3} \\ \omega \leftarrow 2 \cdot \pi \cdot f_s \\ Z1 \leftarrow R4 + \dfrac{1}{j \cdot \omega \cdot C1} \\ Y2 \leftarrow \dfrac{1}{R1} + \dfrac{1}{R2} + \dfrac{1}{R3} + \dfrac{1}{j \cdot \omega \cdot L1} + \dfrac{1}{Z1} \\ Isec \leftarrow \left| Y2 \right| \cdot Vch2 \\ Iprim \leftarrow Isec \cdot Turns \\ \dfrac{Iprim}{Vin} \end{array} \right.$$

$$Yt_s := \frac{Ytest(s)}{\sqrt{2}}$$

Dividing by the square root of 2 converts the Peak-to-Peak Voltages to RMS.

Figure 7.21 Using test data to calculate the Admittance characteristic

The integer variable s is used to identify a particular row in the array. The relationship between the voltage *Vin* induced in the conductor and the voltage at channel 2 of the oscilloscope *Vch*2 is defined by (7.1).

The admittance of the secondary winding of the current transformer is calculated by using the set of equations defined in the subroutine of Figure 7.16.

Multiplying this value by the channel 2 voltage *Vch*2 gives the value of the current *Isec* in the secondary winding. Multiplying *Isec* by the number of turns gives the current in the conductor, *Iprim*.

The ratio of the current in the conductor to the applied voltage *Vin* gives a value for the admittance at the spot frequency f_s. All the values are recorded in the vector *Yt*. Figure 7.22 illustrates the frequency response of this parameter over the range 1–20 MHz.

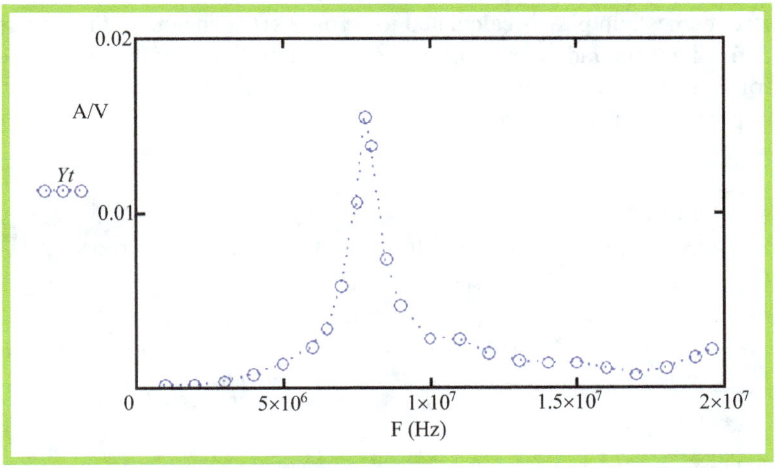

Figure 7.22 Admittance of conductor, derived from test results

As far as the predicted response was concerned, the shape of this curve was extremely encouraging. There is a single peak at just below 10 MHz and an indication that there could be a second peak at just below 30 MHz. This would relate to resonances at the quarter-wave and three-quarter-wave frequencies.

It could reasonably be expected that the dipole model of Figure 5.7 would be capable of simulating the response. Such a task would be just a matter of assigning numerical values to the components, defining the relevant equations, and setting these out in the worksheet.

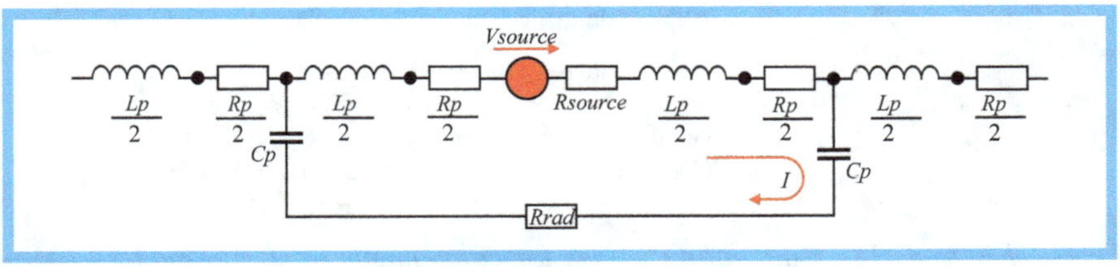

Figure 7.23 General circuit model of an isolated conductor

The general circuit model is reproduced in Figure 7.23. A brief examination of the tabulated data reveals that the quarter-wave frequency *fq* of the conductor-under-review is 7.83 MHz.

Given knowledge of this frequency and the measured length of the cable, it is possible to use (2.35) to calculate the propagation velocity. Since there was no magnetic material in the cable, the value of μ_r can be assumed to be unity. So, (2.36) can be used to derive a value for the relative permittivity.

Equations (2.26) and (2.27) can then be used to calculate theoretical values for the primitive capacitance *Cp* and primitive inductance *Lp*. Since (5.3) gives the theoretical value for the radiation resistance *Rrad* for a dipole, there is now enough information to assign values to the components of the circuit model of Figure 7.23.

Figure 7.24 is a copy of the page of the worksheet which calculates the values of the reactive and resistive parameters of the conductor. The frequency at which skin effect comes into play is calculated to be 6.89 kHz. The only other parameters to be defined are the radiation resistance and the amplitude of the voltage source. The former is set at the same value as a conventional dipole, 73 Ω. The latter is set at unity to allow the response to be defined as amperes per volt.

Worksheet 7.4 Page 3

$\varepsilon_o := 8.854 \cdot 10^{-12}$ F/m $\mu_o := 4 \cdot \pi \cdot 10^{-7}$ H/m $\mu_r := 1$ $c := 3 \cdot 10^8$ m/s

$\rho := 1.7 \cdot 10^{-8}$ Ω/m Resistivity of copper

$l := 7.5$ m Length of monopole

$r := 0.5 \cdot 10^{-3}$ m From micrometre measurement of conductor diameter

$$Fx := \frac{4 \cdot \rho}{\mu_o \cdot \pi \cdot r^2} = 6.89 \times 10^4 \text{ Hz}$$

Crossover frequency
See equation (2.60)

$$Lp := \frac{\mu_o \cdot \mu_r \cdot l}{2 \cdot \pi} \cdot \ln\left(\frac{l}{r}\right) = 1.442 \times 10^{-5} \text{ H}$$

Equation (2.27)

$$\frac{Lp}{2} = 7.212 \times 10^{-6} \text{ H}$$

Inductive component value for model

$fq := 7.83 \cdot 10^6$ Hz From tabulated data

$v := 4 \cdot l \cdot fq = 2.349 \times 10^8$ m/s Equation (2.35)

$$e_r := \left(\frac{c}{v}\right)^2 = 1.631$$

Equation (2.36)

$$Cp := \frac{2 \cdot \pi \cdot \varepsilon_o \cdot \varepsilon_r \cdot l}{\ln\left(\frac{l}{r}\right)} = 7.077 \times 10^{-11} \text{ F}$$

Equation (2.26)

Rrad := 73 Ω Equation (5.3)

Vsource := 1 V Allows admittance value to be calculated

$$Rss := \frac{\rho \cdot l}{\pi \cdot r^2} = 0.162 \ \Omega$$

Steady state resistance. See equation (2.57)

Gp := 0 S Good quality insulation assumed

Rsource := 1Ω Guess-value for impedance of current transformer

Figure 7.24 Calculating values for components of circuit model

Assigning these values to the general circuit model leads to the model of Figure 7.25.

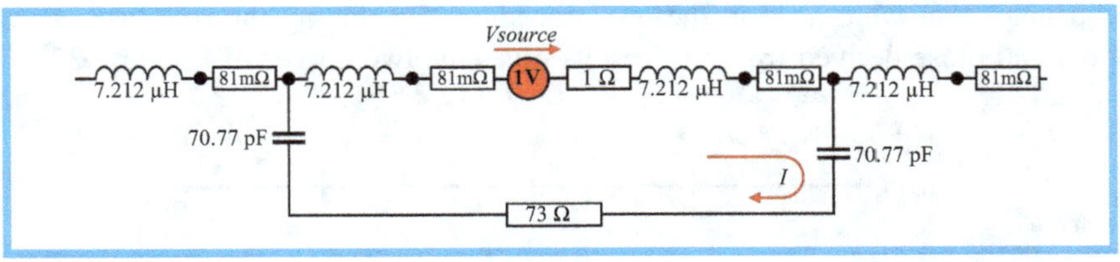

Figure 7.25 Initial circuit model of conductor-under-review

The Mathcad function to calculate the value of the admittance at each frequency is illustrated in Figure 7.26. This is an adaptation of the function *Zbranch(f)* depicted in Figure 4.10.

Worksheet 7.4 Page 4

$$F_i := i \cdot 10^5 \text{ Hz}$$

$$\begin{aligned} Ymodel(i, Vsource) := \;& Rp \leftarrow Rss \cdot \sqrt{1 + \frac{F_i}{Fx}} \\ & \omega \leftarrow 2 \cdot \pi \cdot F_i \\ & \theta \leftarrow \sqrt{(Rp + j \cdot \omega \cdot Lp) \cdot (Gp + j \cdot \omega \cdot Cp)} \\ & Zo \leftarrow \sqrt{\frac{Rp + j \cdot \omega \cdot Lp}{Gp + j \cdot \omega \cdot Cp}} \\ & Z1 \leftarrow Zo \cdot \tanh\left(\frac{\theta}{2}\right) \\ & Z2 \leftarrow Zo \cdot \operatorname{csch}(\theta) \\ & Z3 \leftarrow 2 \cdot Z1 + 2 \cdot Z2 + Rrad + Rsource \\ & \frac{Vsource}{|Z3|} \end{aligned}$$

$$Ym_i := Ymodel(i, Vsource)$$

Figure 7.26 Calculating the admittance of circuit model over a range of frequencies

Distributed parameters *Z1* and *Z2* are determined and then used to calculate the loop impedance *Z3*. Dividing the amplitude of the voltage source *Vsource* by the magnitude of the loop impedance gives a value for the current which will be flowing in the loop.

Since the magnitude of *Vsource* is unity, the output variable of the function *Ymodel(i, Vsource)* is the admittance of the loop shown, at the frequency F_i. The input variable *i* is an integer which points to the relevant row of the frequency vector *F*. The other input variable *Vsource* allows the response to be determined when the source voltage itself is a function of frequency. (Figure 5.20 illustrates how the threat voltage of the incoming wave can vary.)

The vector *Ym* stores the results of the calculations over the same range of frequencies that were used in the test procedure. Displaying the response of this model and those derived from test results gives the two curves of Figure 7.27.

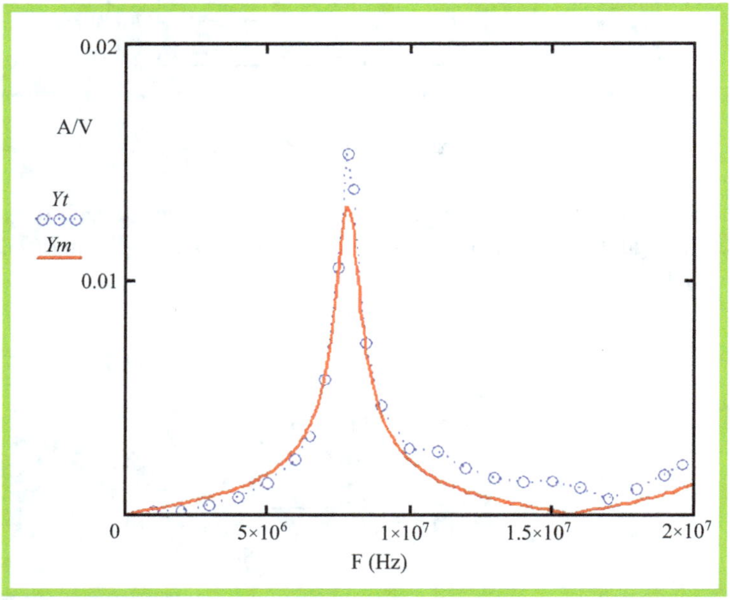

Figure 7.27 Comparing response of the test with that of the initial circuit model

The close correspondence in the region between 1 MHz and 5 MHz indicates that the value calculated for the primitive capacitance *Cp* was fairly accurate. The fact that the peaks of both models occur at the same frequency indicates that the value of *Lp* was also accurate. However, there is an obvious discrepancy between the magnitudes of the two peaks.

It should be remembered that *Rrad* was set at 73 Ω, the theoretical value for a dipole. As this is not a dipole but an isolated conductor, it is fairly easy to derive the actual value of *Rrad* including environmental factors for this and any other isolated conductor simply by adjusting *Vsource* (in Mathcad) until the peak voltages are aligned. Figure 7.28 shows that the peaks are aligned when *Vsource* = 1.176 V.

Since Vsource was initially set at unity, it is now a simple matter to determine the value of *Rrad* by dividing the initial *Rrad* of 73 Ω by the 'offset' *Vsource* of 1.176 V, giving the actual *Rrad* value including environmental factors of 62.075 Ω. Which is basically Ohm's Law since *Rrad* can be manipulated as a simple resistance in such an instance. For any particular cable, the radiation resistance at the half-wave frequency is assumed to be a constant.

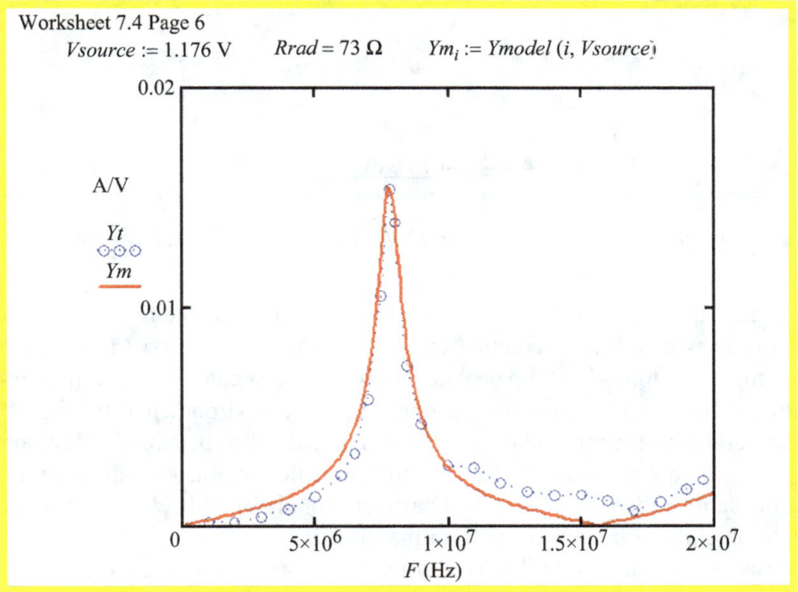

Figure 7.28 Response of model when peaks are aligned

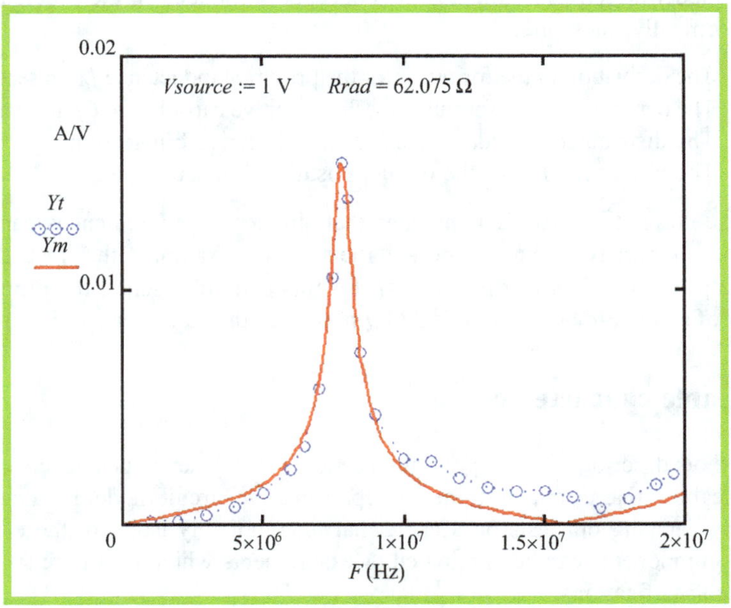

Figure 7.29 Response of model with the radiation resistance calculated to be 62.075 Ω

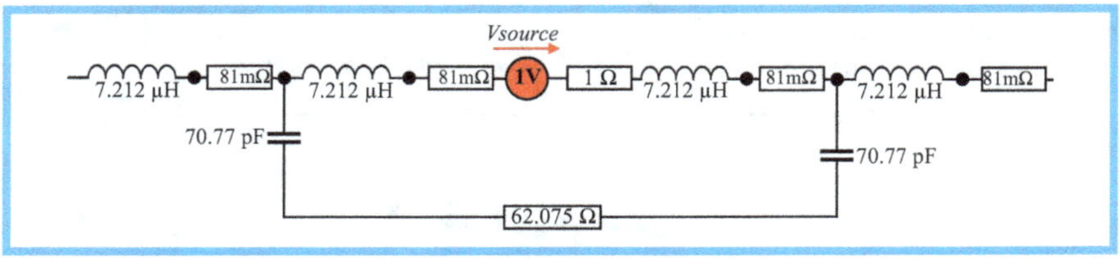

Figure 7.30 Final circuit model of conductor-under-review

Figure 7.29 shows the resultant curve and Figure 7.30 shows the final circuit model. This correlates well with the actual response in the range 0 to 5 MHz, as well as replicating the amplitude of the peak response. This test can be used to measure the radiation resistance of the conductor along with any environmental factors.

Deviations in the region between 10 and 20 MHz can be explained by the fact that, above the quarter-wave frequency, current is flowing backwards and forwards along the cable, rather like waves in a harbour. Simulating this effect would require extra complexity to be introduced into the model.

However, as far as EMC is concerned, the critical frequencies are when the peaks are high. So minor deviations between reality and model at other frequencies are of little concern when assessing the probability that the equipment will pass the formal EMC tests.

The close correlation between the response of the actual system and the theoretical model effectively validates the concepts introduced in the previous chapters. Specifically, these are:

- The derivation of the formulae for the primitive inductance *Lp* in section 2.2.
- The derivation of the formulae for the primitive capacitance *Cp* in section 2.1.
- The distributed impedance circuit model, derived in section 4.1.
- The circuit model for the dipole, postulated in section 5.2.

The close correlation also demonstrates that simple test equipment is capable of providing extremely accurate measurements. For example, the fact that the response peaks at 7.83 MHz means that, for this particular setup, the propagation velocity of antenna-mode current is 234.9 m/μs (see the value for *v* in Figure 7.24).

7.5 Cable characterisation

The method described in the previous section can be developed to characterise specific cables; that is, to determine the representative circuit model for a particular cable. By carrying out tests on a cable that is completely isolated, the effects of external components can be eliminated. Measurements which result are specific to the cable specifications.

In the example described here, a circuit model is derived for a 15 m length of twin-core power cable. Two setups are involved: the first to measure the emitted radiation, and the second to measure the wire-to-wire coupling.

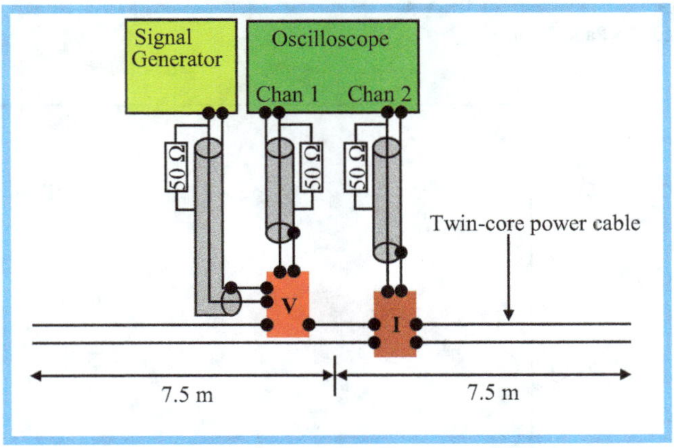

Figure 7.31 Test to measure the emitted radiation characteristics of the cable

Figure 7.31 illustrates the first setup. This configuration is essentially a practical implementation of the general circuit model of an isolated cable, derived from theoretical considerations in Section 5.2: the virtual conductor. Figure 5.13 illustrates that this model and (5.25) and (5.26) define formulae for component values. Since each half of the 15 m cable is represented by a triple-T network, the length l of each network is 7.5 m.

A voltage transformer is used to inject a voltage in series with one conductor of the cable. Both ends of the cable are open-circuit. If these terminations had both been short-circuited, then it would be clear that the transformer was inducing a voltage in the loop formed by the signal and return conductors; a differential-mode voltage. The existence of open circuit terminations does not change the fact that a differential-mode voltage is being injected.

This injected voltage creates a current which flows along the signal conductor. Electromagnetic coupling between the two conductors creates a current which flows in the opposite direction along the return conductor, the differential-mode current. The injected voltage also creates an antenna-mode current which flows along the cable and converts into an electromagnetic wave.

A current transformer clamped around both conductors was used to measure the amplitude of the antenna-mode current.

The range of frequencies was from 1 to 20 MHz and the output displayed in Figure 7.32 is the transfer admittance *YT*, in amperes per volt. The worksheet used to create this graph is shown in Figure 7.33, where there is a major peak at approximately 7.4 MHz. Figure 7.34 illustrates the second setup. Again, the voltage transformer injects a differential-mode voltage into the cable. But this time, the monitored current is the differential-mode current.

In Figure 7.35, there are two peaks, at about 5.67 MHz and at about 17.02 MHz. The worksheet used to create this graph is shown in Figure 7.36.

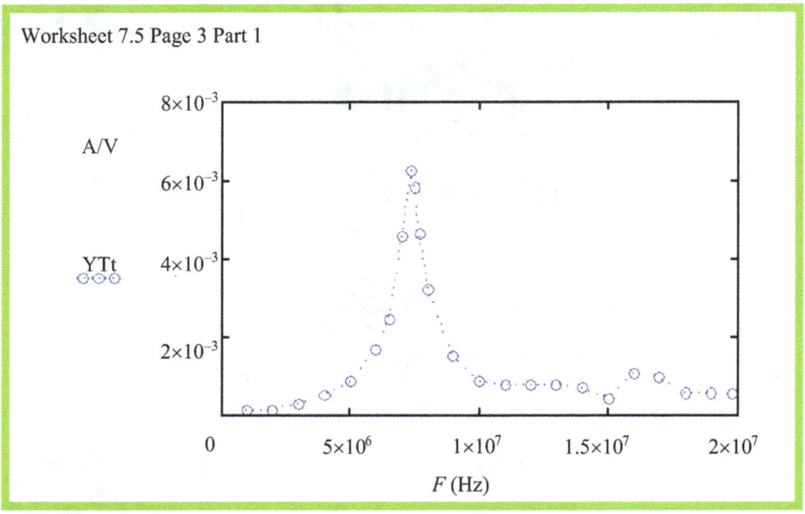

Figure 7.32 Results of radiated emission test

The peak frequencies were simply obtained by printing out Figures 7.32 and 7.35, drawing a vertical line through the peaks to the *x*-axis and measuring the offset from the nearest frequency marker.

The general circuit model for antenna-mode coupling of a transmission line is derived in section 5.2 on the virtual conductor and illustrated by Figure 5.13. Rearranging this in the form of a bridge circuit leads to Figure 7.37. This layout simplifies the definition of the two current loops. In the configuration-under-review, the voltage source is in series with conductor 1, which is reflected in the location of *Vsource* in the diagram. Transforming the circuit components into distributed impedances leads to Figure 7.38.

Component values can be assigned by following a systematic process, starting with the measurement of the conductor radii r_{11} and r_{22}, the spacing between conductors, r_{12}, and the length l of the cable. Using equations available in previous chapters, initial values can be assigned to the resistive and inductive components. This is carried out by Page 4 of Worksheet 7.5, illustrated by Figure 7.39.

Since the conductors of all cabling in electrical systems are sheathed or supported by insulating material, and since this material has a significant effect on the propagation velocities of the signals, it is necessary to determine the value of the relative permittivity of the dielectric before values for the capacitors can be defined. This is done on Page 5 of Worksheet 7.5, illustrated by Figure 7.40.

Having provided initial values for all the components of the model, the next step is to simulate the responses of both tests. This is carried out by the main program, illustrated by Figure 7.41. There are two outputs: the vector *YTm*, which simulates the response of the radiated emission test, and *Ym*, which simulates wire-to-wire coupling.

Worksheet 7.5 Page 1

$$\text{data1} := \begin{pmatrix} 1 & 420 & 0.4 \\ 2 & 420 & 0.4 \\ 3 & 420 & 0.8 \\ 4 & 420 & 1.4 \\ 5 & 435 & 2.4 \\ 6 & 350 & 3.8 \\ 6.5 & 365 & 5.8 \\ 7 & 370 & 11 \\ 7.38 & 370 & 15 \\ 7.5 & 370 & 14 \\ 7.7 & 365 & 11 \\ 8 & 365 & 7.6 \\ 9 & 365 & 3.6 \\ 10 & 360 & 2 \\ 11 & 360 & 1.8 \\ 12 & 360 & 1.8 \\ 13 & 360 & 1.8 \\ 14 & 355 & 1.6 \\ 15 & 360 & 1 \\ 16 & 375 & 2.6 \\ 17 & 320 & 2 \\ 18 & 270 & 1 \\ 19 & 280 & 1 \\ 19.8 & 285 & 1 \end{pmatrix}$$

$R1 := 300 \qquad R2 := 51 \qquad R3 := 50 \qquad R4 := 850$

$L1 := 200 \cdot 10^{-6} \qquad C1 := 60 \cdot 10^{-12} \qquad \text{Turns} := 10$

$s := 1..\text{rows}(\text{data1}) \qquad fl_s := \text{data1}_{s,1} \cdot 10^6$

$$YTt_s := \begin{array}{|l} Vch1 \leftarrow \text{data1}_{s,2} \cdot 10^{-3} \\ Vin \leftarrow \dfrac{51+50}{50} \cdot \text{Vch1} \\ Vch2 \leftarrow \text{data1}_{s,3} \cdot 10^{-3} \\ \omega \leftarrow 2 \cdot \pi \cdot fl_s \\ Z1 \leftarrow R4 + \dfrac{1}{j \cdot \omega \cdot C1} \\ Y2 \leftarrow \dfrac{1}{R1} + \dfrac{1}{R2} + \dfrac{1}{R3} + \dfrac{1}{j \cdot \omega \cdot L1} + \dfrac{1}{Z1} \\ Isec \leftarrow |Y2| \cdot Vch2 \\ Iprim \leftarrow Isec \cdot Turns \\ Yt \leftarrow \dfrac{Iprim}{Vin} \end{array}$$

$$YTt_s := \frac{YTt_s}{\sqrt{2}}$$

Figure 7.33 Using test data to calculate the admittance characteristic

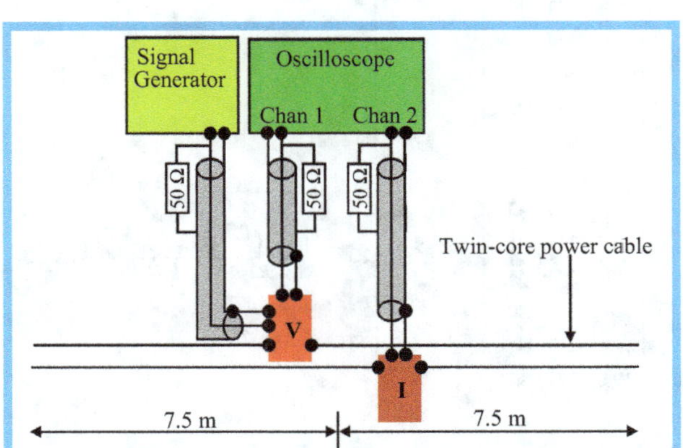

Figure 7.34 Test to measure transmission line characteristics of cable

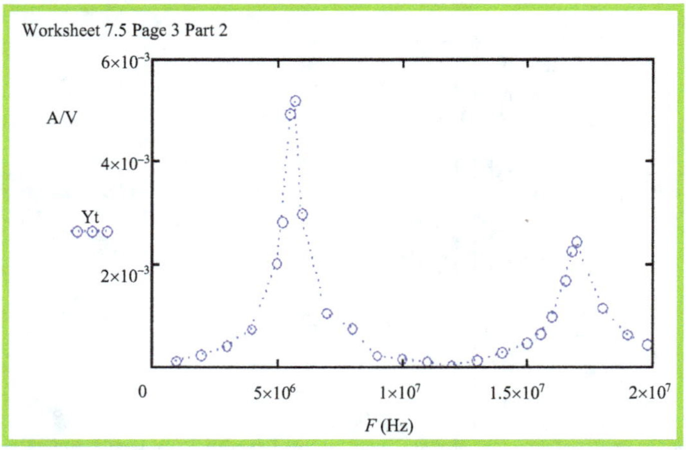

Figure 7.35 Results of wire-to-wire coupling test

Worksheet 7.5 Page 2

$$data := \begin{pmatrix} 1 & 420 & 2.8 \\ 2 & 420 & 6.4 \\ 3 & 420 & 11 \\ 4 & 430 & 20.5 \\ 5 & 460 & 60 \\ 5.2 & 480 & 88 \\ 5.5 & 420 & 135 \\ 5.7 & 325 & 110 \\ 6 & 330 & 63.5 \\ 7 & 370 & 25 \\ 8 & 380 & 18.5 \\ 9 & 375 & 5.2 \\ 10 & 375 & 3.6 \\ 11 & 380 & 1.8 \\ 12 & 375 & 0.6 \\ 13 & 380 & 3 \\ 14 & 380 & 6.6 \\ 15 & 390 & 11.5 \\ 15.5 & 405 & 17 \\ 16 & 415 & 26 \\ 16.5 & 395 & 43 \\ 16.8 & 325 & 47 \\ 17 & 280 & 44 \\ 18 & 270 & 20 \\ 19 & 290 & 11.5 \\ 19.8 & 295 & 8.2 \end{pmatrix}$$

$$s := 1..\text{rows}(data) \qquad f_s := data_{s,1} \cdot 10^6$$

$$Yt_s := \left| \begin{array}{l} Vch1 \leftarrow data_{s,2} \cdot 10^{-3} \\ Vin \leftarrow \dfrac{51+50}{50} \cdot Vch1 \\ Vch2 \leftarrow data_{s,3} \cdot 10^{-3} \\ \omega \leftarrow 2 \cdot \pi \cdot f_s \\ Z1 \leftarrow R4 + \dfrac{1}{j \cdot \omega \cdot C1} \\ Y2 \leftarrow \dfrac{1}{R1} + \dfrac{1}{R2} + \dfrac{1}{R3} + \dfrac{1}{j \cdot \omega \cdot L1} + \dfrac{1}{Z1} \\ Isec \leftarrow |Y2| \cdot Vch2 \\ Iprim \leftarrow Isec \cdot Turns \\ Yt \leftarrow \dfrac{Iprim}{Vin} \end{array} \right.$$

$$Yt_s := \frac{Yt_s}{\sqrt{2}}$$

Figure 7.36 Using test data to calculate the wire-to-wire coupling

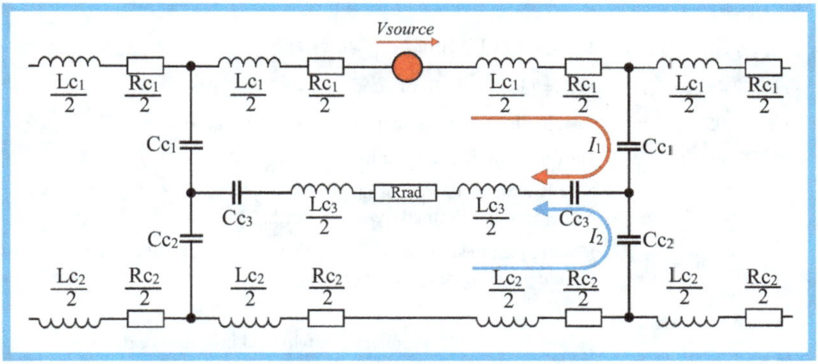

Figure 7.37 General circuit model of an isolated cable

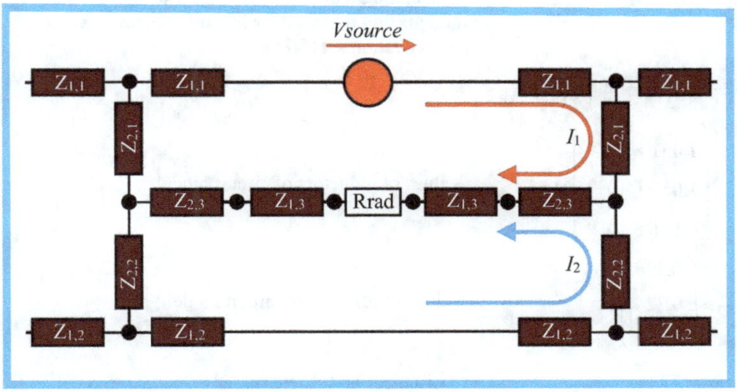

Figure 7.38 Distributed parameter circuit model

Response characteristics of the model are compared with those of the actual tests in the final page of the worksheet and are displayed here as Figures 7.42 and 7.43.

We now have a 'base-line' that will allow us to adjust the model to reflect the 'real-world' parameters, which provides the lossless theoretical model with values that enable the model to accurately represent the real-world characteristics of the cable-under-test.

We will start with the response of the radiated emission test. On examination of Figure 7.42, we see that both the frequency and amplitude of the respective peaks are out of alignment.

With regard to the wire-to-wire model, it can be seen from Figure 7.43 that although the frequency of the peaks is in alignment the amplitude of the peaks is not and increasingly deviates from the test plot as the frequency increases.

Worksheet 7.5 Page 4

$\rho := 1.7 \cdot 10^{-8}\ \Omega/\text{m}$	$\mu_o := 4 \cdot \pi \cdot 10^{-7}$ H/m $\quad \mu_r := 1$
$r11 := 0.48 \cdot 10^{-3}$ m	$r22 := r11$ Radii of conductors
$r12 := 1.95 \cdot 10^{-3}$ m	Spacing between conductors. See Figure 2.17
$l := 7.5$ m	Half the length of the cable
$Ra := 0.648\ \Omega$	Steady state resistance of length $l \times 4 = 30$ m $= 2 \times 15$ m Twin-core cable. Initially defined by equation (2.57)
$Rss := \begin{pmatrix} Ra \\ Ra \\ 0 \end{pmatrix}$	Placing steady state resistance values in the vector Rss. Conductor 3 represents the environment
$Fx := \dfrac{4 \cdot \rho}{\mu_o \cdot \pi \cdot r10^2}$ $\quad Fx = 7.476 \times 10^4$ Hz	Frequency at which skin-effect starts to modify resistance. See equation (2.60)
$Lc_1 := \dfrac{\mu_o \cdot \mu_r \cdot l}{2 \cdot \pi} \cdot \ln\left(\dfrac{r12}{r11}\right)$ $Lc_2 := \dfrac{\mu_o \cdot \mu_r \cdot l}{2 \cdot \pi} \cdot \ln\left(\dfrac{r12}{r22}\right)$ $Lc_3 := \dfrac{\mu_o \cdot \mu_r \cdot l}{2 \cdot \pi} \cdot \ln\left(\dfrac{l}{r12}\right)$	Calculating inductance values for the circuit model and placing these values in a three-element vector. See equation set (5.25)
$\dfrac{Lc}{2} = \begin{pmatrix} 1.051 \times 10^{-6} \\ 1.051 \times 10^{-6} \\ 6.191 \times 10^{-6} \end{pmatrix}$ H	Values of inductors of circuit model
$\dfrac{Rss}{2} = \begin{pmatrix} 0.324 \\ 0.324 \\ 0 \end{pmatrix}\ \Omega$	Values of resistors of circuit model
$Rrad := 73\ \Omega$	Radiation resistance. To be determined as per Section 7.4

Figure 7.39 Calculation of values for resistors and inductors

The first step is to adjust the model to account for all the parasitic 'stray' capacitance inherent in the setup. This is easily done by adjusting the value of fqb until the peaks align on the horizontal axis as shown in Figure 7.44. The second step is to find the value of Rrad using the method described in Section 7.4, this aligns the amplitude of the peaks as shown in Figure 7.45.

The third step is to adjust the model to account for the Environmental Frequency Losses. We do this in a crude fashion simply by introducing a constant 'EFL' and multiplying it by each successive index in the Ym_i loop, thus increasing the impedance of the model as the frequency increases.

The EFL value given in Figure 7.46 is the value obtained after adjustment of the EFL constant to align the model as also shown in Figure 7.46. This allows for Step 4 to be undertaken without needing too many iterations to vertically align the peaks.

So, having accounted for EFL, we now have to account for Environmental Line Losses 'ELL', which we then add to the Ym_i loop. As can be seen in Figure 7.47, the ELL constant value shown vertically aligns the peaks.

Step 5, the final step accounts for the environmental parasitic capacitances inherent to the twin-core cable. We do this by adjusting the value of FQA until the peaks align on the horizontal axis as shown in Figure 7.47. It should be noted that both steps 4 and 5 are shown on the same plot as the FQA adjustment required is so small and thus is hard to notice visually!

Worksheet 7.5 Page 5

Note: fqa and fqb values obtained graphically

$\varepsilon_o := 8.854 \cdot 10^{-12}$ F/m $\qquad c := 3 \cdot 10^8$ m/s

$fqa := 5.67 \cdot 10^6$ HZ — The frequency at which the peak occurs in test of wire-to-wire coupling

$va := 4 \cdot l \cdot fqa = 1.701 \times 10^8$ m/s — Velocity of propagation of electromagnetic wave along transmission line. See equation (2.35)

$\varepsilon_{ra} := \left(\frac{c}{va}\right)^2$ — Relative permittivity of dielectric of cable, acting as a transmission line. See equation (2.36)

$\varepsilon_{ra} = 3.111$ F/m — Value of relative permittivity to be used to define capacitor values for conductors

$fqb := 7.4 \cdot 10^6$ Hz — The frequency at which the peak occurs in radiated emission test

$vb := 4 \cdot l \cdot fqb = 2.22 \times 10^8$ m/s — Velocity of propagation of electromagnetic wave along cable. See equation (2.35)

$\varepsilon_{rb} := \left(\frac{c}{vb}\right)^2$ — Relative permittivity of dielectric of cable when it is acting as an aerial. See equation (2.36)

$\varepsilon_{rb} = 1.826$ F/m — Value of relative permittivity to be used to define capacitor value for monopole

$$Cc_1 := \frac{2 \cdot \pi \cdot \varepsilon_o \cdot \varepsilon_{ra} \cdot l}{\ln\left(\frac{r12}{r11}\right)}$$

$$Cc_2 := Cc_1$$

$$Cc_3 := \frac{2 \cdot \pi \cdot \varepsilon_o \cdot \varepsilon_{rb} \cdot l}{\ln\left(\frac{l}{r12}\right)}$$

Calculating capacitance values for the circuit model and placing these values in a three-element vector. See equation set (5.26)

$$Cc = \begin{pmatrix} 9.258 \times 10^{-10} \\ 9.258 \times 10^{-10} \\ 9.23 \times 10^{-11} \end{pmatrix} \text{F}$$

Values of capacitors of circuit model

Figure 7.40 Calculating values for the capacitors

Worksheet 7.5 Page 6

$i := 1..200 \quad Fi := i \cdot 10^5$ Defining the frequency range for the model

$$Zbranch(s) := \left| \begin{array}{l} \omega \leftarrow 2 \cdot \pi \cdot F_s \\ \text{for } k \in 1..3 \\ \quad \left| \begin{array}{l} Rc_k \leftarrow Rss_k \cdot \sqrt{1 + \left(\dfrac{F_s}{F_x}\right)} \\ \theta \leftarrow \sqrt{(Rc_k + j \cdot \omega \cdot Lc_k) \cdot j \cdot \omega \cdot Cc_k} \\ Zo \leftarrow \sqrt{\dfrac{Rc_k + j \cdot \omega \cdot Lc_k}{j \cdot \omega \cdot Cc_k}} \\ z_{1,k} \leftarrow Zo \cdot \tanh\left(\dfrac{\theta}{2}\right) \\ z_{2,k} \leftarrow Zo \cdot \mathrm{csch}(\theta) \end{array} \right. \\ z \end{array} \right.$$

Copy of function introduced in Figure 4.10

Equations for loop impedances, derived from inspection of Figure 7.38

$$Zloop(s) := \left| \begin{array}{l} Z \leftarrow Zbranch(s) \\ Z11 \leftarrow 2 \cdot (Z_{1,1} + Z_{2,1} + Z_{1,3} + Z_{2,3}) + Rrad \\ Z12 \leftarrow -2 \cdot (Z_{1,3} + Z_{1,3}) - Rrad \\ Z22 \leftarrow 2 \cdot (Z_{1,2} + Z_{2,2} + Z_{1,3} + Z_{2,3}) + Rrad \\ \begin{pmatrix} Z11 & Z12 \\ Z12 & Z22 \end{pmatrix} \end{array} \right.$$

$$V := \begin{pmatrix} 1 \\ 0 \end{pmatrix}$$

$$YTm_i := \left| \begin{array}{l} Z \leftarrow Zloop(i) \\ I \leftarrow \mathrm{lsolve}(Z, V) \\ \left| I_1 - I_2 \right| \end{array} \right.$$

Calculating response of transfer admittance for radiated emission from the cable

$$Ym_i := \left| \begin{array}{l} Z \leftarrow Zloop(i) \\ I \leftarrow \mathrm{lsolve}(Z, V) \\ \left| I_2 \right| \end{array} \right.$$

Calculating response of admittance of open-circuit transmission line

Figure 7.41 The main programme, calculating the response of the circuit model

The final component values obtained by this procedure are shown in Figure 7.48 and can now be assigned to the general circuit model of Figure 7.37, producing the representative circuit model for the configuration-under-review as shown in Figure 7.49.

The ability to create a circuit model of such a cable leads to several advantages:

- The basic feature of a circuit model of a length of cable is that component values are proportional to length of the cable. Per-unit-length parameters can be derived, and later used to define models for any length of cable used in the system-under-review. (It is important that the cross-section of the cable remains uniform.)

- Measurements can be carried out on long cables at relatively low frequencies and used to predict the performance of shorter cables at higher frequencies.
- Having determined a circuit model for a particular cable, the file can be stored in a library, similar to that used by SPICE software for devices such as Operational Amplifiers or Schmitt triggers, though a database of the pertinent values would allow for the characteristics of cables to be more easily compared and evaluated for potential use in varied applications.

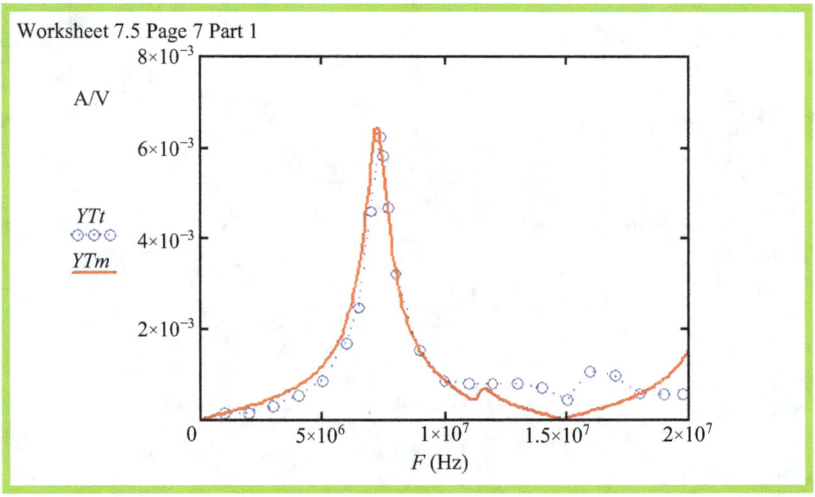

Figure 7.42 Radiated emission. Responses of cable and model

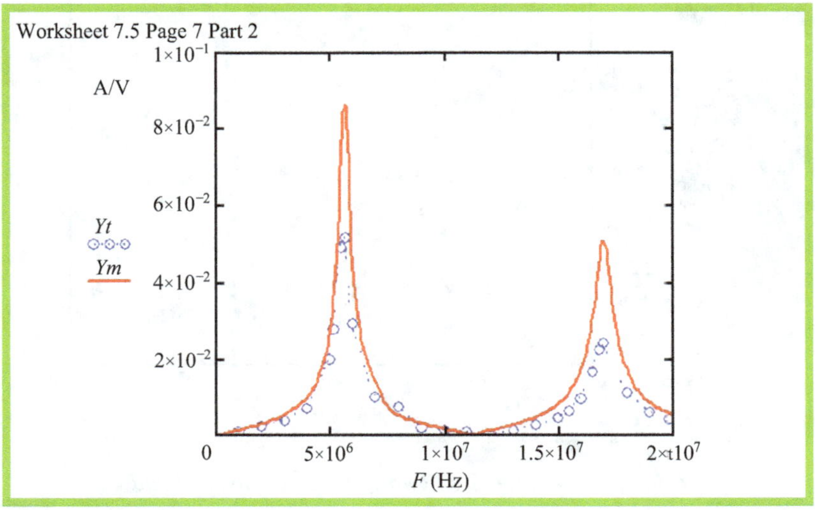

Figure 7.43 Transmission line responses of cable and model

- Combining the models of the interface equipment with that of the cable will allow the radiated emission of the system-under-review to be simulated.
- It is also possible to simulate the response of the system to external interference.

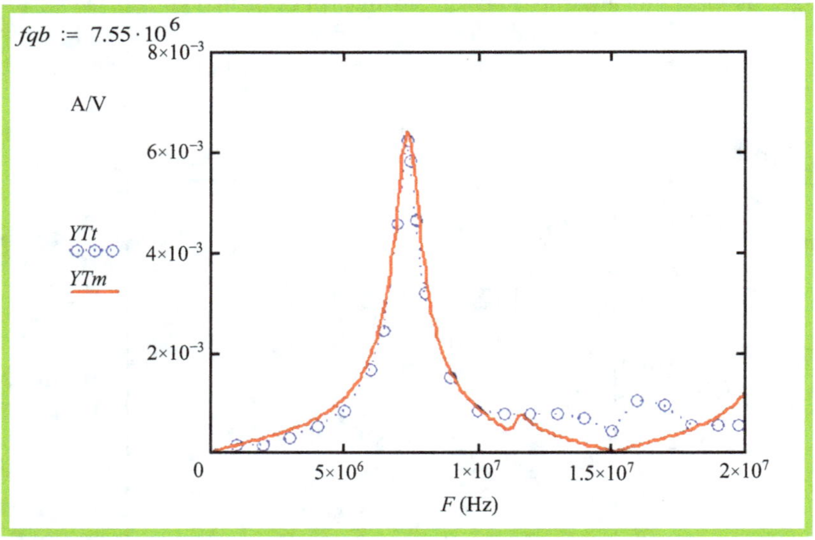

Figure 7.44 Step 1 Aligning the peaks using fqb

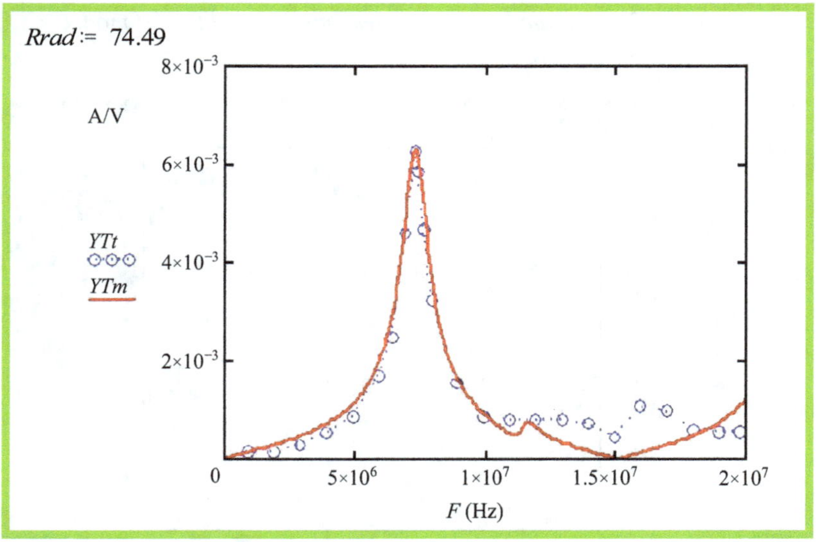

Figure 7.45 Step 2 Aligning the peaks using Rrad

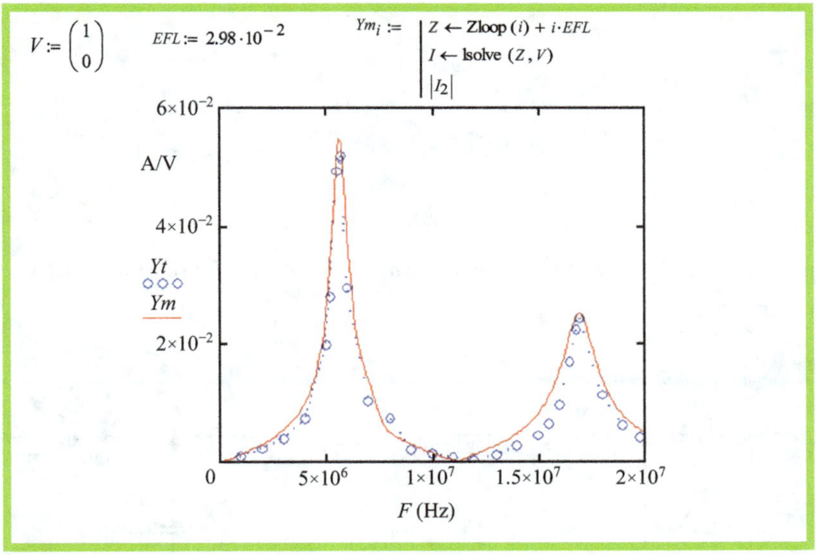

Figure 7.46 Step 3 Aligning the peaks using EFL

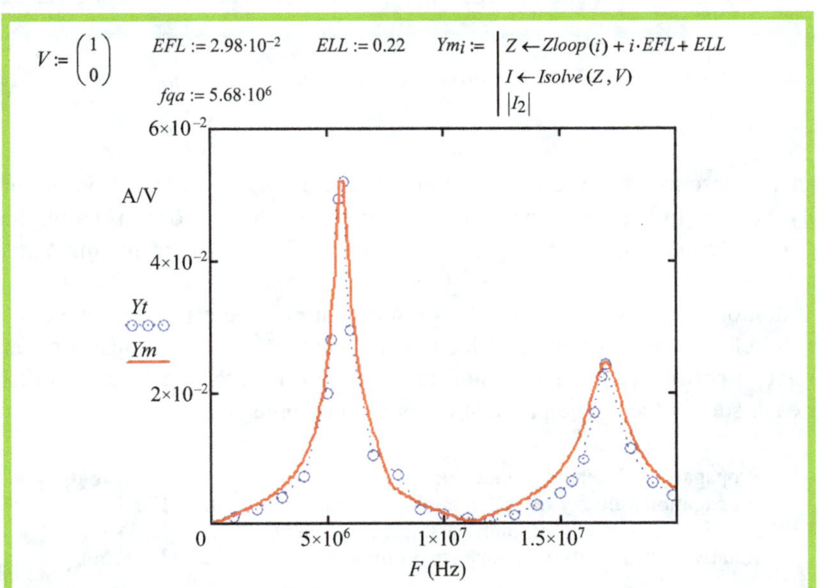

Figure 7.47 Steps 4 and 5 Aligning the peaks using ELL and fqa

$$\frac{Rss}{2} = \begin{pmatrix} 0.324 \\ 0.324 \\ 0 \end{pmatrix} \Omega \qquad \frac{Lc}{2} = \begin{pmatrix} 1.051 \times 10^{-6} \\ 1.051 \times 10^{-6} \\ 6.191 \times 10^{-6} \end{pmatrix} H \qquad Cc = \begin{pmatrix} 9.226 \times 10^{-10} \\ 9.226 \times 10^{-10} \\ 8.867 \times 10^{-11} \end{pmatrix} F$$

$$Rrad := 74.49$$

Figure 7.48 Final component values for general circuit model

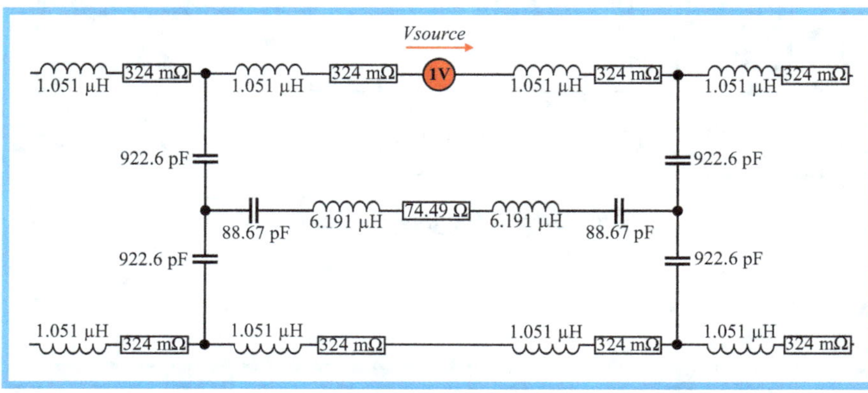

Figure 7.49 Representative circuit model of cable under test

It can be reasoned that the field pattern of antenna-mode propagation is almost totally external to the cable and so includes more air than it does insulating material. The differential-mode propagation is more confined to the insulating material, so it travels more slowly.

Additionally, this set of tests does more than provide a reason for the difference. It defines the propagation velocities and provides an actual measurement of the relative permittivity. These values are obtained from Page 5 of the Worksheet after each step in the above procedure has been completed.

Propagation of antenna-mode current,	vb	226.5 m/μs
Propagation velocity of differential-mode current,	va	170.4 m/μs
Relative permittivity of cable, acting as an antenna,	ε_{rb}	1.754 F/m
Relative permittivity of transmission line,	ε_{ra}	3.1 F/m

7.6 Cable transients

Since the signal generator used with the setup of Figure 7.31 was capable of generating square waves as well as sine waves, it seemed logical to check the response of the cable to transient signals. However, the waveform supplied by the output of the voltage transformer deviated significantly from the ideal square wave. As such, it was decided to deliver the input signal to the cable via a resistor network. The setup is illustrated by Figure 7.50.

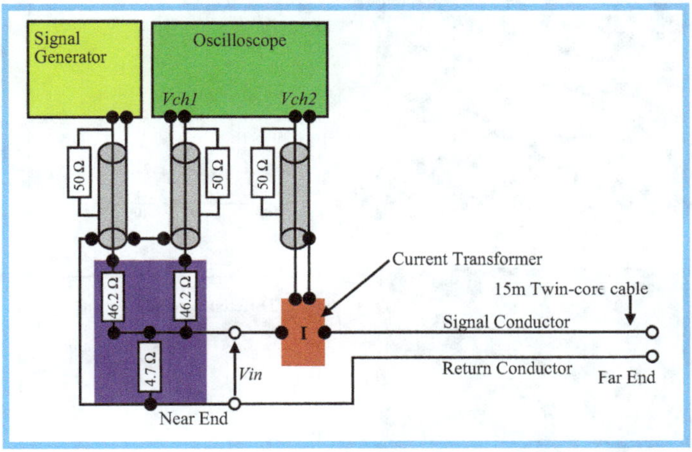

Figure 7.50 Setup for radiated transient test

This configuration allowed the cable to be analysed as a transmission line. Since the far end was open-circuit and the resistance at the near-end was very low, it was to be expected that multiple reflections would take place and that the monitored current would provide some insight into what was happening along the cable.

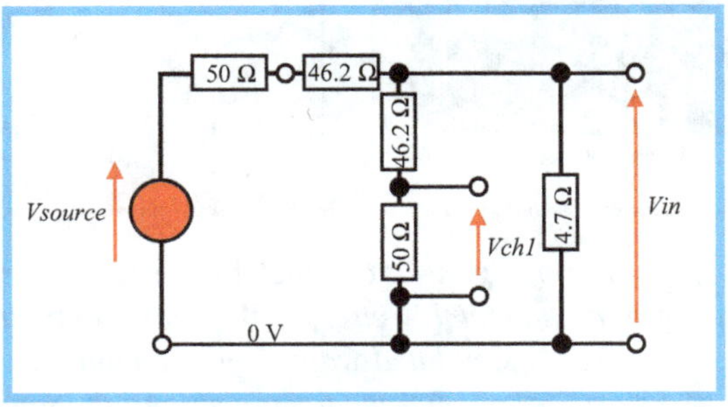

Figure 7.51 Circuit model of test setup

Figure 7.51 illustrates how the voltage monitored at channel 1 input terminals is related to the voltage delivered to the cable. Inspection of this model gives:

$$Vch1 = \frac{50}{96.2} \cdot Vin \tag{7.9}$$

Figure 7.52(a) is a photograph of the actual waveform monitored by channel 1. The current delivered to the line causes minor perturbations to the input voltage for about 2 μs after each step change, after which *Vin* reaches its open-circuit value. The period for the square wave was about 8 μs. That is, the oscillator frequency was set at about 125 kHz. The program described in section 6.6 was used to create the waveform of Figure 7.52(b). Input variables were set to simulate an initial falling edge at 150 ns and a rising edge at 4.1 μs. The output variable was selected to be *Vch1*.

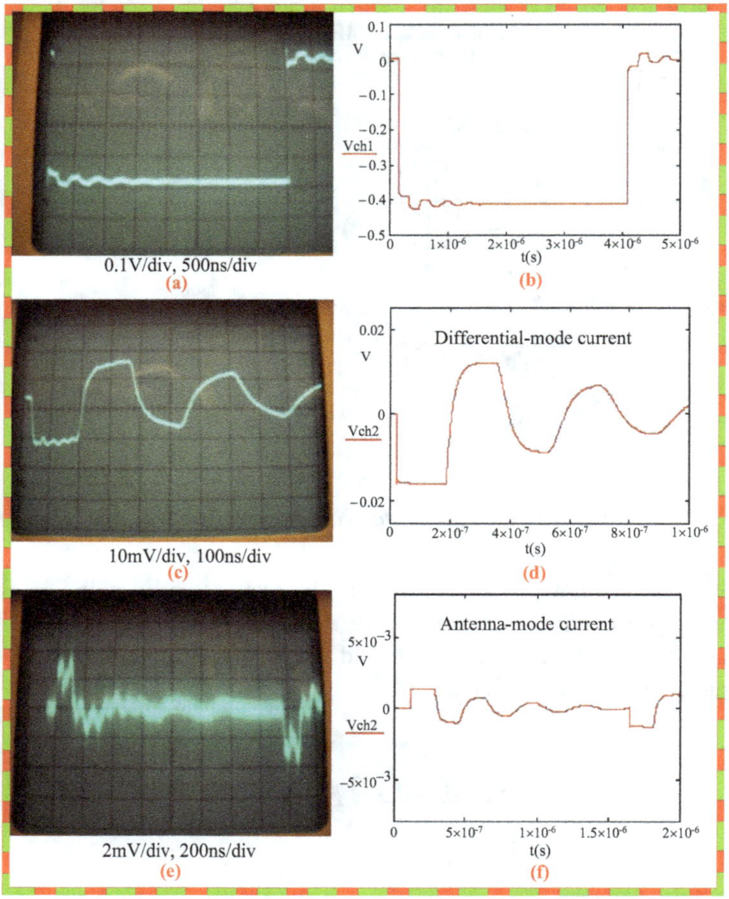

Figure 7.52 *(a) Waveform of signal at channel 1 of oscilloscope, (b) simulated waveform at channel 1, Vch1, (c) Waveform at channel 2 of oscilloscope, (d) simulated waveform at channel 2, Vch2, (e) Waveform of antenna-mode current and (f) Simulation of antenna-mode current*

Figure 7.52(c) is a photograph of the waveform displayed by channel 2 of the oscilloscope. The oscillations during the first 200 ns could be due to reflections emanating from the ground conductors of the test equipment. They could also be due to cable or connector discontinuities. The program of section 6.6 was run again, with input variables set to simulate an initial falling edge at 20 ns and a sweep time of 1 μs, and the output variable was selected to be *Vdiff*. Figure 7.52(d) was the result. The period between the first and second edges of the waveform is twice the differential-mode propagation delay of the 15 m cable – the time taken for the reflected wave to return to its source.

The setup of Figure 7.50 was modified to route both conductors of the cable through the core of the current transformer. Antenna-mode current was now being monitored. This is shown in Figure 7.52(e). Running the simulation program a third time, with the time of the initial rising edge set at 120 ns, the time of the falling edge set at 1.65 μs, and with *Vrad* selected as the output variable, gave Figure 7.52(f). The rise and fall times of the model were set by the value of *dt*. From the worksheet of Figure 6.27:

$$dt = \frac{T}{N} = \frac{83 \cdot 10^{-9}}{100} = 830 \text{ ps}.$$

Comparing Figure 7.52(c) with Figure 7.52(d) reveals that the simulated waveform of the differential-mode current *Idiff* is very similar to the actual waveform. As expected, the waveform gradually changes from a square wave to a sine wave as the higher frequency components of the square wave are degraded and 'lost' to 'ground' and the environment via both intrinsic and parasitic capacitances of the cable and test equipment.

There are some deviations, however:

- The leading edge of the step on the observed signal is not quite as fast as the theoretical step. This is almost certainly due to the finite response times of the signal generator and oscilloscope.
- There is a ripple on the actual waveform which lasts for about 200ns. This is probably due to reflected current returning from the conductors of the structure.

The similarity between Figure 7.52(b) and Figure 7.52(a) is just as significant. These compare the monitored voltage at channel 1 with the simulated voltage *Vch1*. The ripple during the first 2 μs following each step change is due to the voltage drop in the source resistance *Rg* caused by the current delivered to the cable.

Most notable is the fact that the corners of the leading and trailing edges are rounded. If the loading effect had been due solely to the current waveform displayed in Figure 7.52(b), then the corners would not have been much sharper. The fact that they are rounded is a clear demonstration that the source resistance is carrying antenna-mode current as well as differential-mode current.

There is one significant difference between Figure 7.52(a) and Figure 7.52(b):

There is a sharp 'spike' after each step change of the monitored waveform. This is due to capacitive coupling between the output of the signal generator and the input to channel 1 of the scope.

The two terminals connecting co-axial inner conductors to the 46 Ω resistors shown in Figure 7.50 were very close to each other.

However, the existence of this spike does not obscure the fact that the ends of each step-change are rounded.

Figures 7.52(e), and 7.52(f), also display significant similarities. These compare the waveform of the antenna-mode current carried by the cable with the waveform generated by the simulation. It is quite clear that the waveform resulting from the simulation follows the same basic pattern as that observed on the oscilloscope; that is, a damped oscillation which is a miniature version of *Idiff*. There is also a higher frequency component superimposed on top of the damped oscillation. This lasts for many more cycles than the underlying waveform of Figure 7.52(f), indicating that the damping it experiences is insignificant. The probable cause is a current component that is oscillating backwards and forwards between the cable and the structure. Damping is much less because this current component does not flow in the source resistance *Rg*.

The fact that the test results and the simulated waveforms are closely correlated provides a fair degree of confidence in the reliability of the model. Figure 7.52(d) is certainly more representative of the actual response than the simple delay-line model developed in section 6.2. A point worth remembering is that although the capacitors and inductors of that simple model have been replaced by the parameters of propagation time and characteristic impedance, these properties continue to exist. (Section 6.3 covers this aspect in more detail.)

Since the delay-line model has introduced the parameter of time into the computations, it becomes possible to think in terms of the energy stored in the line. That is, the static energy of electric charge and the kinetic energy of moving charges. In fact, (6.5) for an RLC circuit is very similar to the equation of motion of a mass on a spring.

The inclusion of the network of *Crad*, *Ro* and *Inf* of Figure 6.34 enables the effect of energy storage caused by antenna-mode current to be separated from the effect of storage of differential-mode energy in the delay line. When a step voltage is applied to the open-circuit line, the differential-mode current delivered to that line is constant for a period equal to twice the propagation time and so is the antenna-mode current. After that period, the capacitor *Crad* has been charged up and contains a lot of potential energy. The initial flow of current into *Crad* represents the kinetic energy required to deliver that charge.

The fact that the current waveform reverts rapidly from a square wave to a sinusoidal wave at the natural frequency of the assembly shows that the transient energy of the initial step has been converted into the dynamic energy of an LC circuit. From Figure 7.52(c), the time for one cycle of this waveform is about 0.34 μs, corresponding to about 2.9 MHz. That is, a frequency which is close to the quarter-wave frequency of a 15 m cable acting as a monopole antenna.

Figure 7.32 shows that, at the frequency of resonance, the radiation emitted by the assembly reaches a peak value.

Since the amplitude of this sinusoidal wave of Figure 7.52(c) decays rapidly, it is reasonable to assume that this loss of energy is mostly due to radiated emission. The inclusion of *Ro3* in the model of 6.34 allows the amplitude of the antenna-mode current itself to be evaluated.

The setup of Figure 7.50 offers another way of deriving values for cable parameters, simply by using data from the transient response. Since the voltage and current are almost constant during the first 166 μs, the value of the characteristic impedance *Ro* can be calculated quite quickly. Since the transit time is also known, the loop inductance *La* and loop capacitance *Ca* of the 15 m cable can also be derived, as illustrated in Figure 7.53.

From the waveform of Figure 7.51, the amplitude of the step voltage measured by Channel 1, at the end of the first exponential rise, is: Vchl := 0.39 V

The voltage applied to the input terminals of the line is related to Vch1 by equation (7.9)

$$Vin := \frac{96.2}{50} \cdot \mathrm{Vch1} = 0.75\ \mathrm{V}$$

From the waveform of Figure 7.52 (c), the voltage measured by Channel 2 is: Vch2 := 0.016 V

The ratio of the amplitude of Vch2 to the amplitude of the current monitored by the current transformer is given by equation (7.7)

$$RT := 2.27\ \Omega$$

Invoking equation (7.6)

$$Iout := \frac{\mathrm{Vch2}}{\mathrm{RT}} = 7.048 \times 10^{-3}\ \mathrm{A}$$

The ratio of Vin to Iout gives a measure of the characteristic impedance of the 'loss-less' cable:

$Ro := \frac{Vin}{Iout} = 106\ \Omega$. This can then be divided by $\sqrt{2}$ to give $\frac{106}{\sqrt{2}} = 74.953\ \Omega$,

which is the RMS value, and is very close to a cable whose nominal characteristic impedance is rated at 75 Ω, a very common and widely used characteristic impedance.

The waveform of Figure 7.52 (c) also gives a measure of the time taken for a current step to propagate to the far end and return to the near end. This is the time difference between the first and second edges of the waveform. This gives:

$$T := \frac{166.10^{-9}}{2}\ \mathrm{s}$$

The capacitance and inductance of the cable can be determined by invoking equations (6.45) and (6.44).

$$C_{Trans} := \frac{T}{Ro} = 780\ \mathrm{pF}$$

$$L_{Trans} := T \cdot Ro = 8.8\ \mu\mathrm{H}$$

These results can be compared with those obtained from cable characterisation tests. From Figure 7.49.

$$C_{Char} := 2 \cdot \frac{922.6}{2} \cdot 10^{-10} = 922.6\ \mathrm{pF}$$

$$L_{Char}\ 8 \cdot 1.051 \cdot 10^{-6} = 8.408\ \mu\mathrm{H}$$

Figure 7.53 Calculating component values using transient and frequency response data

Figure 7.53 also derives values for these reactive components, using data from the frequency response tests on the same cable.

	Transient analysis	Frequency response analysis
Capacitance between conductors	780 pF	922.6 pF
Loop inductance	8.8 μH	8.408 μH

The reasonably close correlation between these results shows that it is possible to obtain a first estimate of the values of both the inductance and the capacitance of a cable, just by using data from a couple of transient waveforms.

It is useful to summarise the reasoning involved in the transient analysis and to identify its significance in the design of electronic equipment.

Any step voltage between two terminals of a cable will cause a transient wavefront to propagate along the cable at a velocity somewhat less than the speed of light. During the transit time, some of the energy will be radiated into the environment and some will be stored temporarily in the form of static charge.

The static charge between cable and environment will disappear in the form of a high-frequency signal which rapidly decays in amplitude; see Figure 7.52(e). This means that both the static energy in the charge and the dynamic energy used to create that charge are converted into radiated interference. The charge between the conductors will eventually settle down to a steady-state value, but not before some of the energy is converted into differential-mode radiation; see Figure 7.52(c).

This will happen with every step-change in voltage. Since any signal is effectively a continuous series of step changes, it is inevitable that a significant portion of the signal on a twin-conductor cable will radiate away in the form of EMI.

If the separation between send and return conductors is uncontrolled, then most of the transient energy will be converted into unwanted radiation. This is very useful for covert surveillance, but not much else.

The best ways of minimising the amount of electromagnetic energy in a transmission line which is lost to the environment are to use either a co-axial cable, a screened pair, or a waveguide.

If an unscreened twin-conductor cable is used, the easiest way of reducing the level of radiated energy is to absorb it in a resistive component. However, the additional heat generated in the resistive component will also need to be dissipated. Sections 8.5 and 8.6 describe a range of design techniques which enable this objective to be achieved.

7.7 Capacitor characterisation

It is entirely possible to use the technique of circuit modelling to create representative circuit models of components such as inductors, capacitors and filters. The setup would be as illustrated in Figure 7.54.

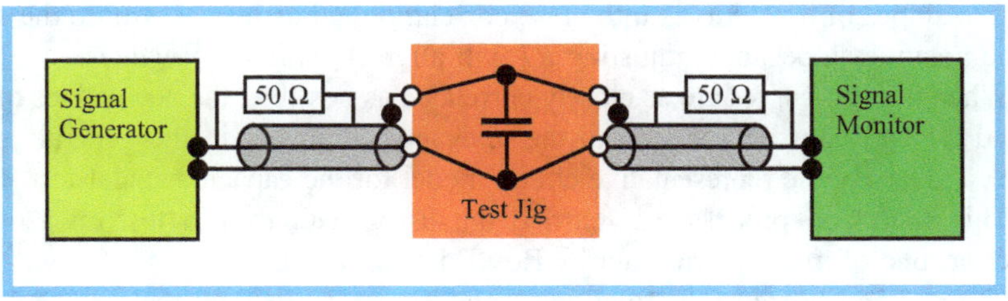

Figure 7.54 Test setup for characterising components

In this instance, the component-under-test is a capacitor, mounted in the test jig. The jig is coupled to a signal generator using a 50 Ω co-axial cable. A similar cable is used to route the output signal to a signal monitor. A transfer function analyser was used to provide the input signal and to monitor the output. Alternatively, the source could be a signal generator and the monitor could be an oscilloscope or a spectrum analyser.

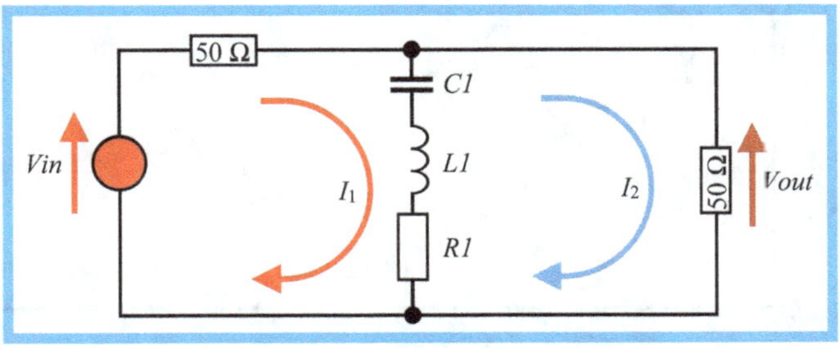

Figure 7.55 General circuit model for capacitor characterisation

Figure 7.55 shows the general circuit model for this setup. Since the output impedance of the generator is 50 Ω and the characteristic impedance of the co-axial cable is 50 Ω, the cable between the source and *jig* is effectively transparent. The same applies to the output link, that between *jig* and signal monitor.

Since all capacitors possess inductive and resistive properties, the simplest model for this particular component-under-test is the series LCR circuit. For this type of test, the relevant parameter is the transfer impedance.

$$ZT = \frac{Vout}{I_1} \tag{7.10}$$

So, the test procedure would be to apply a constant-voltage variable-frequency to the test jig and note the amplitude of the output signal at a set of spot frequencies.

This gives the frequency response of the transfer impedance *ZTt*; that is, the transfer impedance derived from test results.

It is a simple matter to write a program to calculate the frequency response of the transfer impedance *ZTm* of the circuit model. Both responses can then be plotted on the same graph. Initial guess values are assigned to *C1, L1* and *R1*, and these parameters are then treated as input variables. Adjusting *C1* will bring the negative slopes of both curves into close correlation, adjusting *L1* will do the same for the positive slopes, and adjusting *R1* will align the minimum values.

When the two curves are as closely correlated as possible, the new values of *C1, L1* and *R1* can then be assigned to the relevant component in the general circuit model. This gives the representative circuit model for the capacitor-under-test.

This analytical process was carried out using data from a test on a 60 nF capacitor, one of many carried out by Roy Ediss ediss-electric.com.

Figure 7.56 illustrates the final graph and Figure 7.57 shows the representative circuit model. A copy of the Mathcad worksheet used to create the graph is provided by Figures 7.58 and 7.59.

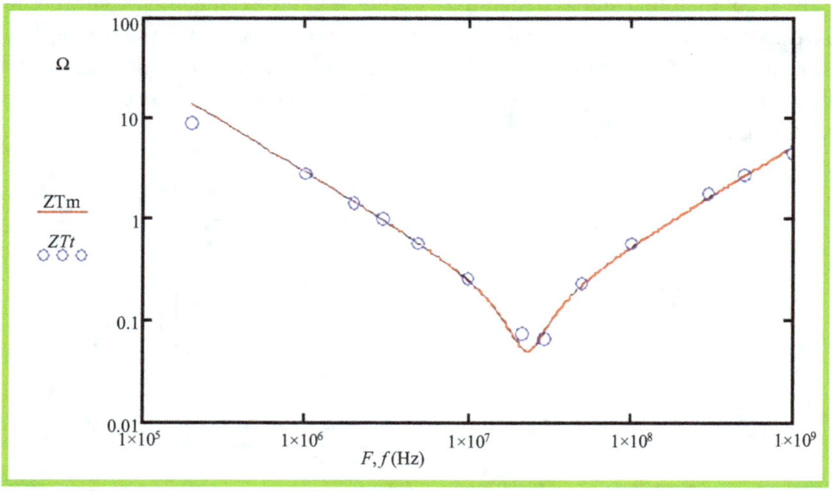

Figure 7.56 Transfer impedances of test setup and circuit model

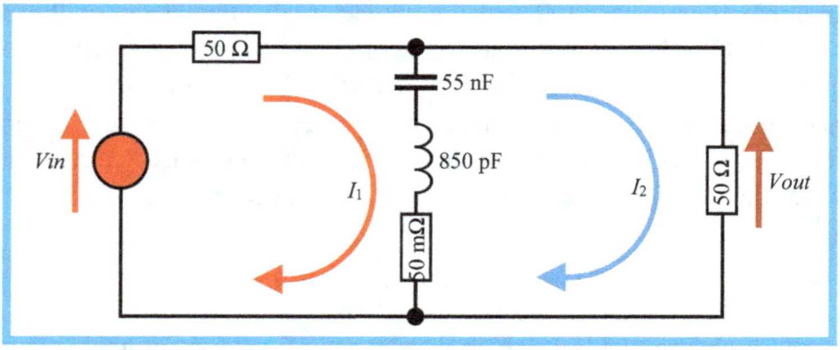

Figure 7.57 Representative circuit model of Murata 100 nF capacitor-under-test

Worksheet 7.7. Page 1

$$D := \begin{pmatrix} 0.2 & 430 & -9020 & \text{"58 nF"} \\ 1 & 190 & -2700 & \text{"53 nF"} \\ 2 & 93 & -1400 & \text{"53 nF"} \\ 3 & 71 & -975 & \text{"54 nF"} \\ 5 & 30 & -554 & \text{"54 nF"} \\ 10 & 30 & -251 & \text{"62 nF"} \\ 21.6 & 49 & -53 & \text{"0"} \\ 30 & 30 & 60 & \text{"396 pH"} \\ 50 & 27 & 221 & \text{"696 pH"} \\ 100 & 79 & 553 & \text{"877 pH"} \\ 300 & 400 & 1670 & \text{"881 pH"} \\ 500 & 944 & 2500 & \text{"793 pH"} \\ 1000 & 2500 & 3700 & \text{"586 pH"} \end{pmatrix}$$

Column 1: frequency, MHz
Column 2: resistance, mΩ
Column 3: reactance, mΩ
Column 4: text

$s := 1..\text{rows}(D)$

$f_s := D_{s,1} \cdot 10^6$

$R_s := D_{s,2} \cdot 10^{-3}$

$X_s := D_{s,3} \cdot 10^{-3}$

$ZTt_s := \left| R_s + j \cdot X_s \right|$

Data copied from: ediss-electric.com/bypass_caps_pdf/1_100n0201X5R_murata.pdf

Figure 7.58 Calculating frequency response of transfer impedance from test data

Worksheet 7.7. Page 2

See Figure 7.55

$R1 := 0.050\ \Omega$ $\quad C1 := 55 \cdot 10^{-9}\ \text{F}$ $\quad L1 := 850 \cdot 10^{-12}\ \text{H}$

$Rn := 50\ \Omega$ $\quad Rf := 50\ \Omega$ $\quad V := \begin{pmatrix} 1 \\ 0 \end{pmatrix} \text{V}$

$i := 2..10000$ $\quad F_i := i \cdot 10^5$

$$ZTm_i := \left| \begin{array}{l} \omega \leftarrow 2 \cdot \pi \cdot Fi \\ Z11 \leftarrow Rn + R1 + \dfrac{1}{j \cdot \omega \cdot C1} + j \cdot \omega \cdot L1 \\ Z12 \leftarrow -\left(R1 + \dfrac{1}{j \cdot \omega \cdot C1} + j \cdot \omega \cdot L1 \right) \\ Z22 \leftarrow \text{Rf} + R1 + \dfrac{1}{j \cdot \omega \cdot C1} + j \cdot \omega \cdot L1 \\ Z \leftarrow \begin{pmatrix} Z11 & Z12 \\ Z12 & Z22 \end{pmatrix} \\ I \leftarrow \text{lsolve}(Z, V) \\ Vout \leftarrow |I2 \cdot Rf| \\ Iin \leftarrow |I1| \\ \dfrac{Vout}{Iin} \end{array} \right.$$

Equation 7.10

Figure 7.59 Calculating the frequency response of the transfer impedance of the circuit model

7.8 When theory meets reality: analysis

In this chapter, the theory and models outlined and developed in the previous chapters have been adapted to create simple models which can be tested in the real world of bench testing. Two primary models have been developed and tested, those being cable characterisation and transient response. Though both models have been developed from the 'primitives' each model differs in its utilisation and it is worth noting that the transient model only requires the RMS value of the current to calculate the characteristic impedance of the cable, whereas the cable characterisation model requires all currents to be manipulated using their RMS values, since it retains the reactive components in the model.

Starting with the cable characterisation model of Chapter 7.5, it can be seen that the 'ideal' response of the model depicted in Figures 7.42 and 7.43 requires that the model be modified to account for parasitic capacitance and environmental factors. From what I have been able to establish, the bench testing was carried out in an atmosphere with a temperature range between 21°C and 25°C, in high humidity conditions.

The step-by-step procedure detailed in Chapter 7.5 thus modifies the model to compensate for these environmental factors. However, if a cable is to be used in any environment which is not within the narrow environmental range given here, the cable should be tested over the full environmental conditions in which it is to be used. Ideally, a cable should be tested over a wide range of temperatures in varying mediums (Vacuum, Atmosphere, Earth and Fluids) and where applicable humidity and pressure levels. Preferably, this should be done in a Faraday cage where the frequency and magnitude of electromagnetic waves of interest can be introduced into the environment as either source or threat. The data obtained for each cable tested in such a fashion can then be added to a database and it will be possible to determine how such a cable will perform in any given environment. Thus, with a database of multiple cable characteristics, the most appropriate cable for any given application can be selected. Another environmental condition which should be characterised is gravity, which acts upon all physical bodies and will therefore alter the characteristics of each and every cable dependent upon the cable's composition, mass and length. Unfortunately, for such a characterisation to be conducted, a space station or orbital platform will be required.

The transient response model differs significantly from the cable characterisation model in that the model represents a 'loss-less' cable, and as such the reactive components present in the cable characterisation model are not required, with the model dependent upon the test results to formulate the capacitive and inductive values of the cable. As such, although there is a close correlation between the values obtained by the two models, it is not possible to compensate for the environmental factors associated with the reactive components within the transient model, which accounts for the slight difference in values depicted by Figure 7.53.

Additionally, a third model has been provided in Chapter 7.7, which is able to characterise all passive components. Given that all the models presented and tested in this chapter are able to provide results that closely correlate with the test data and with each other. It can be said with a high degree of confidence that the primitives, theory and models developed in the previous chapters have the capacity to generate accurate and meaningful results, which can be used to simulate how the properties of a cable will affect the environment, and how susceptible a cable is to EMC/EMI.

Chapter 8
Practical design

There are a number of design concepts and techniques in existence which are aimed at improving EMC. Many of them are supported by the analytical approach. But some are not.

A concept which can be traced back for many decades is that of the 'equipotential ground plane'. There is no such thing. Since the 'method of images' of electromagnetic theory is often quoted as the basis for this concept, it is useful to identify the fallacies underlying the correlation. Section 8.1 does just that. Designers should resist the temptation to treat the structure as a convenient return path for signals and power. **The most effective use of the conducting structure is as a shield.**

One deduction which is continually reinforced during the process of creating circuit models is the fact that the signal and return conductors should be as close together as possible.

Section 8.2 shows that this requirement is implemented to good effect in the design of printed circuit boards by routing the signal tracks as close to the ground plane as possible. It also shows that, for signals between boards and between equipment, this requirement is best achieved by allocating a return conductor to every signal or power conductor.

In terms of EMC, one of the most counterproductive concepts is that of the 'single-point ground'. Section 8.3 illustrates how this technique guarantees the creation of intractable problems. In contrast, the existence of a multitude of 'ground loops' causes a significant reduction in the level of interference.

Improvements achieved by the inclusion of dedicated return conductors can be nullified by the inappropriate design of the interface circuitry of equipment units. Section 8.4 identifies the need for current balance in the interconnecting cables and illustrates a few methods of achieving this objective.

Section 8.5 focuses on the fact that optimum transmission is achieved by terminating transmission lines with a resistance equal in value to the characteristic impedance. If the signal is transmitted efficiently, it follows that emissions due to interference are minimal. Circuit modelling confirms that a configuration which exhibits low emission is less susceptible to external interference.

Whether the common-mode loop is open-circuit at one end or short-circuited at both ends, it is inevitable that there will be reflections. Electrical energy will be stored temporarily in this loop and will quickly depart into the environment in the form of interference. It is possible to absorb this energy and damp down the oscillations. Section 8.6 describes ways of doing this.

Section 8.7 deals briefly with the subject of equipment shielding and identifies the basic requirements. Measures used to protect buildings and equipment from the indirect effects of lightning are also described.

Every technique described in this chapter defines the interface circuitry at both ends of the signal link. This provides visibility of both the signal loop and the common-mode loop. Without such visibility, there is no point in even trying to assess the interference coupling characteristics, let alone analyse them.

By defining the interface circuits at both ends of the link, an important feature of printed circuit board design is identified; the interface circuits provide a buffer between the internal wiring of each board and the cabling used to carry signals between boards.

This means that the interface circuits can be designed to handle the higher levels of interference which are bound to exist on the longer conductors.

The length of the signal link is an important parameter. The longer the link, the lower the frequency at which interference reaches its highest peak. Since the power available is inversely proportional to the square of the frequency, the longest links deliver the highest level of interference energy. The buffer circuits on the printed circuit boards should be able to absorb whatever unwanted energy arrives via the signal link.

8.1 Grounding

Reliance on the use of the conducting structure as the universal return path for all signals and all supplies is probably the most prevalent cause of EMC problems. This could be due to the widespread belief in the existence of the equipotential ground plane. There is no such thing.

Ground planes are an extremely useful design feature of printed circuit boards and integrated circuits. But this does not mean that they are equipotential surfaces. Nor does it mean that a conductor designated as 'ground' or 'earth' is automatically a zero-voltage reference point for all signals in the system.

The concept of a surface at which all points are at zero voltage probably arises from a misunderstanding of the method of images used in electromagnetic field theory. So, it is a useful exercise to review the reasoning behind the technique.

The image solution method is usually illustrated by considering an infinite length of line charge parallel to a grounded conductor of infinite length [16]. It is possible to simulate the electric field distribution by replacing the ground plane with an insulating surface and placing an image conductor the same distance below the plane as the actual conductor is above the surface.

Figure 8.1 illustrates a configuration where two conductors are routed over a plane surface. A steady voltage is applied between the near end of conductor 1 and the near end of the ground plane (conductor 3). The far end of conductor 1 is open-circuit. Conductor 2 is short-circuited to ground at both ends. Contour lines, depicting intermediate voltages, are also shown on the figure.

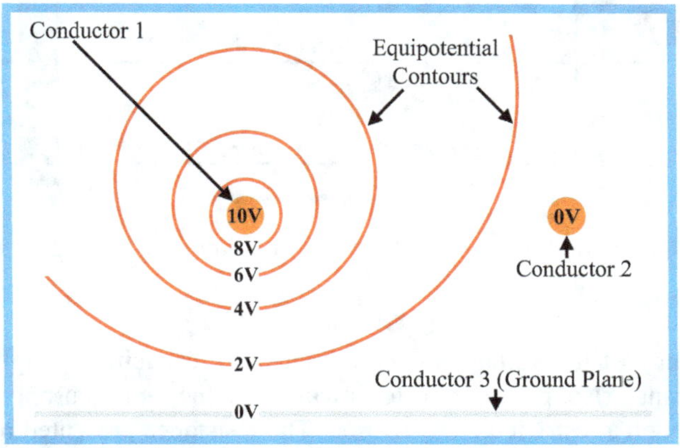

Figure 8.1 Two conductors over a ground plane

Since the voltage applied between conductor 1 and the ground plane is constant, no current is flowing. Under these circumstances, the ground plane is indeed an equipotential surface. But with no current flowing, there can be no signal and no interference. It is also useful to note that the contour lines do not intersect conductor 2. The field pattern is asymmetrical.

If now, the far end of conductor 1 is short-circuited to the ground plane and a current at constant rate-of-change is caused to flow in that conductor, returning via the ground plane, then, although the actual voltages would be different, the pattern of contour lines would remain exactly the same. However, there would be a voltage difference between each cross-section.

If the method of images had been used to create this field pattern, it would have been assumed that the image of conductor 1 had been carrying exactly the same current.

This means that the voltage developed along conductor 1 would be balanced by a voltage along the image conductor. Since this image conductor represents the effect of the plane, it follows that the voltage developed along the ground would be the same magnitude as the voltage along conductor 1.

The coupling mechanism can be analysed by assuming that a low-frequency sinusoidal waveform is applied to the near-end terminals of conductors 1 and 3, and that the short-circuit link between the near-end of conductor 2 and ground is replaced by a voltmeter. Figure 8.2 is a circuit diagram depicting the inductive properties of the three conductors. A sinusoidal voltage $V2$ would be developed between the near-end terminals of conductor 2 and ground. Since the impedance of a voltmeter is high, minimal current would flow in conductor 2, and $V2$ would be a measure of the voltage along $L3$, the inductance of the ground plane.

If a voltage is developed along the ground plane, then that conductor cannot possibly be regarded as an equipotential surface.

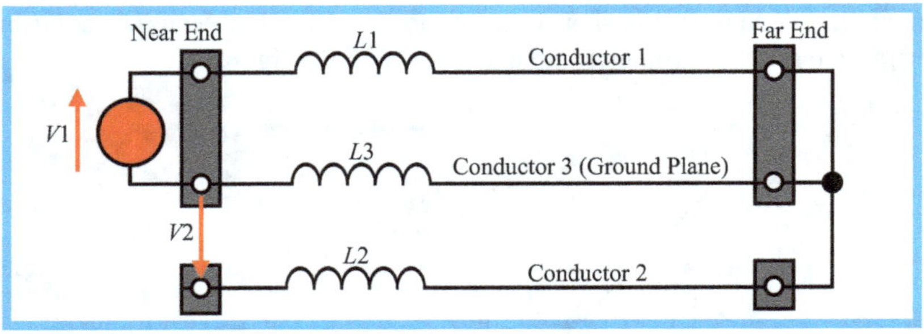

Figure 8.2 Inductive properties of two conductors over a ground plane

If the frequency of the voltage source is increased, capacitive coupling becomes more significant. Also, proximity effects will cause the return current in conductor 3 to flow through a restricted cross-section. The resistance presented by the ground will also increase. The end result is that the simple model of Figure 8.2 can best be modelled by the triple-T network of Figure 2.19. The ground plane can be a source of both electric and magnetic field coupling.

Acceptance of the concept of the equipotential ground has inevitably led to further misconceptions. Notable among these is the belief that all connections made to the 'ground' conductor are automatically held at zero voltage.

In circuit diagrams, it is often desired to define that a connection exists between two separate points without the need to draw a line between those points. The normal convention is to identify those points with a unique symbol. This avoids the use of 'tramlines' crisscrossing the diagram and allows the reader to focus attention on the functional purpose of the circuit.

The most frequently used symbols are for 'ground' or 'earth'. From a functional point of view, it is assumed that any voltage difference between such points can only have a marginal effect on circuit function. It is then assumed that, for all practical purposes, there is no voltage difference.

However, such an assumption is only valid if the area enclosed by the relevant circuit loop is extremely small. This means that there are a few problems when the convention is used with dual-layer or multilayer boards. In these assemblies, the traces used to link components are very close to the ground plane. Hence, the area enclosed by the differential-mode loop is always very small. The ground plane provides a return path for the current, just where it is needed. Section 8.3 provides more information on this aspect of design.

But that is not the case with single-sided boards. When such a board is used, great care is needed to ensure that the loop area involved is as small as possible. To do this, it is necessary to use solid lines to depict conductors on the board. Symbols indicating the presence of hidden connecting links have no place on such a diagram.

If it transpires that the conductive trace linking two points on a single-sided board follows a long circuitous route, then there is an increased probability that the signal in question will be prone to interference problems. If the designer does not have visibility of this track, then the existence of an interference problem will come as a surprise.

At the time when electronic circuitry was based on the use of valves, the components were usually mounted on an aluminium chassis and the circuit diagram depicted this conductor as a solid line at the bottom of the page. The power supply rail was also included at the top of the page, identifying the *R-C* networks used to minimise the interaction between each stage of the supply from the next. Visibility of every circuit path was maintained. This allowed the designer to minimise the area of every circuit loop, whether that loop carried power supply current, signal current or both. Care was taken to ensure that high current in the output stages did not affect the sensitive circuitry at the front end.

If this disciplined approach were utilised during the design of single-sided printed circuit boards, then a significant improvement in the EMC of that particular circuit function could be achieved. In the case of single-sided boards, such an approach might be desirable. In cases where the connecting links are between boards of an equipment unit, this approach would be extremely important. In cases where the link is between equipment units, such an approach is essential.

Assumption of the existence of an equipotential conductor is evident in every circuit diagram which includes the ground symbol or any of its variants. This may be reasonable if the purpose is to describe the function of the circuit, but totally counter-productive if the objective is to analyse EMC. The existence of these symbols means that the return path for every signal current is undefined. The same applies to power supply currents.

If half the circuitry is undefined at the initial design stage, then subsequent stages such as detailed drawing, component manufacture, wiring and assembly are left to the discretion of whoever is carrying out the task in question.

The EMC of the system is effectively out of control. By the time the completed system is submitted for EMC testing, the only thing to do is cross one's fingers and hope for the best.

If the physical relationships between the signal and return conductors are defined at the initial stage of the project, then the designer is able to use all the analytical tools described in the preceding chapters and all the techniques described in the following sections. Visibility of this relationship can be maintained at all subsequent stages of development.

It can be concluded that the conductors of the structure should not be used as a convenient return path for signals and power. Any current flowing along the structure will create a voltage along that length and any voltage between different locations on the structure is a source of interference.

Up to now, the focus of this section has been on what the structure should not be used for. To identify just what it should be used for, it is necessary to go back to Figure 8.1. Here, it shows that neither the equipotential contours of an electric field nor the magnetic lines of force of a magnetic field penetrate the ground plane. Any circuitry on the underside of this barrier is shielded from the worst effects of the electromagnetic field above. From the point of view of EMC, this is a desirable design feature.

Another significant aspect of the magnetic field distribution emerges. The lines of force do not link with conductor 2. This is because a loop has been formed between this conductor and ground. Transformer action ensures that the current in this loop creates an opposing magnetic field which precisely balances the field emanating from conductor 1. That is, this secondary loop tends to act as a shield which minimises the penetration of the magnetic field to the right of conductor 2.

This action is put to good use in the design of Faraday cages and the assembly of conductors used to protect buildings from the worst effects of lightning. So, it can also be reasoned that, even where the conductors of the structure do not present an impenetrable surface, they still perform a very useful shielding function.

If the structure is designed to form a lattice network of conducting loops, then any interference will create a circulating current in each loop, and the field created by these circulating currents will tend to balance the incoming interference. Such a structure behaves as a shield which radiates unwanted electromagnetic fields back into the environment.

8.2 Conductor pairing

As indicated in the previous section, the concept of the equipotential ground conductor is extremely attractive to those involved in the design of systems which are installed on a conductive framework such as on vehicles, aircraft, or spacecraft.

It is reasoned that, since the framework provides a very convenient return path for all currents in the system, there is no need to install unnecessary return conductors which cost money and add mass.

This impression is reinforced by the known effectiveness of the ground plane in multi-layer printed circuits, where it does seem to behave as an equipotential surface. Since the reason for this is not immediately apparent, it is useful to review the behaviour of the ground plane in a little more detail.

Figure 8.3 illustrates part of a printed circuit board assembly where two adjacent tracks are connected to two voltage sources and the far ends are short-circuited to the ground plane.

Using the technique of composite conductors described in Chapter 3, it is possible to determine the distribution of current in a cross-section of the printed circuit board.

If it is assumed that the voltage sources apply step voltages of 1 V of opposite polarity simultaneously to both tracks, then the rate of change of current in all three conductors can be calculated using the process described in Chapter 3. The resultant distribution in all three conductors is illustrated by Figure 8.4.

It is evident that the current in the left-hand track is positive and that in the right-hand track is negative. Also, the rate of change of current in each track is unevenly distributed, with the greatest rate of change occurring at the outer edges.

However, the most significant feature of the illustration is the fact that most of the current in each track returns via the area of the ground plane that is immediately adjacent. Effectively, the ground plane provides a return path for the current precisely where it is needed.

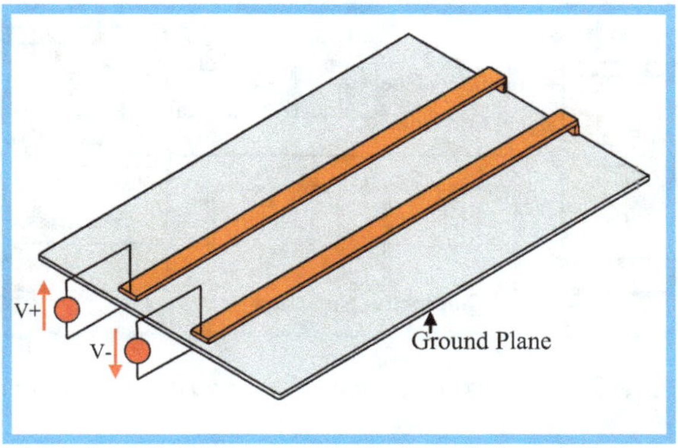

Figure 8.3 Two printed circuit tracks over a ground plane

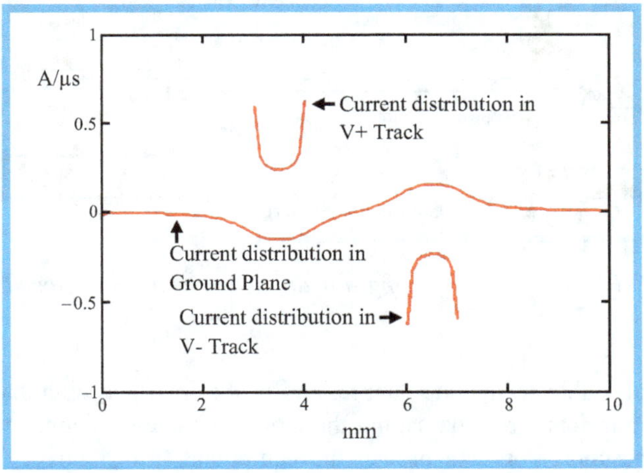

Figure 8.4 Distribution of current in cross-section of printed circuit board

This effect is also present in systems where the supporting framework is constructed of conducting material. In such a system, it is possible to route interconnecting cables along that framework to utilise the structure as a return conductor. Figure 8.5 illustrates such a setup.

However, it is never possible to match the close coupling achieved on a printed circuit board. It is inevitable that wide gaps will occur between cable and structure, and the area enclosed by the current loop can become quite large. Stray coupling is increased. Reducing the level of this unwanted coupling can only be achieved by minimising the separation between cable and structure.

Adding cable trays or cable conduits as shown in Figure 8.5 will help toward this objective. Effectively, the structure is being brought closer to the cable.

Even so, the signal link is still quite susceptible to interference transients. The longer the cable, the higher the level of interference.

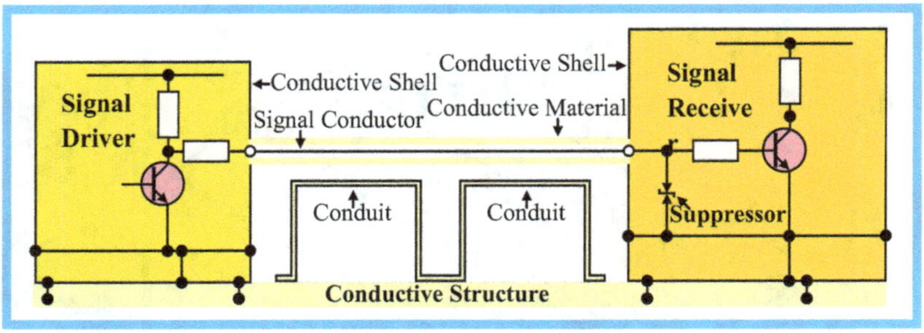

Figure 8.5 Using the framework as a return conductor

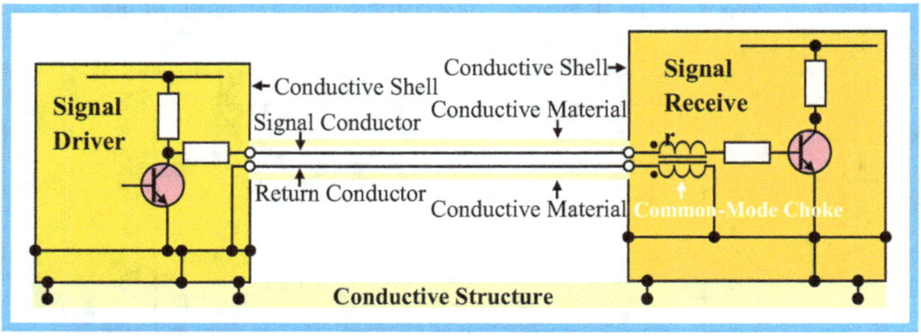

Figure 8.6 Allocating a dedicated return conductor

It becomes advisable to include suppressors at the interface circuitry to prevent high voltage transients from damaging the more sensitive semiconductor devices.

Although suppressors can protect the equipment from damage, they cannot prevent interference pulses from getting through to the processing circuitry. Incorrect messages can be received. Data can be corrupted.

If, at the initial design stage, it is decided to allocate a return conductor to every signal conductor, then this will create two separate loops: the differential-mode loop and the common-mode loop. Analysis of this configuration is provided by Section 2.8, where it is concluded that the signal and return conductors should be held as closely together as possible along the entire route from driver to receiver. It is much easier to hold two conductors together than to minimise the spacing between cable and structure.

Given this configuration, there are many ways of minimising interference coupling. For example, the suppressor can be replaced by a common-mode choke, as shown in Figure 8.6. This significantly reduces the amplitude of interference pulses in the common-mode loop whilst having minimal effect on performance.

Also, it is very helpful to reduce the area of the common-mode loop by routing the signal/return pair very close to the structure. It generally reduces the antenna efficiency of that loop.

8.3 Ground loops

The single-point ground is another concept that is well past its sell-by date. Such a configuration is shown in Figure 8.7, where three different printed circuit boards are used to process the signal. Each board is supplied with a steady-state voltage of 15 V, derived from a stabilised supply module.

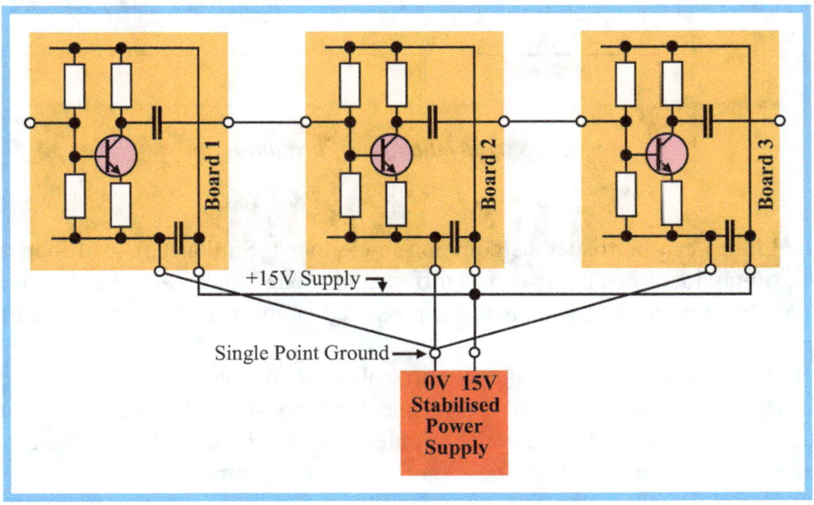

Figure 8.7 Single-point ground

The intent of the single-point ground is to prevent the supply return currents from sharing a common conductor. It is reasoned that if there is no common conductor along which a voltage can be developed, then there will be no interference. This could possibly be the case if the current drawn by each printed circuit board is constant in value. However, this never happens in practical situations as the current in every conductor is continuously changing.

Loop areas enclosed by the signal currents are extremely large. Moreover, the loops formed by signal and supply currents share a common area. Magnetic flux will link with both loops. Transformer action will ensure that high levels of self-induced interference exist between the signals carried between boards.

That is, intra-system interference is likely to cause significant problems. Nor does such a high level of sensitivity bode well for the time when the completed equipment is subjected to formal EMC testing.

Figure 8.8 illustrates how the performance can be dramatically improved. Changes to the configuration are twofold:

- The supply conductors are kept close together.
- A return conductor is allocated to each signal conductor.

Comparing Figures 8.7 and 8.8 enables the essential difference between these two configurations to be identified. The latter circuit contains ground loops.

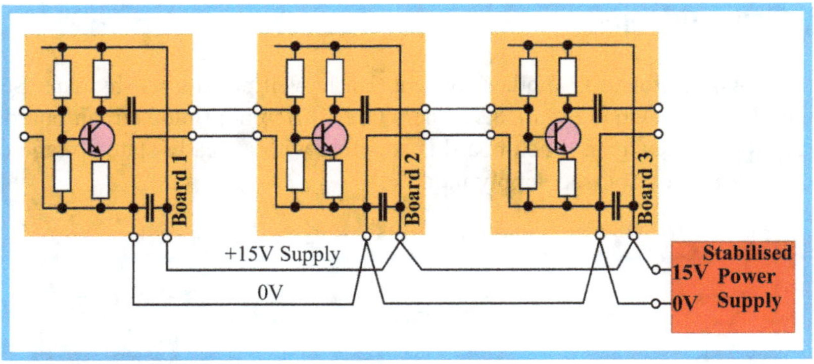

Figure 8.8 Introduction of ground loops

Ground loops can be treated as common-mode loops. Similarly, signal loops and power supply loops can be treated as differential-mode loops. An important feature of this configuration is the fact that it is possible to minimise the loop area of every differential-mode loop.

The way is now open for the circuit analysis techniques described in the previous chapters to be employed. Assessment of the signal links in the light of Section 4.3 should allow potential problems to be identified. Common-mode rejection can be included in the design of the interface circuitry.

It is also worth noting that on a multilayer circuit board, the ground plane carries the return current for every signal current carried by the copper tracks on the adjacent surface. This gives rise to a large number of circulating currents. Figure 8.4 illustrates the fact that the ground conductor does indeed carry currents which flow in opposite directions, and the only way for this to happen is for a loop current to exist in the flat conducting surface. The ground plane is effectively a large number of ground loops.

The protection afforded by ground loops is even more important at the system level, where cables are used to route signals between equipment units. If each unit of equipment is shielded, as illustrated in Figure 8.25, and if every cable is shielded by a screen braid, and if those braids are bonded at each end to the units they interconnect, then the entire system will be shielded. Circuitry within the system will be protected from the worst effects of external fields, and radiated emission will be attenuated by the shield barrier. In such a configuration, there will be a multitude of ground loops.

It does not matter whether the link-under-review is on a printed circuit board, interconnects two boards, or interconnects two units of equipment; the existence of the ground loop is an essential part of good EMC design.

In large systems, it is also good practice to route the cables as close to the conducting structure as possible. Any high-amplitude transient in a conducting member of the structure, such as those due to a lightning strike, will create an electromagnetic field which will couple with every conductor routed alongside that member. This will induce essentially the same voltage in series with every conductor of the cable. Differential voltages will be minimal, and interference will be reduced.

8.4 Common-mode rejection

The configuration illustrated by Figure 8.9 is perfectly adequate for the vast majority of signal links inside units of equipment. However, there could be significant interference problems when the links are between widely separated units of equipment. There are also situations where unwanted transient signals appear inside a single unit. Such interference manifests itself as large spikes on an oscilloscope screen. The amplitude of these spikes can usually be reduced by the introduction of common-mode rejection circuits.

8.4.1 Differential amplifier

Figure 8.9 illustrates one of the simplest methods, the differential amplifier. This is an operational amplifier with the feedback and input networks organised such that:

$$V3 = V1 - V2 \tag{8.1}$$

That is, the output voltage of the amplifier in the signal receiver is a function of the voltage difference between the two input terminals. Equally important is the relationship between the signal and return current. Ideally, this is:

$$Isignal = Ireturn \tag{8.2}$$

If this condition is achieved, then any voltage developed along the signal conductor is precisely balanced by an equal and opposite voltage developed along the return conductor.

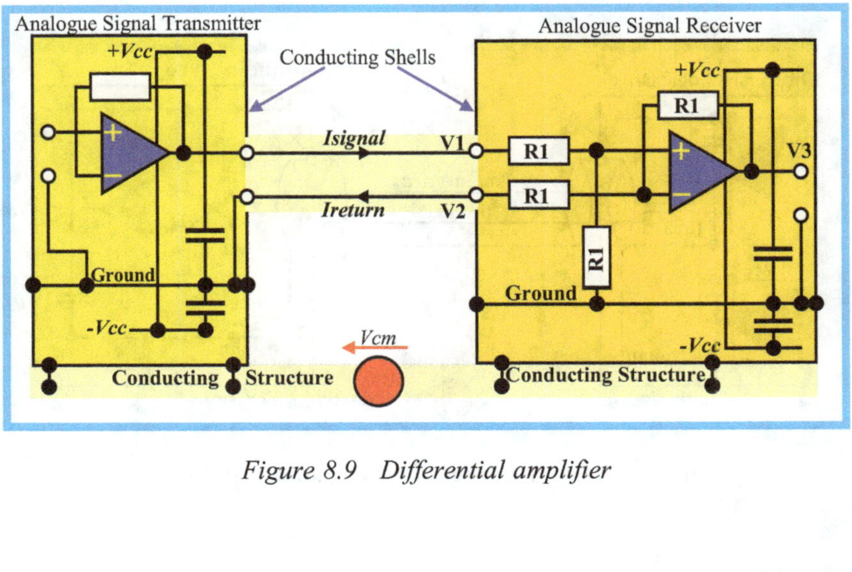

Figure 8.9 Differential amplifier

In theory, this means that the signal passed between the transmitter and the receiver is immune to any interference signal appearing in the common-mode loop. That is, the source *Vcm* has no effect on the transmitted signal.

In practice, this is not the case. Several factors are involved. All four resistors should be the same value, within a close tolerance. To minimise reflections, the value chosen for R1 should be the same as the characteristic impedance of each conductor. At high frequencies, the level of rejection also depends on the length of the two cable conductors being identical.

If the amplitude of the interference exceeds the linear range of the amplifier, signal distortion will occur. Common-mode rejection is also limited by the bandwidth of the amplifier.

However, these considerations do not really diminish the usefulness of the differential amplifier. Protection from common-mode interference is provided over a wide range of frequencies. The useful bandwidth is from zero frequency to the upper frequency limit of the amplifier, and some amplifiers have a very fast response indeed.

Such a configuration is amenable to analysis using the techniques described in the previous chapters. Figure 2.19 illustrates how the connections between the two units can be modelled.

8.4.2 Differential logic driver

Figure 8.10 illustrates one way of transmitting logic signals between widely separated units of equipment. A screened pair is used to shield the signal and return conductors.

The amplifiers in the differential logic driver are fed by complementary inputs from a bi-stable circuit. When the non-inverted signal is 'high', the inverted signal is 'low', and vice versa. These two inputs are fed to the comparator at the differential logic receiver.

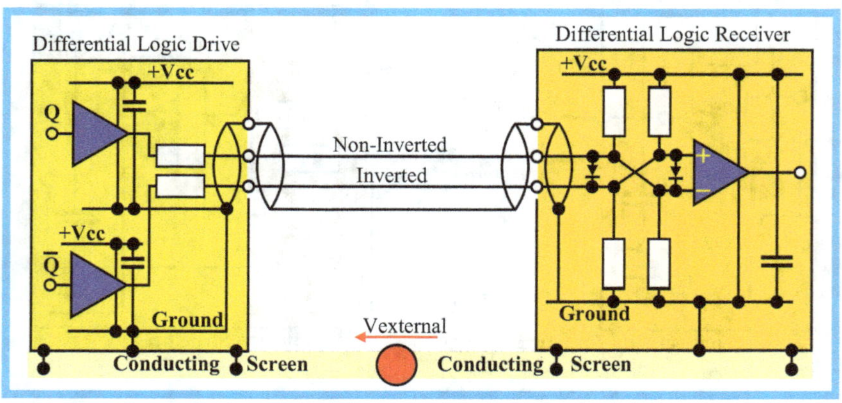

Figure 8.10 Differential logic signal transmission

In the absence of any connection to the receiver, the two input terminals of the comparator would be held at half the supply voltage, Vcc, by the two potential dividers. (This means that some hysteresis needs to be built into the amplifier.) When a connection is made to the transmitter, there will be current flowing from the transmitter to the receiver down one conductor, and an equal amount flowing from receiver to transmitter down the other, assuming that the supply voltage is the same in both items of equipment.

Electromagnetic coupling between the conductors will maintain this balance during transient conditions. As with analogue signal transmission, the response will be limited only by the capabilities of the active devices.

Since both the 'inverted' and 'non-inverted' signal voltages are referenced to 'ground', a connection is needed between the units to carry any spurious transient current during a change in the state of the logic signal. A three-conductor cable is needed. This requirement is best met by the installation of a screened twisted pair.

It will be noted that the screen is connected to the shield of both units of equipment as well as to the grounded conductors. Since this balance depends on the cable conductors being of equal length and the values of the resistors matching precisely, there will always be some coupling of common-mode voltage into the output signal.

8.4.3 Differential analogue driver

The screened pair can also be used to carry an analogue signal. In this case, the driver would be as shown in Figure 8.11. The advantage of using this configuration is the fact that the voltage output of the driver is balanced with respect to local ground, as is the input current to the receiver.

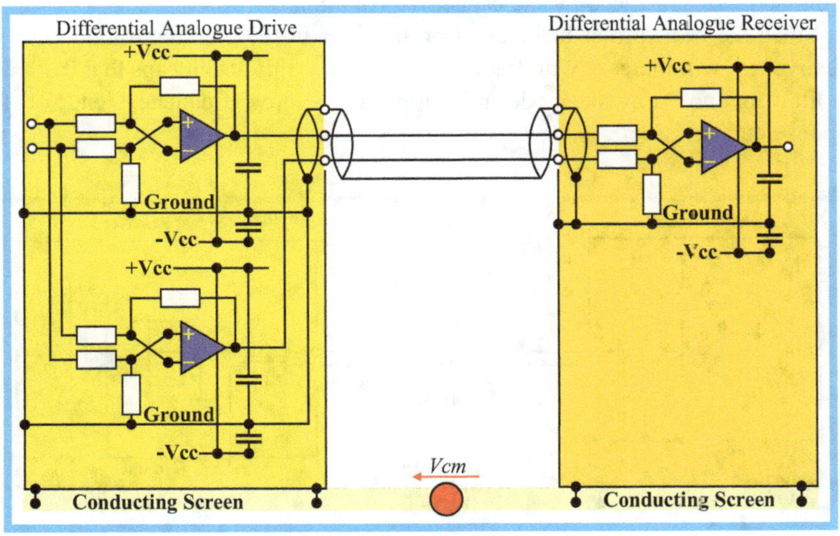

Figure 8.11 Differential analogue driver

At high frequencies, there will be significant common-mode coupling with the screen. Since both the differential-mode voltage and differential-mode current are balanced with respect to the screen, the voltages due to common-mode coupling will tend to balance each other out. Since perfect balance is unachievable, there will always be some common-mode current in the screen.

If the units are not shielded, then the cable screen will still connect the power supply returns.

8.4.4 Common-mode choke

Figure 8.12 illustrates the use of the common-mode choke. Transformer T1 consists of a number of turns of twin-conductor cable, bifilar-wound onto a ferrite core. As far as the common-mode current is concerned, the transformer behaves as an inductor.

With a differential-mode signal, the currents in the two conductors are equal in value but opposite in direction. Since the inductances of the two conductors are almost identical, the voltages induced along the cable will tend to cancel out. As far as the differential mode signal is concerned, the impedance presented by the transformer is negligible. However, some minimal coupling between common-mode and differential-mode signals is bound to exist.

Since the transformer is transparent to differential-mode signals, its effect on the common-mode loop can be modelled by inserting an inductance in series with the structure.

There will be a small amount of flux linking each conductor which does not link the other. In the classical model of a transformer there would be a primary inductance and a secondary inductance. It is important that a bifilar winding is used because this will minimise the primary and secondary inductances. Also, the balancing action is enhanced by the capacitive coupling.

Since the transformer behaves as an inductor in the common-mode loop, its effectiveness will increase with frequency. However, this also means that there will be little or no common-mode rejection at the low frequency end of the response curve.

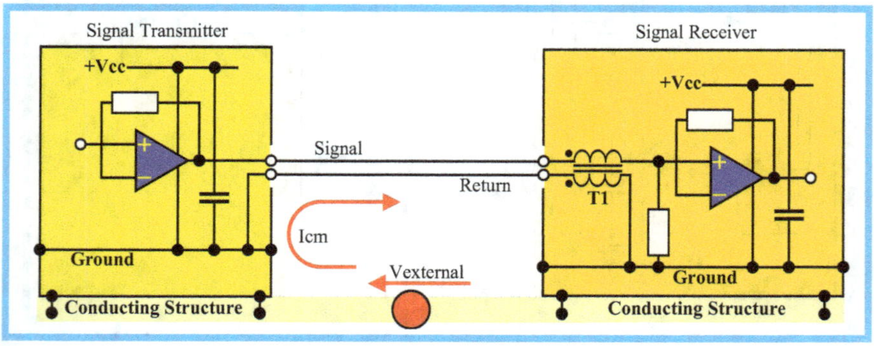

Figure 8.12 The common-mode choke

At the upper end of the response, capacitive effects come into play, and these will limit the level of rejection offered to unwanted signals. There is also a distinct possibility that resonances will occur; a transformer acts as an open circuit at high frequencies. 'Soft' ferrites which behave resistively can help to ameliorate this effect.

Hence, it is a useful exercise to identify the frequencies at which the common-mode rejection is minimal. Section 4.3 helps with this task.

The common-mode choke can sometimes be useful in minimising troublesome effects after the equipment has been manufactured. Split-core ferrites are available, and these can be used to clamp round cables.

8.4.5 Transformer coupling

Another way of achieving a current balance between the signal and return conductors is to break the connection between the return conductor and ground at the receiving end of the line.

With analogue signals, a transformer could be used, as illustrated by Figure 8.13.

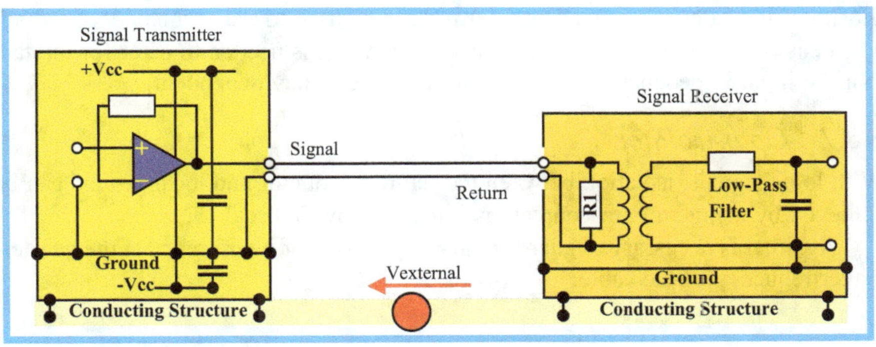

Figure 8.13 Floating transformer at the receiving end of the line

The problem with transformer coupling is that the bandwidth of the signal is limited. The purpose of the resistor R1 across the primary winding is to increase the useful bandwidth. However, if the purpose of the link is to transmit a narrow-band signal, then the transformer is the obvious choice.

This configuration gives excellent common-mode rejection at low frequencies. However, the immunity to interference reduces with frequency, due mostly to the inter-winding capacitance of the transformer. Resonance can occur at the quarter-wave frequency of the line, and the interference can actually be amplified.

The response to interference in the common-mode loop is very similar to the curve of Figure 4.12. To avoid problems with high-frequency coupling, it is necessary to ensure that the upper frequency of the signal is significantly less than the quarter-wave resonance of the cable.

Also, it is a good idea to include a low-pass filter at the receiver to limit its bandwidth to that of the expected signal.

8.4.6 Centre-tapped transformer

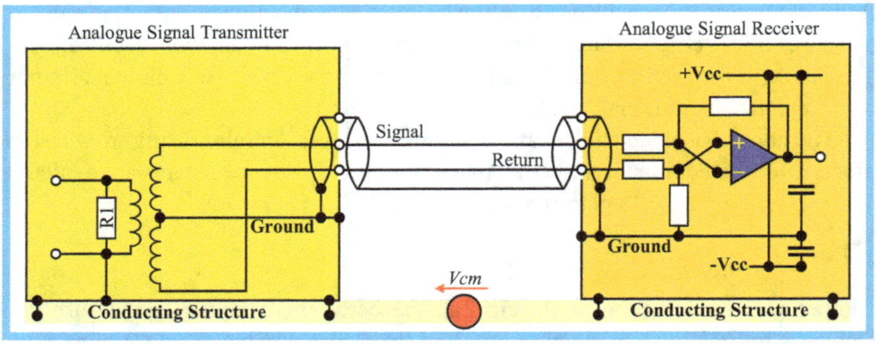

Figure 8.14 Transformer driver

A transformer can also be used in the signal-transmitting circuitry. Figure 8.14 illustrates how a centre-tapped transformer can be used to create a balanced output. The resistor R1 in parallel with the primary winding is needed to cater for the fact that the input impedance of the transformer is frequency dependent.

8.4.7 Opto-isolator

With logic signals, isolation between the return conductor and local ground can be achieved by using an opto-coupler, as illustrated by Figure 8.15.

Again, it is necessary to guard against high levels of interference at the quarter-wave frequency of the cable.

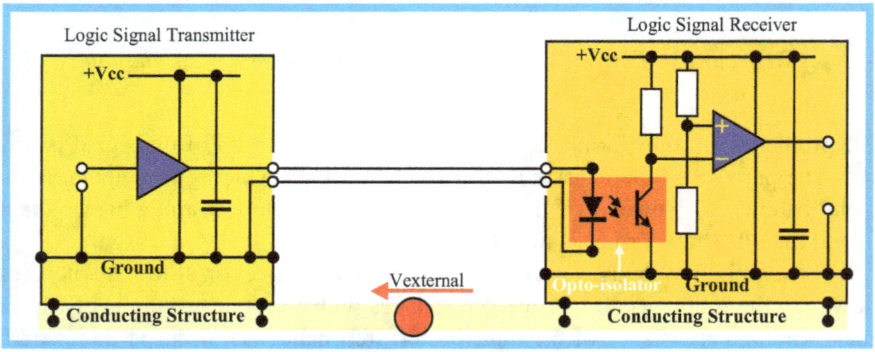

Figure 8.15 Opto-isolator for logic signals

8.5 Differential-mode damping

Although transmission line equations are not often used in the design of low-frequency circuits, they still apply at those frequencies. There is no cutoff point in the spectrum below which transmission line theory ceases to be valid.

One important concept is that of characteristic impedance. Optimum transmission is obtained when the input impedance of the receiver is equal to the characteristic impedance of the line. There is no reflection of energy. If the source resistance is also equal to the characteristic impedance, then there are no reflections at either end. This is standard practice when dealing with radio-frequency signals. It is worthwhile considering the use of the characteristic impedance to terminate the lines in the configurations of Figures 8.9 and 8.10.

If the line is operating at maximum efficiency, then it must also be experiencing minimum loss. Since much of the loss is due to radiation, it follows that by terminating the line correctly, minimum radiation will occur.

It is shown in Section 2.9 that the transfer admittance for conducted susceptibility is exactly the same as for conducted emission. Hence, this configuration also experiences minimum susceptibility.

8.5.1 Transient damping

By its very nature, logic circuitry creates a continuous succession of transient steps in the current delivered by the power supply rails on a printed circuit board. These transients have minimal effect on the supply voltage because of the smoothing capacitors positioned at strategic locations on the board. Even so, there are still transient current demands on the cables delivering power from the stabilised supply module.

This causes transient voltages to be developed along the cable conductors.

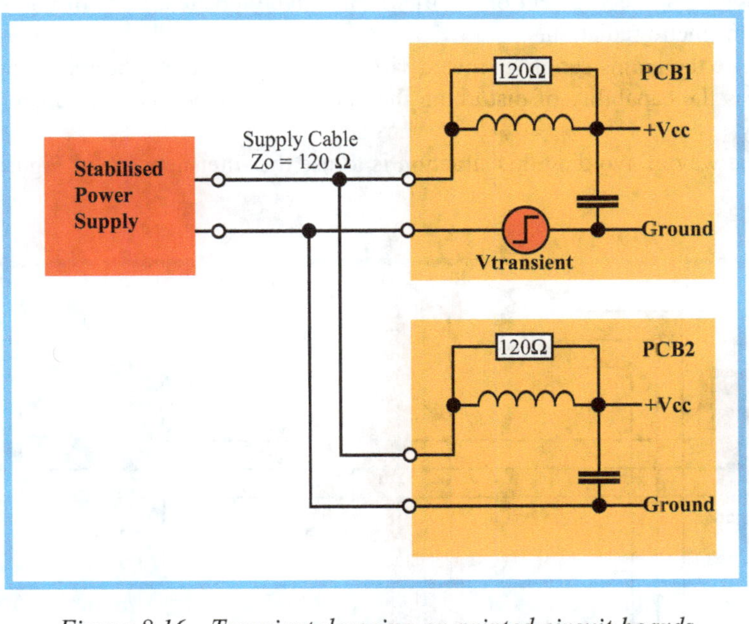

Figure 8.16 Transient damping on printed circuit boards

In Figure 8.16, it is assumed that PCB 1 handles relatively high current surges compared to PCB 2. The resultant voltage transients can be simulated by the source, *Vtransient*, shown in the figure. As far as transients are concerned, the inductor behaves like an open circuit and a capacitor behaves like a short circuit. So, the voltage transient is delivered to the cable via the 120 Ω resistor. Similarly, the step voltage arriving at PCB 2 'experiences' a resistive load of 120 Ω.

If the characteristic impedance of the supply conductors is 120 Ω, then there will be no reflections at PCB 2. If there are no reflections, then there will be no 'ringing' transients in the supply cable. Removing these ringing pulses will remove a potential source of interference.

8.5.2 Mains filtering

The fact that a capacitor behaves like a short circuit to transients means that any filter capacitor between live and neutral of the 220V∼240V 50Hz (UK mains supply) will reflect all of the energy of an incident pulse back up the line. When an equipment unit is switched on, all of the transient energy is reflected back into the mains supply.

From the point of view of the equipment-under-review, this is a 'don't care' situation. The capacitor has provided an effective block to the incoming interference. If every equipment unit is protected in this way, then the transient pulse will simply reverberate backwards and forwards along the supply cable. This appears as a ringing transient, which eventually decays to zero. Since the decay is due to the energy being converted to electromagnetic radiation, it means that the act of switching any equipment on or off will inevitably create a burst of high-energy electromagnetic radiation.

Since the mains supply is routed virtually everywhere in a building, this transient has the capability of disturbing the operation of any other equipment in the vicinity.

One way of avoiding this situation is to use the filter network of Figure 8.16.

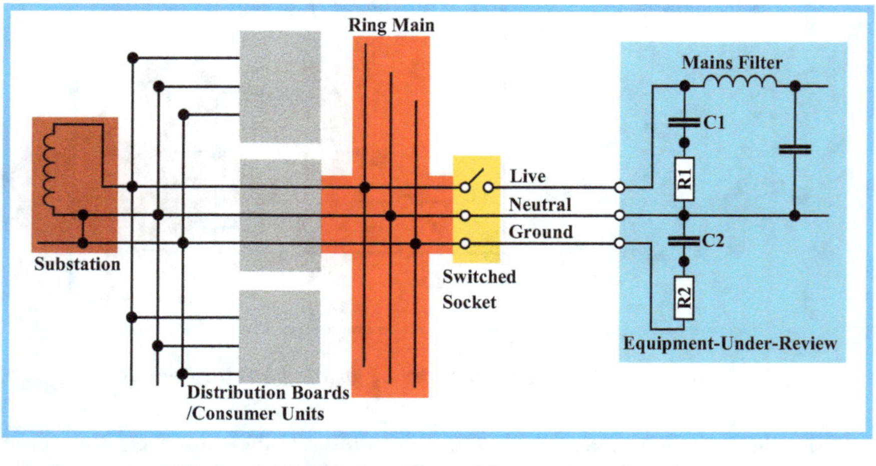

Figure 8.17 Mains filter with transient damping

Another method is illustrated by Figure 8.17. Here, the live and neutral conductors can be viewed as a transmission line supplying differential-mode power. The neutral and earth conductors would then be carrying common-mode loop current.

Any fast transient current arriving from the supply is routed via the capacitor C1 into the load resistor R1. If R1 is equal in value to the characteristic impedance of the supply conductors, then there will be no differential-mode reflection. Similarly, the resistor R2 absorbs any transient power in the common-mode loop.

The use of resistors in the mains filters makes it possible to reduce the level of high-frequency interference emanating from the power supply cables, at least in the immediate vicinity of the equipment-under-review. Although perfect impedance matching is unlikely, such terminations offer a dramatic improvement on filters which reflect all the energy.

Although the mains supply conductors are not designed as controlled-impedance transmission lines and they have many stubs and branches along them, the fact remains that there are usually several metres of cable between the equipment and the mains socket, and tens of metres of three-core cable between the mains socket and the local switchboard. To a first approximation, these cables can be represented by a triple-T circuit model such as the one illustrated by the general circuit model of Figure 4.6. (If there was no point in trying to simulate the characteristics of the mains supply, then Line Input Simulation Networks (LISNs) would not exist.)

The tests described in Section 7.5 show that a two-core mains cable has characteristics which can be closely defined. The same is true of three-core cable. It is entirely possible for designers to obtain a length of such cable and carry out measurements which define the inductance, capacitance and resistance of each conductor of that cable, and calculate the values of the characteristic impedances.

Given knowledge of the values of the characteristic impedances of the line, it is possible to define the optimum values for the resistors for the filter.

If such resistors were to be included, they would damp down the transients in the length of cable which supplies the equipment-under-review. They would also attenuate interference caused by 'broadband over power-lines'.

8.5.3 Solenoid switching

Another rich source of transient pulses is the action of solenoids. A fair amount of energy is stored by the current in the winding. When the current source is switched off, the energy in the magnetic field is converted to a high transient voltage, and this acts as a step voltage input to the supply conductors. All of the energy is then converted to a ringing transient in the cable. The amplitude of the transient dies down fairly quickly because power has radiated away in the form of an electromagnetic field.

Figure 8.18 illustrates how this transient energy can be safely absorbed. When the switch SW1 closes, a step voltage is propagated down the line formed by the supply and return conductors. If the resistor R2 is equal to the characteristic impedance of the line, then there will be no reflection from the initial edge. If the value of the capacitor is selected such that the current waveform is critically damped, then the wave-shape of the current in the cable will be a single pulse, decaying to a constant value. High-frequency ringing should be minimal.

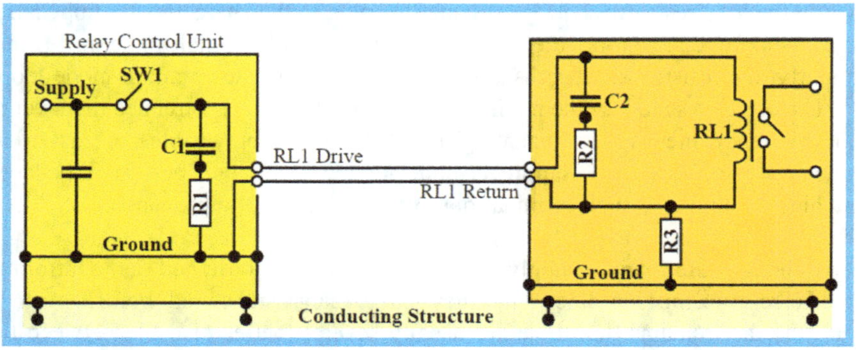

Figure 8.18 Relay transient damping

More importantly, when the switch opens, most of the energy stored by current in the inductance of the solenoid will be absorbed by R2. Most of the energy stored in the capacitance of the line will be absorbed by R1.

It is also a fact that when SW1 is closing, the contacts will bounce several times. The input supply to the solenoid is not a clean step function. Rather, it is a series of pulses of varying mark/space ratio. There is no way of avoiding the creation of some level of interference. However, the amplitude of the radiated field can be minimised by using a conductor pair to carry the supply current and by using damping resistors at each end, as Figure 8.18 illustrates.

If the structure had been used to carry the return current, then the amplitude of the radiated field would have been much larger.

8.5.4 Commercial filters

It is useful for designers to be aware of the fact that commercial filters can cause unexpected problems.

These devices are characterised by subjecting them to a standard test. A signal generator is used to apply a voltage to the input terminals of the device and the output voltage is measured. The output impedance of the signal generator is 50 Ω and the load presented by the voltage measuring equipment is 50 Ω. Maintaining a constant voltage and varying the supply frequency allows a response curve to be derived. This is the definitive curve recorded on the manufacturer's data sheets.

These curves usually exhibit a flat response at low frequencies, then form a downward slope which indicates that as the frequency rises, then the attenuation increases. It should not be assumed that when the filter is installed in an item of equipment, the attenuation it provides will match that of the manufacturer's specification.

Usually, the filter presents a capacitive load at the input terminals. This capacitor resonates with the inductance of the line to provide a peak in the response. This peak indicates that there is a voltage gain, and sometimes the gain can be as high as 20 dB.

If the purpose of the filter is to remove high-frequency content from the signal arriving at the equipment input terminals (or from the signal delivered to the output terminals), then it is necessary to include the cable parameters in the simulation as well as the circuitry at the far end, in analysing its performance.

8.5.5 Use of carbon

Carbon is a material worth considering when selecting which components to use, since it has a relatively high resistance which can be used to damp down reflections.

For example, carbon-composite resistors have minimal inductance and capacitance, making them very useful as damping resistors.

The high-voltage co-axial cables used on the photon detector assembly, part of the Space Telescope, have a layer of carbon composite material between the inner conductor and the insulation, and a graphite layer between the insulation and the shield. It is possible for current transients to occur in the high-voltage supplies due to switch-on and switch-off conditions as well as to partial discharges. Fast transients are rich in high-frequency components. Skin effect will ensure that these components of the current flow in the adjacent conducting surfaces. Since these surfaces are composed of carbon material, then the transient current will flow along a highly resistive path.

Carbon tips on the static dischargers on aircraft present a high resistance in series with the discharge path, minimising the amplitude of each transient.

8.6 Common-mode damping

As indicated in Section 4.4, there are always peaks in the response of the transfer admittance, whether a grounded or a floating configuration is adopted. It is usually possible to select which of these is the best choice for a particular application. Sometimes neither configuration provides the required level of common-mode rejection and it is necessary to damp down the peaks in the response.

The only way of reducing the level of these peaks is to absorb the unwanted energy. Although it is possible for a soft ferrite to absorb some of the energy in the resistive region of its frequency range, it is much better to be able to insert a pre-defined value of resistance into the common-mode loop over a pre-defined range of frequencies.

8.6.1 Common-mode resistor

One method of including a resistor in the common-mode loop would be to add a third winding to the common-mode choke and connect a low-value resistor between the terminals, as illustrated by Figure 8.19. Since the bifilar winding carries the common-mode current, there will be a voltage induced in this third winding proportional to the rate of change of this current. The 'choke' now really needs to be viewed as a transformer. Since the third winding is loaded by the resistor $R1$, the effect is to insert a resistance in series with the common-mode loop. If the third winding has the same number of turns as the bifilar winding, then the value of the resistance inserted into the common-mode loop will be R1.

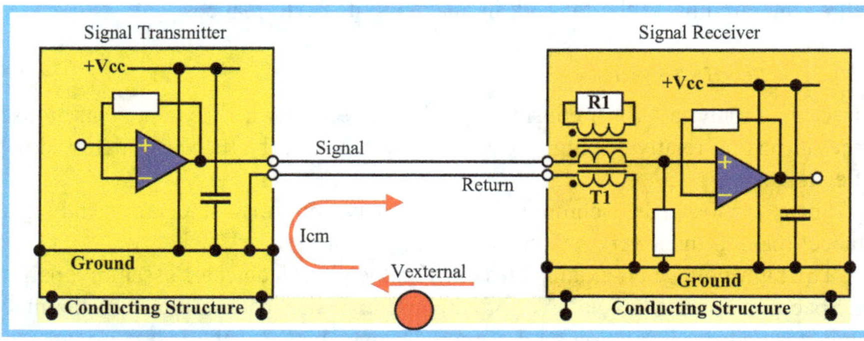

Figure 8.19 The common-mode resistor

At all frequencies, this resistor is transparent to the signal current. But at the frequencies for which the transformer is effective, the device acts to absorb the unwanted energy in the common-mode loop. The value of the resistor is determined by the characteristic impedance of that loop. Damping action is effective at all frequencies where the impedance of the winding is greater than the resistance. Figure 7.5 illustrates the fact that the frequency range can be quite wide.

One benefit from this technique is that the value of the common-mode resistor can now be defined by the designer. The other is that it can be used at relatively low frequencies.

8.6.2 Minimising pickup

Another possible use for the common-mode resistor is to prevent radio-frequency signals from being picked up by a microphone lead.

It is not at all obvious that this would be a problem with public address systems which restrict their bandwidth to the audio-frequency band.

However, common-mode signals arriving at the pre-amplifier can be converted to differential-mode signals by the mechanisms described in previous chapters.

Any non-linearity in the circuitry of the mixer can result in demodulation of the radio-frequency signal. The demodulated signal will then be processed in exactly the same way as the intended audio signal.

To deal with this problem, a ferrite core can be clamped around the cable and a short piece of wire can be threaded through the core to act as the third winding and connected to a resistor of defined value (say 100 Ω). This acts to provide a resistive load in series with the antenna-mode current. It is effective over the entire bandwidth at which the toroid acts as a transformer. Figure 8.20 illustrates the setup. This may not eliminate the problem of inadvertent demodulation, but it will reduce the interference level to a point where the signal is inaudible.

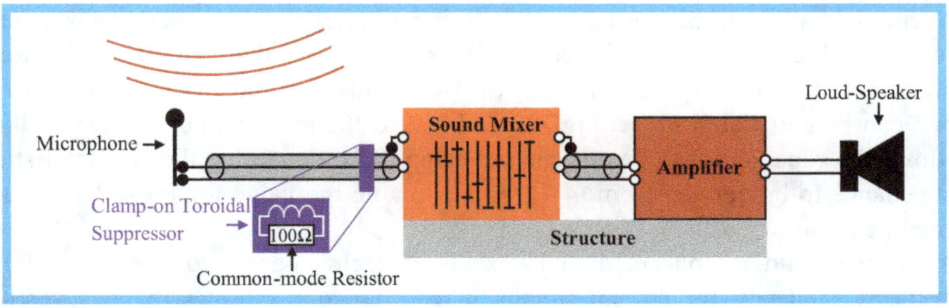

Figure 8.20 Antenna-mode damping

It is important that the device be fitted as close to the grounded end of the cable as possible, because it is at this end, the current will reach a maximum value. Transformer action depends on the rate of change of current. At the microphone end of the cable, there is no antenna-mode current. If the device were to be located near the microphone, it would be totally ineffective.

An alternative solution to such a problem would be to use a shorter length of microphone cable to change its quarter-wave frequency. This was the solution adopted for a public address system in a Church, when it had been possible to hear faint voices coming from the speakers during quiet times of an evening service.

8.6.3 *Tri-axial cable*

Another method of absorbing power in the common-mode loop is illustrated by Figure 8.21. This method is particularly useful in situations where connections are made between the bench test equipment and the equipment-under-test. In such setups, it is inevitable that the interconnecting cables will be routed in close proximity to each other and that cross-coupling will occur. Even with co-axial cables, the effect of this type of interference can be quite noticeable, particularly under resonance conditions. The need arises for extra shielding to be provided and for this shielding to be free from resonances.

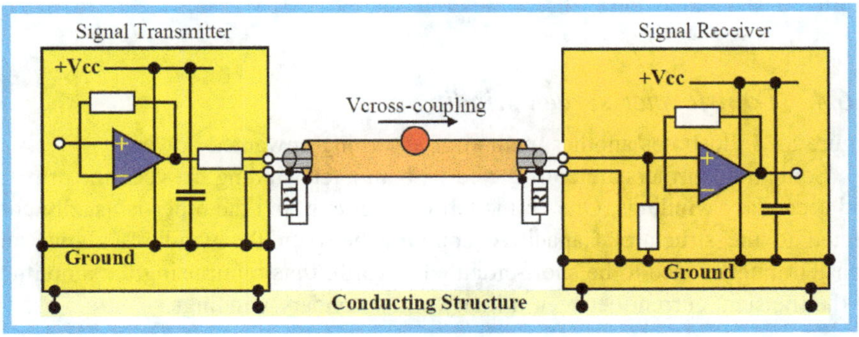

Figure 8.21 Tri-axial cable

A tri-axial cable can be assembled by sliding a braided sheath over an insulated co-axial cable. This assembly then consists of an inner and an outer transmission line, where the outer line is formed by the two braids. The inductance of this outer line can be calculated by setting *r*23 and *r*33 to the same value in (2.93). This allows the capacitance to be determined using (2.28), enabling the characteristic impedance to be derived from (4.7), thus allowing the value for R1 to be calculated as equal to *Zo*.

If the braid is connected to the shell of each co-axial connector via R1, then any tendency for current in this inner loop to resonate will be severely damped.

To achieve good performance at very high frequencies, several resistors would need to be assembled in an annular ring at each termination to minimise the inductance of the resistive components.

The setup of Figure 8.21 can first be viewed as a source of interference. Any high-frequency signal passing between these two units will flow along the inner conductor of the coaxial cable, returning via the screen. A small voltage will be developed along this screen, causing current to flow in the loop formed by the screen and the outer shield. The resistors R1 will reduce the amplitude of this current, as well as minimising reflections in this outer loop.

With any setup linking the test equipment to the equipment-under-test, there are bound to be at least three such cables. It is inevitable that these will be routed in close proximity in some cases even touching. Although an outer woven braid of insulating material will prevent direct contact, electromagnetic coupling cannot be avoided.

If the setup of Figure 8.21 is now viewed as a victim of such stray coupling, then the effect of this coupling can be represented by a source voltage *Vcross-coupling* in series with the outer shield. This will cause a current to flow in the outer loop. Since the resistors R1 limit the amplitude of this current, the voltage developed along the length of the inner screen will be extremely small.

This means that the interference caused by the coupling of signals between cables of the test setup will be extremely small. The introduction of damping resistors into the screen/shield loops represents an improvement on the level of shielding that can be provided by a configuration where the outer shields are simply shorted to the connector shells.

8.6.4 Transformer screen-winding

Figure 8.22 illustrates another method of absorbing unwanted power.

Some transformers are constructed with an inter-winding between the primary and secondary windings. One end is left open-circuit and the other is usually connected to the structure. Capacitive coupling between the windings allows any transient interference to be short-circuited to earth, thus minimising the amplitude of the transient current between primary and secondary windings.

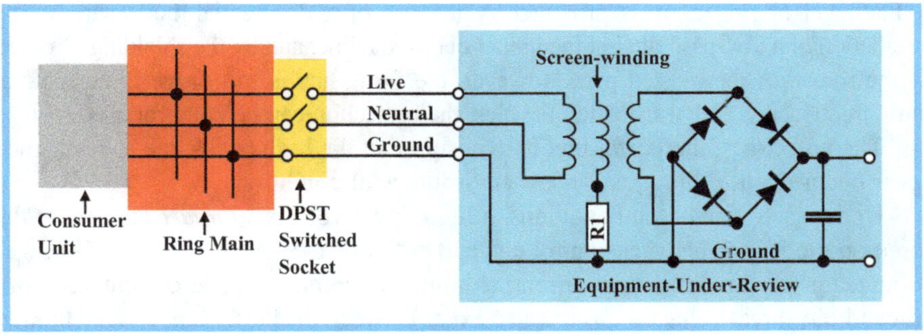

Figure 8.22 Absorbing power via screen-winding

However, the existence of a capacitor at the end of a transmission line means that it is effectively short-circuited. The line is certain to resonate. Interference current at these frequencies will be amplified, just like in a tuned circuit.

Such a situation can be avoided by inserting a resistor in series with the inter-winding. This will damp down any resonant peaks.

8.6.5 Common-mode filter

Figure 8.23 illustrates a method of absorbing radio-frequency power picked up by long cables. It uses a firing line for an electro-explosive device (EED) as an example. Under normal operation, a steady state voltage is applied to the firing line. Since the EED is usually a low-resistance device, a high current will flow and the device will heat up sufficiently to cause ignition.

Under all other conditions, it is essential that a safety margin exists between any interference current and the no-fire current of the device. It could be that the system is intended to operate under conditions where the electromagnetic environment is extremely harsh, such as when it is located near a high-power transmitter.

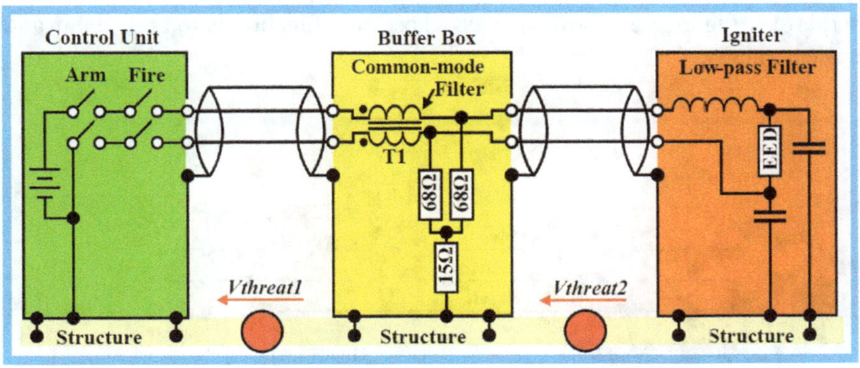

Figure 8.23 The common-mode filter

A fair level of protection is provided by the use of a differential-mode supply, the installation of a low-pass filter at the igniter end of the line, and by the shielding afforded by the screen of the twisted pair. However, the firing line is still prone to resonate at some frequencies, and at these frequencies, the protection can be compromised.

The purpose of the buffer box is to ensure that the high peaks due to resonance never occur even in the most severe environmental conditions.

Under such worst-case conditions, a high threat voltage *Vthreat1* appears in the common-mode loop between control unit and buffer box. Since transformer T1 acts as an inductor to common-mode current, it limits the common-mode current very significantly. A threat voltage *Vthreat2* also exists between the buffer box and the Igniter.

This creates a current in the loop formed by the screen and structure. Current in the screen develops a voltage along its resistance, and this small voltage appears as a voltage source in the common-mode loop formed by the screen and the firing line. Since the resistors in the buffer box match the characteristic impedances of the conductors they terminate, there is no resonance.

Also, the resistor network acts as a bridge circuit to balance the current in the two conductors of the firing line. **The result is a firing circuit which is safe to use in very hostile environments**.

8.6.6 Transformer-coupled resistors

Apart from lightning pulses, the most powerful source of interference is the main power supply. Although the power is delivered at a very low frequency, the transients created when loads are switched on and off make it a rich source of high-frequency electromagnetic fields.

When any load is switched on, there is an initial current transient due to the charging current in the stray capacitors (C1 and C2). This is reflected back up the line by the motor inductance and reflected further at all discontinuities along the way. There is nowhere for this unwanted energy to go, except into the environment. When a machine is switched off, the energy stored in the inductors and capacitors can only expend itself in the form of a burst of interference.

It is in such situations that the transformer-coupled resistance can be used to good effect. Figure 8.24 shows a filter whose sole function is to absorb unwanted transient energy.

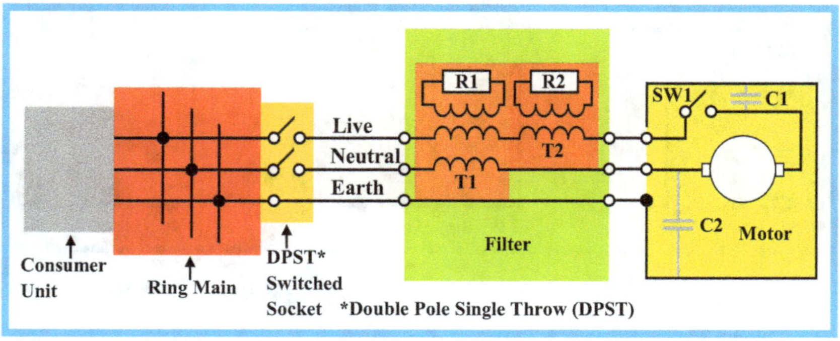

Figure 8.24 Transformer-coupled resistance

This energy can emanate from the motor switching current or from the transients generated by other switched loads on the supply. One resistor absorbs differential-mode interference, whilst the other minimises the common-mode current.

Transformer T1 is tri-filar wound and T2 is bi-filar wound. The resistor R1 is approximately equal in value to the characteristic impedance associated with the common-mode loop, whilst R2 is the same as the characteristic impedance of the differential-mode loop. The mutual inductance of each transformer is determined by the lowest frequency at which the resistor is effective.

Such a circuit can help to reduce the level of environmental pollution caused by switching transients in heavy-duty loads.

It could be used to prevent the nuisance tripping of the residual current detector in the mains switch box. Such tripping can occur at switch-on if there is a large value of capacitance between the live and earth conductors at the load end.

8.7 Shielding

Since every book on EMC carries a section on shielding, there is no need for an extensive treatment of the subject in these pages. However, it is useful to apply the lessons learnt in the previous chapters to review the shielding properties of the structure.

The action of any electromagnetic field on a conductor is to induce a current along that conductor. If the frequency of the radiation is high, then most of this current will be concentrated on the surface of the conductor; the skin effect being predominant.

With a dipole antenna, this current is routed to a receiver via a transmission line. If the conductor is a square, flat surface such as an equipment panel, then the current travels to the edge of the square. Since this represents a discontinuity in the current flow, some of it is reflected back. Some of it continues to flow down the edge and back along the opposite face of the panel.

Any transient current flowing in a conductor will emit a radiated field. That is, the panel acts as a repeater and radiates a significant proportion of the transient energy back into the environment. Since current is flowing on both sides of the panel, radiation is in all directions. As far as shielding is concerned, such a panel is highly inefficient.

If the shield were to be spherical, then there would be no direct path between the inner and outer surfaces. At the outer surface of the conductor, part of any external field would be reflected back into the environment, but some would be absorbed into the material. At the inner surface of the conductor, some energy would be reflected back to the outer surface and some would propagate into the volume enclosed by the sphere. The ratio between external and internal power levels is described as the shielding effectiveness of the material.

This shielding action is equally useful in preventing internally generated field from radiating out into the environment. This means that a conducting surface can provide an effective barrier to the propagation of electromagnetic energy at all frequencies, provided that there are no gaps in the surface.

8.7.1 Equipment shielding

In practical situations, the construction of the enclosure is more akin to a box-like structure than a spheroid. There are joints between panels and framework, holes in panels for ventilation, gaps to allow visibility of meters, indicator lamps and screen displays; and more holes for potentiometer spindles, switches, push-buttons and keyboards.

This means that any current flow will be subject to multiple discontinuities and multiple reflections. Since current density will increase at the discontinuities, 'hot-spots' can develop in the conducting surface immediately adjacent to slots. Under resonance conditions, these slots can act as dipole antennae.

Methods of dealing with gaps between structural components are to use crimped seams or conductive gaskets. Cut-outs used to provide visibility of meters and displays can be covered with conductive glass with the conductive surface of the glass RF-bonded to the shield all around the edge. Holes for access shafts can be protected by a waveguide below cut-off. All these techniques are described in detail by [21].

The purpose of most units of electronic equipment is to process signals which arrive and depart via connectors and cables. If the cables are unscreened, then any interference they pick up will be routed directly into the shielded enclosure. This will render the shield ineffective.

One option would be to filter the signals at the shield wall.

Another option would be to sheath the cable with a screen braid and RF–bond this braid directly to the shield wall, either directly or via the shells of suitable connectors. Such a construction ensures that most of the interference current flows via the cable screen to the outer surface of the enclosure. Methods of maintaining the integrity of any shield are illustrated in Figure 8.25.

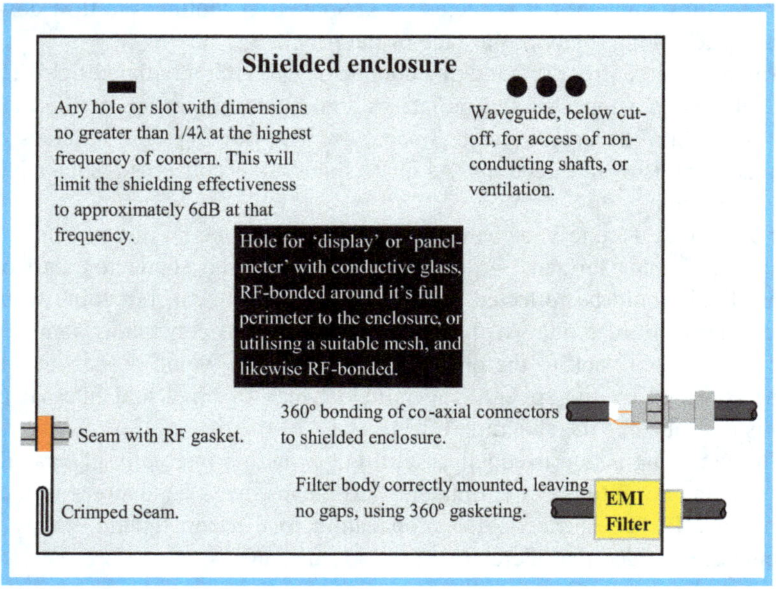

Figure 8.25 Maintaining shield integrity of the equipment unit

It is possible to design enclosures which have a shielding effectiveness of 100 dB or more. However, this can only be confirmed by testing. In the absence of such tests, it is better to assume a limit to the frequency at which shielding is effective.

This can be determined by following the guidelines provided by the Ordnance Board [22]. Essentially, the procedure is to examine the equipment and measure the length of the largest gap in the shielding, *Lgap*. The cutoff frequency is then:

$$f_{limit} = \frac{300 \cdot 10^6}{4 * Lgap} \tag{8.3}$$

Below this frequency, it can be assumed that the shielding effectiveness is 20 dB; that is, a voltage or current ratio of ten to one. Above this frequency, it is better to assume that the shielding effectiveness is zero. When assessing the EMC of any particular signal link, the approach could be to implement the analytical techniques described in Chapters 2 to 6, and then to invoke the shielding factor as described above. This would give a worst-case figure.

In cases where the structure does not meet the rigorous guidelines of Figure 8.25, it still provides some measure of shielding by virtue of the fact that there will be circulating currents in the conducting loops which absorb the energy of the incoming radiation; in much the same way that a breakwater protects boats inside a harbour. Chapter 3 describes a method which has been successfully used by researchers to locate regions in aircraft structures where the indirect effects of a lightning strike are least severe. This identifies routes along which cables can safely be laid. Such an approach can apply to any structural assembly.

Carrying out tests of actual performance, as described by Chapter 7, would allow a more accurate figure to be assigned to the shielding provided to any particular signal link.

Other techniques for calculating shield effectiveness can be found in [5].

8.7.2 Shielding of buildings

Protecting buildings and structures from the effects of lightning presents a more daunting task, but the approach is essentially the same.

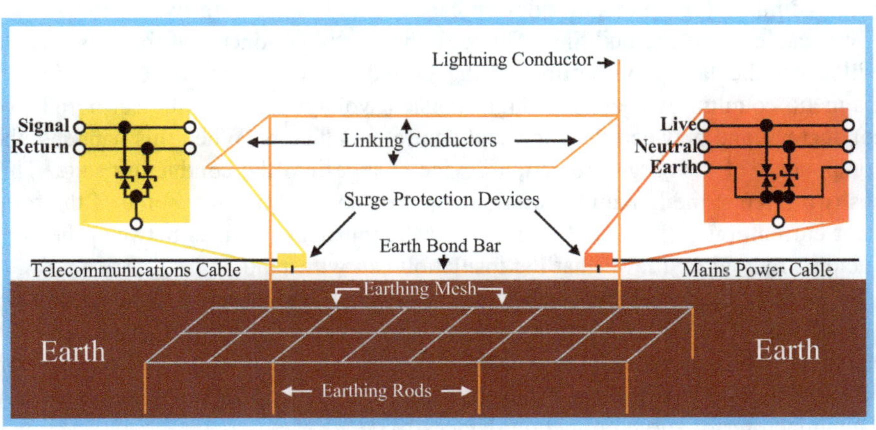

Figure 8.26 Lightning protection of building

Figure 8.26 illustrates the protection measures which have evolved. Instead of a continuous conducting surface enclosing the building, the outer shield consists of a cage-like network of conducting links. Lightning conductors on the roof provide attachment points for any direct strike as well as a vertical conducting path down the outer walls to a grid-like network of conductors buried in the earth. Linking conductors provide horizontal paths between the verticals.

The nature of antenna-mode transient current is that it flows on the outer surface of conductors. The purpose of the horizontal links is to enable the downlinks to share the current. This minimises the amplitude of the transient current in any particular conductor, and minimises the level of pulsed magnetic field it creates.

The presence of these links ensures that the lightning transient flows on the outer surface of the building. That is, the assembly of lightning conductors acts in the same way as a shield. Most of the current flows down the lightning conductors into the earth. It will also flow away from the building via any other conductor. So, it can be expected that a proportion will be carried by the earth conductor of the power line. High transient current in the conductors inevitably creates high transient voltages between these and any other conductors in the near vicinity, with the risk of arcing. This can happen at ground level as well as at height.

To prevent damage to telecommunications cables, surge protection devices can be fitted at the point of entry to the building. These are usually assemblies of metal oxide varistors.

At voltages below a pre-determined threshold, a varistor behaves like a pair of back-to-back Zener diodes. If the threshold voltage is exceeded, the current increases rapidly, and avalanche breakdown occurs. Some devices are capable of handling extremely high currents, of the order of 40 kA, albeit for a short period, though long enough to survive a lightning current surge. During a lightning transient, the reaction time is fast enough to prevent the voltage transient from exceeding the voltage tolerance capability of the equipment being protected.

Current into the telecommunications line is limited by the inductance and resistance of the line when it is acting as an antenna. The cable acts to carry a small proportion of lightning current away from the building, and, hopefully, continues to serve its normal function after the event.

A similar surge protection device can be fitted at the point where the service power cable enters the building. Although the 'earth' conductor of the power cable will carry the larger proportion of the current flowing into the cable, electromagnetic coupling will create a high transient voltage between the earth and live conductors, and between the earth and neutral conductors. When this happens, the components of the surge protection device change into the conducting state. This has two effects: the live and neutral conductors now act to carry some of the transient current away from the building, and the transient voltages between the conductors are held to a level that the insulation can withstand.

If lightning strikes either the telecommunications cable or the power cable outside the building, the relevant surge protection device will act to ensure that the electrical equipment inside the building remains safe. A more detailed description of the method used to protect buildings from the effects of lightning is provided by a manufacturer of surge protection devices [23].

The measures illustrated by Figure 8.26 can provide protection from the worst effects of lightning, including injury to humans, fire, and structural damage.

Preventing damage to electrical and electronic systems requires further measures, such as other surge suppression devices between the main power distribution board and the equipment. Lightning protection zones, such as screened rooms, can also be installed.

It should be noted that surge protection devices can only provide protection from damage. They cannot prevent interference from upsetting the normal functioning of the equipment.

To prevent upset due to interference, one or more of the other techniques illustrated in this chapter need to be included in the system design.

8.7.3 Use of carbon

If the surface is highly conductive, some of the radiation is reflected, but most is conducted into spaces where its presence is undesirable. If the surface contains resistive material such as carbon, much of the unwanted electromagnetic energy is converted into heat.

Examples of the usage of carbon for this purpose are:

- Blocks of carbon composite material are used to line anechoic chambers.
- A bitumen layer is added to the walls of secure buildings.
- Carbon composite used in the structure of aircraft absorbs much radiated energy.
- Static dischargers have carbon tips to create a voltage drop between the structure and the environment. This limits the discharge current to safe levels.
- Fuel pipes made of carbon composite material prevent the build-up of static voltages.
- Carbon in vehicle tires, making contact with tarmac, prevents vehicles acquiring astatic charge.

It is a useful exercise to investigate the materials planned for use in the equipment under development, to check whether the properties of carbon can be exploited.

8.8 Filling in a few blanks and clarification of Practical Design using PCB example

Throughout this chapter and indeed this book and underlying all the design considerations espoused in the previous sections; **'Is the fact that there is no such thing, nor can there ever be such a thing as an "equipotential" earth and most certainly not in the "real" physical world which we are but a very small part of'**. However, a better, much less binary question would be: How close to an 'equipotential' ground can we get? The answer to that is very much materially, physically and environmentally specific. With this in mind we are able to answer a question which is frequently asked about SPICE simulation of EMC/EMI. The answer to that question is yes, it can be used to simulate EMC/EMI conditions but with one very important caveat. The accuracy of the simulated response is wholly dependent upon the physical implementation of the circuit-under-review.

Figure 8.27 depicts two PCBs, each having four layers with the left (Host) PCB and the right PCB being the (Device). Since the system was designed to operate as an asynchronous one-to-many, many-to-one configuration then both the Host and Device

PCBs incorporate a voltage regulator based 3.3 V power supply to power both the SOC (System on Chip) a combined microcontroller and a 2.45 GHz (GFSK) Gaussian frequency-shift keying modulated differential-mode transceiver running from a 16 MHz clock and the output stage switched transmit/receive common-mode RF amplifier. A simple Balan is used to connect the SOC transceiver to the RF amplifier. The two tracks used for the microcontroller switching of the RF amplifier each incorporate a 1 KΩ resistor which is used to protect the microcontroller from transient reflections and also eliminate/dampen any transients which otherwise would be transmitted into the environment creating intermittent transmit and receive errors.

In addition to the 2.45 GHz wireless communication components, the Device PCB also incorporates a 5.0 V voltage regulator to supply a differential driver chip (DDC), whose output is set to the DMX512 Standard, 44 Hz, 512 channel, by the microcontroller.

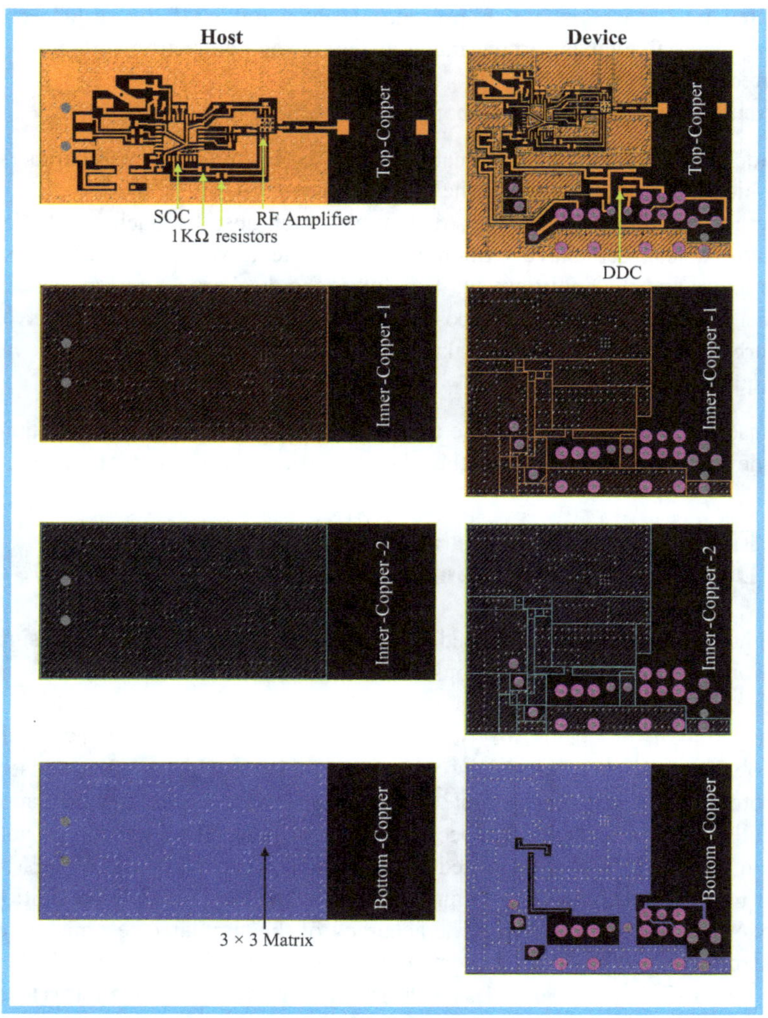

Figure 8.27 Transceiver PCB layouts

Looking at the Top and Bottom Coppers of both the Host and Device circuits, a SPICE model would indicate, characteristic impedance of the chip antenna tracks notwithstanding, that the circuits would operate as intended. Although both circuits would indeed work, due in large part, to the through hole plating which effectively stitches the bottom copper 'ground planes' to the top copper ground zones which provide shielding for the top copper tracks via the ground plane, the ground plane would not be sufficient (big enough) to allow the RF amplifier to operate at its optimum power level. In fact, with the antenna matched using a VNA (Vector Network Analyser), the line-of-sight maximum range was measured to be 48 m. However, when the two Inner Coppers were introduced and the range was remeasured, a range of 157 m was obtained. There are no functions in SPICE which will allow the simulation of such a design process, which effectively triples the physical size of the effective ground plane and simultaneously reduces the distance between the top copper tracks and the ground plane as described in Section 8.2.

Using the through-hole plated stitching method depicted in Figure 8.27, EMC can be all but totally eradicated and susceptibility to EMI can be greatly reduced. This should come as no surprise, since it is basically a macro representation and implementation of the ground loops described in Section 8.3. Additionally, it should be remembered that 'heat' is also electromagnetic radiation, which if not dispersed, can severely affect the signal-to-noise ratio of a signal as the more heat that's present, the higher the noise base level of the system becomes. The 3×3 matrix of through-hole plates is located directly below the RF amplifier, thus providing not only a grounding point but also an adequate heatsink, as the increased size and therefore the mass of the electrically conductive ground plane of a multi-layer PCB is able to 'sink' substantially more thermal noise than a double-sided PCB.

So, you may well be asking yourselves, what the hell has mass to do with EM Theory? The answer to that question is beyond the scope of this book to vindicate in full but will be comprehensively shown and validated in, should circumstances permit, **Concept AE_{42} 'A Gravity Centric approach to Grand Unification'**. I can, however, leave this section with a little teaser. It is basically a reasoning and a logic exercise. To begin with, every accredited physics book states that a Photon is massless and Energy is dimensionless. Now, given the newly acquired data from the James Web Space Telescope (JWST), we have empirical data which supports the concept of gravitational lensing. This would mean that either photons do indeed have mass and/or that Energy is by its very nature attractive. Given such an observation and the fact that both gravity which acts upon mass, and the field strength of electromagnetic radiation, follow an inverse square law, then unifying these two fields shows that mass is indeed intrinsic to EM theory.

Chapter 9
System design

Chapters 2 and 3 have used the concepts of electromagnetic theory to create circuit models which allow electromagnetic coupling between adjacent circuits to be simulated using circuit theory. The resultant simplification in mathematics comes at a price. Action and reaction anywhere in the system are assumed to be instantaneous. This means that the wavelength of the maximum frequency of the simulation must be greater than ten times the maximum dimension of the assembly-under-review.

Chapter 4 shows how this limitation can be relaxed by invoking the concept of distributed parameters. Even so, the limitation is that the wavelength of the maximum frequency of the simulation must be greater than ten times the maximum size of the cable section. However, this means that the maximum frequency of the simulation is no longer limited by the length of the cable.

Chapter 5 extends the application to analyse the coupling between cable and environment. Here, the simplifications are achieved by assuming worst-case analysis. This is acceptable when analysing EMC, where the criterion is to ensure that the actual interference is always less than that predicted.

The transient analysis of Chapter 6 reveals many features of electromagnetic coupling which are not generally known and which provide a much-improved understanding of the phenomena.

The development of the modelling process was made possible by the fact that many tests were carried out on many cable assemblies. Chapter 7 describes some of them. Correlating the result of each test with those of the model allowed defects in the model to be corrected. This was a more unforgiving approach than one based purely on theory. Even so, every equation is traceable back to formulae derived in textbook theory.

Chapter 8 has identified several techniques which can be used to minimise the level of interference coupled into and out of a signal link. It is possible to glean a few simple guidelines which identify the basic concepts used in all these techniques, and these are listed in Section 9.1. If implemented at the initial stage of a project, these guidelines should ensure that the EMC of the system can be achieved in a cost-effective manner.

In all the previous chapters, the approach has been to build up circuit models from basic building blocks and to develop those models to deal with increasingly complex coupling mechanisms. Having developed these models to a point where they provide a reasonably accurate simulation of the performance of actual hardware, it is possible to reverse the process and describe a top-down approach. This is the objective of Section 9.2.

Section 9.3 identifies the function of the interface circuits as buffers between the signal processing functions on printed circuit boards and the signal distribution function of the wires and cables. Chapter 8 provides details of a wide range of such interface circuits.

Having established a clear set of relationships between currents and voltages in the signal link and the electric and magnetic fields in the near vicinity, it is possible to extend those relationships to include the EMC requirements.

That is, the design of the system can be tailored to meet those requirements. Section 9.4 establishes a relationship between the threat environment and the level of interference experienced by the victim circuit.

Section 9.5 relates the design of a potential source of interference to the maximum level of radiated interference, as specified by the formal requirements.

One of the responsibilities of every engineer is to ensure adequate preparations are made before embarking on a design project. Section 9.6 identifies those aspects of the planning necessary to implement the approach described here. In particular, it recommends that enough bench test equipment be provided to meet the needs of the project, and that Mathematical software be installed on the personal computers. It goes on to describe methods of identifying those signal links which could pose problems, and of characterising those links during the initial development of the system. It concludes by stating that the best way of dealing with an EMC problem is to create a circuit model and review it in light of the guidelines of Section 9.1.

9.1 Design guidelines

Although the approach has been to create circuit models which simulate the behaviour of electromagnetic field coupling, the fundamental parameters used in the construction of every model have been derived from electromagnetic theory and all the equations used in the analyses of Chapters 2 to 6 are traceable back to that theory. Chapter 7 shows that there is a clear correlation between the actual performance of a physical assembly and the response of its circuit model.

It is possible to distil a few simple guidelines from the analyses and assessments of the previous chapters. If implemented, they will ensure that the system under development has the best possible chance of meeting all the EMC requirements.

9.1.1 Structure as a shield

Probably, the most important conclusion of the previous chapter is that designers should avoid the temptation to use the structure as a convenient return path for signals and power.

Every conductor of the structural assembly possesses the property of inductance. Any changing current will develop a voltage along the length of that conductor. If voltages can exist between different locations on the structure, that structure cannot be regarded as an equipotential surface. Every signal and every power line which uses the structure as a return conductor is guaranteed to interfere with every other signal in the system.

Section 8.1 explains this in more detail and goes on to reason that it is better to design the structure as a network of conducting loops. Any interference field will cause current to circulate in each loop, and the field created by the circulating current will tend to balance the effect of the incoming radiation. Such a design will create a structure which behaves essentially as a shield. Section 8.7 provides more information on the design of shields.

9.1.2 Return conductors

Every conductor acts as an antenna and any electrical signal can be treated as a sequence of small (or large) step pulses. A transient step pulse travelling along any conductor will create an electromagnetic wave which spreads out in all directions. If another conductor is routed alongside, then the electromagnetic wave will create a current which flows in the opposite direction along that second conductor. That is, the second conductor acts as a receiving antenna. The closer the two conductors are to each other, the more current is captured.

If the second conductor is assigned to carry the return current, then the conductor pair is acting as a transmission line. It appears as though an electromagnetic wave is travelling along the intervening space, guided by the conductor pair. The tests described in Sections 7.5 and 7.6 illustrate this effect.

If the return conductor follows a different route, then the coupling between send and return conductors is reduced, whilst the coupling between differential-mode current and common-mode current is increased. The signal link will emit more interference and its susceptibility to interference will increase. Section 2.8 shows how the parameter values of the circuit model are affected by the relative positions of the conductors in the cross-section of the cable assembly.

Two important objectives are achieved by keeping the send and return conductors of the signal link as close together as possible. The efficiency of the link is improved (there are fewer losses), and the interference level is minimised. Improved EMC and improved efficiency go hand-in-hand.

Since the transfer admittance of the cable as an emitter is the same as the transfer admittance which defines its susceptibility, then the action of routing the two conductors will also reduce the susceptibility of the signal link.

9.1.3 Ground loops

In the vast majority of signal links, it is simply not practical to use any of the techniques described in Section 8.4 to achieve common-mode rejection. Both the output signal and the input monitor are referenced to the zero-volt conductor of their respective printed circuit boards. Rather than use the conducting path via the structure or the single-point ground of Figure 8.7, it is good practice to connect the return conductor to the local structure at both ends of the line. This will form a multitude of ground loops (or earth loops) which each carry common-mode current. An electromagnetic field generated by common-mode current caused by external interference will link both conductors of any signal link and will balance out the interfering signal.

With the configuration illustrated by Figure 8.8, the field generated by the signal conductor will induce a current which flows in the opposite direction along the return conductor. The tests described in Section 7.5 illustrate this fact. As the frequency increases, less and less current will flow in the common-mode loop.

In printed circuit boards, the ground plane acts as a set of return conductors, each located precisely under the associated signal conductor. Figure 8.4 illustrates the fact that current can flow in opposite directions in the ground plane. For this to happen, current must be flowing in a loop in the ground plane. The ground plane acts as a multitude of ground loops.

If the link-under-review is intended to carry a narrow-bandwidth, low-frequency signal, then it is sometimes better to isolate the return conductor from the structure at the receiving end. However, the floating configuration acts to amplify interference levels at the quarter-wave frequency. In such configurations, it is necessary to ensure that the interface circuitry does not respond to frequencies in this range. This can be done by the use of bandwidth filtering.

If the single-point-ground configuration is implemented, then the system will exhibit the worst EMC characteristics possible. Section 8.3 identifies the essential difference between a configuration which uses ground loops and one which is based on the concept of the single-point ground.

9.1.4 Current balance

The existence of a return conductor adjacent to the send conductor provides an opportunity to balance the currents. If the current in the return conductor is equal in magnitude but opposite in direction to that in the send conductor, then the total current at the interface will be zero; that is, the common-mode current will be zero and there will be no interference.

Such a situation is never achieved in practice. But it is something to aim for. Circuit configurations described in Section 8.4 help to minimise interference.

9.1.5 Differential-mode damping

It is good practice to minimise differential-mode reflections at the interface circuitry. If differential-mode energy is reflected back into the transmission line, the only way it can disappear is via electromagnetic radiation. In other words, matched transmission-line design is best for low emissions.

Since the transfer admittance for conducted susceptibility is the same as that for conducted emission, it follows that a matched transmission-line design is best for low susceptibility.

Ideally, the terminations at each end of the line should be the same as the characteristic impedance of the line. This will generally be somewhere in the range of 50 Ω to 200 Ω. Open circuit and short circuit terminations will cause 100% reflections of all signals. Inductive or capacitive loads will cause 100% reflection of transient signals. Section 8.5 describes various ways of damping differential-mode reflections.

9.1.6 Common-mode damping

Viewed from the perspective of conducted susceptibility, current in the common-mode loop is undesirable, since some of it will inevitably appear in the differential-mode loop. Viewed from the perspective of conducted emission, current in the common-mode loop is undesirable, since all of it will inevitably re-appear as interference.

The only way of preventing unwanted energy from radiating out into the environment or coupling into the common-mode loop is to absorb it; that is, to convert it into heat energy.

Components capable of absorbing electrical energy are soft ferrites (especially when operated in frequency ranges where they are most resistive), steel-cored transformers and chokes, and other ferromagnetic materials which exhibit heating losses. Nickel is sometimes used to coat copper conductors to make them more lossy.

It is possible to insert a defined value of resistance in the common-mode loop or the antenna-mode loop using one of the techniques described in Section 8.6. These can be especially useful in reducing the amplitude of high-current long-duration transients in power supplies; the kind usually dealt with by surge suppressors.

Carbon is a material worth considering when selecting components and materials. Sections 8.5.5 and 8.7.3 provide examples of its use.

9.1.7 System assessment

The modelling techniques described in the previous pages describe how to create a circuit model to simulate the interference coupling mechanisms of any signal link.

In the majority of configurations, it is possible to assess the adequacy of the design merely by examining the circuit model from an engineering point of view. If any risks are identified, the design can be modified at an early stage in the development process.

Critical signal links can be identified at a very early stage of product development. One of the actions of such a feasibility study would be to identify tasks which should be carried out in the subsequent stages to define the detailed design and analyse the performance of these links. A method of performing such an assessment is described in Section 9.2.

9.1.8 Bench tests

Having identified critical signal links and carried out a prediction of the performance of each, the next step is to check the actual performance by subjecting it to bench tests.

Chapter 7 is devoted to the subject of bench testing, in terms of the method of checking and in terms of the method of analysis. A representative circuit model can be created of the link-under-review, and this model can be used to predict the results of formal testing in an EMC Test House.

Reports of such analyses will be a significant feature of design reviews.

9.2 Relating the diagrams

Any plan to control the EMC must necessarily involve a top-down view of the system; and this requires the systematic use of diagrams. Since the creation of block diagrams, circuit diagrams and wiring diagrams are part of the normal design process, it is logical to extend the use of the technique to assess the EMC characteristics of the system.

The purpose of this section is to provide an overall view of the process and indicate some of the design considerations involved in meeting EMC requirements. Since there are many books which provide detailed guidance on the design of electronic equipment, such information is not included here.

9.2.1 Circuit diagrams

One such diagram is the circuit illustrated in Figure 9.1. The push–pull emitter follower identifies the module as the output stage of an audio amplifier, the stage which supplies drive current into a loudspeaker. The focus is purely on the circuitry of the module, with minimal information about the interconnecting cables and no information about the circuitry in other modules.

No component values are shown in this particular figure because the diagram is for illustrative purposes only. If component values had been defined, it would have been possible to analyse the performance of this module. Simulation programs with integrated circuit emphasis (SPICE) are designed specifically to deal with such a task. All that the equipment designer needs to do is draw the circuit on a computer screen, define the component values, add a voltage source to the input terminals, and invoke one of the analysis options. Very significantly, SPICE software is capable of analysing both the transient behaviour and the frequency response of the system.

Although SPICE can be used to simulate the behaviour of extremely complex circuitry on printed circuit boards, that facility comes at a price. The previous chapters have shown that the analysis of interference coupling is based on the assumption that voltages exist between 'reference' or 'ground' conductors.

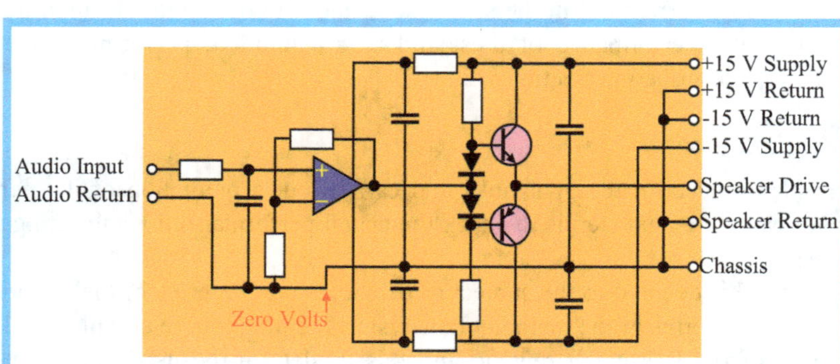

Figure 9.1 Circuit diagram – audio power amplifier

SPICE analyses of the functional performance of complex assemblies are based on the assumption that all points connected to the reference conductor are at zero volts. Given such an assumption, there is no point in attempting to analyse interference coupling.

Section 8.2 shows that, although voltages exist along the surface of a ground plane, the simultaneous existence of circulating currents allows that plane to be regarded as an equipotential surface. Hence, SPICE analysis of complex circuitry on a printed circuit board with a ground plane can provide a reliable simulation of the behaviour of that circuitry. This makes the software a valuable analytical tool. The words 'integrated circuit emphasis' indicate the intended application of the software, that being, the analysis of small-sized modules.

In this particular example, the signals being processed are in the audio range up to about 20 kHz. At this frequency, the quarter-wavelength is more than 3.5 km. Since the largest dimension on the printed circuit board is of the order of 10 cm (corresponding to a quarter-wave frequency of 7.5 GHz), it is not likely that antenna-mode interference will have any direct effect on the board-mounted circuitry.

Although it is possible for high levels of radio-frequency (RF) interference to arrive via the input (or output) connectors, the design of the interface circuitry should be such that all signals outside the operating range of the amplifier are severely attenuated. This is probably the most important precaution.

It is worth noting that non-linearity in the relationship between voltage and current in transistor and diode junctions can cause the modulation of an RF signal to be picked up. The same is true if the centre frequency of the RF signal lies on the sloping part of the frequency response curve of a filter. Care should be taken to select transistors and diodes which are not capable of responding to high-frequency RF signals outside the bandwidth of the signal being processed.

It is also advisable to keep the signal loops on the board as small as possible, either by using a ground plane, or by routing the tracks carrying each signal and its return current closely together.

Given such precautions, it is reasonable to assume that there will be no internal interference between components on the board. By ignoring effects which are judged to be irrelevant, the computational power of SPICE software can be put to good use in analysing the behaviour of extremely complex circuit boards.

Even though the main purpose of SPICE analysis is to simulate system function, it is still possible to assess and analyse the interference coupling characteristics of critical signal links on the board by reviewing them in the light of the guidelines of Section 9.1.

9.2.2 Wiring diagrams

However, this assumption that all connections to the reference node are at zero voltage ceases to have any validity when the inter-module wiring of the equipment is considered. The wiring diagram of Figure 9.2 shows how this particular printed circuit board could be interconnected with the other modules of the equipment and how equipment units are assembled to create a functioning system.

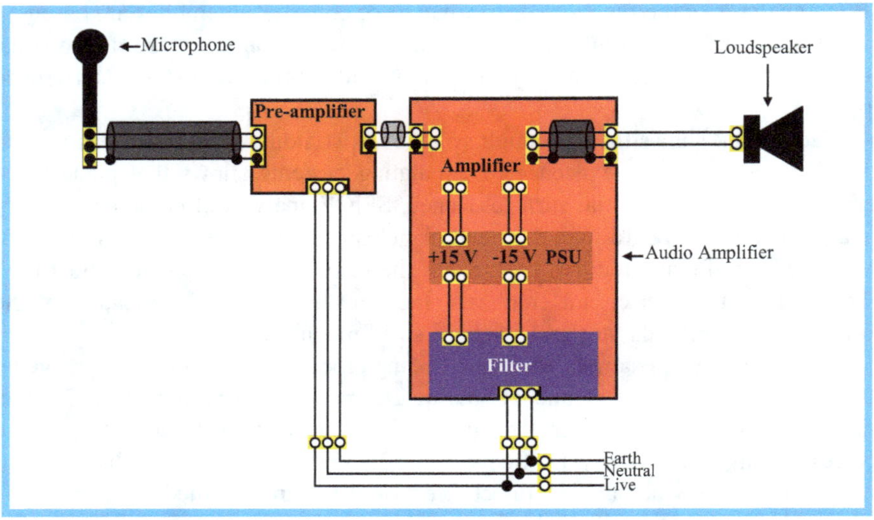

Figure 9.2 Wiring diagram – public address system

In this system, the cables carrying signals and power between the various units are much longer than the printed circuit traces connecting different components on the printed circuit boards. Since the interface circuitry for each signal link is designed to provide a buffer between the interconnecting cable and the signal processing circuitry on the board, it is possible to assume that interference picked up by the cable can be treated separately from that picked up by the traces on the printed circuitry.

The analysis of Section 5.3 shows that the maximum threat voltage which can be induced in a cable of defined length is proportional to the inverse of the frequency.

Given that, the longer the cable, the higher the RF energy which can be induced, and also the fact that the frequency at which maximum RF energy can be induced reduces as the length of the cable increases as illustrated by Figure 5.20.

Then, for a long cable, the design of the interface circuitry should ensure that the signal passed on to the processing circuitry due to the threat voltage is within acceptable limits. This design would almost certainly include a low-pass filter, ensuring that the level of attenuation increases as the frequency increases. This being so, the focus of attention shifts to the design of the interconnecting cables.

Any external electromagnetic field will link with the cables and will cause antenna-mode current to flow in the outer surface of every exposed conductor. Equally, any antenna-mode current created by differential-mode current in the cables will radiate interference away from the system.

For such a system, the assumption that every point on a conductor is at the same voltage is no longer valid. It becomes necessary to represent the coupling to and from the signal loop as a triple-T network, as illustrated by Figure 2.19.

9.2.3 Block diagrams

It is best to approach the problem in a systematic manner. Since most electronic systems are much more complex than the simple public address system depicted here, it is normal practice to illustrate the function of each equipment unit by means of a block diagram, as illustrated by Figure 9.3.

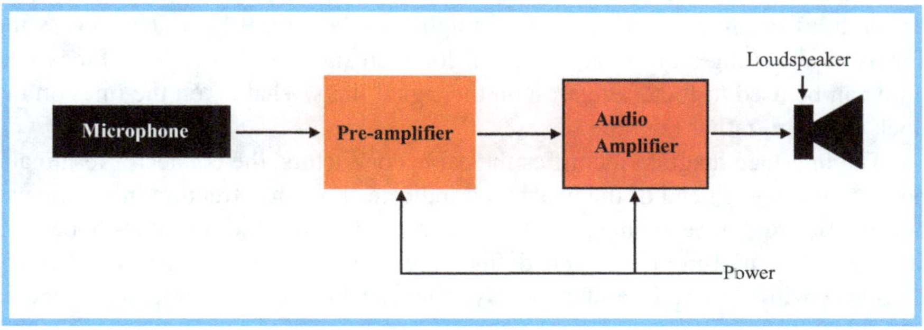

Figure 9.3 Block diagram – simple public address system

Such a diagram gives the designer an overall view of the relationships between the different units, and is usually created during the initial feasibility study of the project. At this stage of development, many details of the design have yet to be defined, and this includes those details which control the EMC of the system.

9.2.4 Interface diagrams

Given the existence of a block diagram which identifies every signal transmitted between every processing module, it becomes possible to focus attention on each signal link, and this function is performed by an interface diagram.

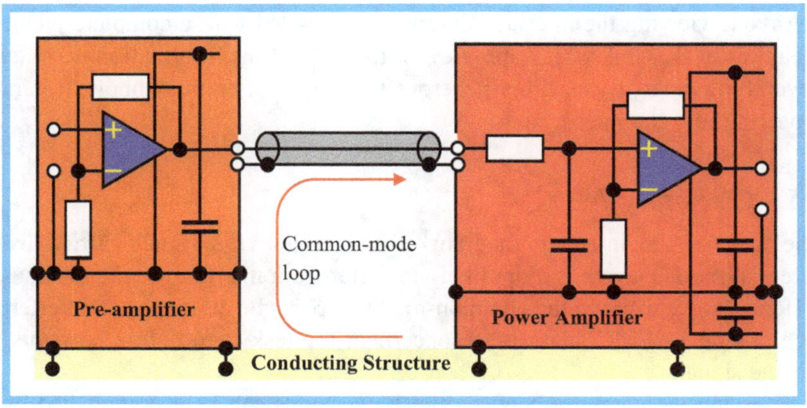

Figure 9.4 Interface diagram of the link between pre-amplifier and power amplifier

Figure 9.4 is just such a diagram. It identifies all the conductors and all the components involved in transmitting a signal from the pre-amplifier to the power amplifier.

By limiting the scope of the diagram to the components involved in the coupling interference between a signal and its environment, unnecessary complication is avoided.

The normal function of block diagrams is to allow attention to be focused on the modules (such as the audio power amplifier of Figure 9.1) which process the signals. By creating an interface diagram for each signal link, the same block diagram can be used to focus attention on the signal links. That is, on the lines on the block diagram, rather than the boxes.

The interface diagram identifies the cable conductors, the connector terminals, the circuitry at each end of the cable, and indicates how the structure interconnects the 'ground' reference conductors at each end. It identifies the common-mode loop and the differential-mode loop. It defines precisely how the transmitter circuitry interfaces with the cable and precisely how the buffer at the receiving end is designed.

This diagram illustrates clearly that three conductors are involved in carrying a signal from one location in the system to another. In this case, it is the inner conductor of the co-axial cable, the screen of that cable, and the structure. Differential-mode current is carried by a loop formed by the signal and return conductors, in this case, the core and screen of the co-axial cable. Common-mode current is carried by the loop formed by the return conductor and structure. In this context, the structure represents the combined effect of all other conductors linking the two equipment units.

This diagram is arguably the most important from the point of view of EMC design. The designer is able to define what sort of cable to use: wire pair, co-axial cable, screened pair, multi-conductor assembly, or something else. He or she is free to use any of the interface circuits described in Chapter 8, or to define a completely new interface. It is also possible to decide whether or not to enclose the cable with an overall screen or to route it along conduits.

Anyone viewing the interface diagram is provided with a complete picture of the features of that link which are relevant to EMC. If the information is contained in several drawings, each with a generous sprinkling of earth symbols, it is rather like peering into the mist.

9.2.5 Circuit models

In the case of the interface diagram of Figure 9.4, the signal carried by the differential-mode current is more likely to suffer from interference than to cause it.

Hence, any current in the common-mode loop can be classed as an interference source. As far as EMC analysis is concerned, the task is to define how this will affect the signal.

Assessing the coupling between these two loops can be accomplished by the use of the circuit model of Figure 9.5. So, the creation of the circuit model becomes the next step in the process.

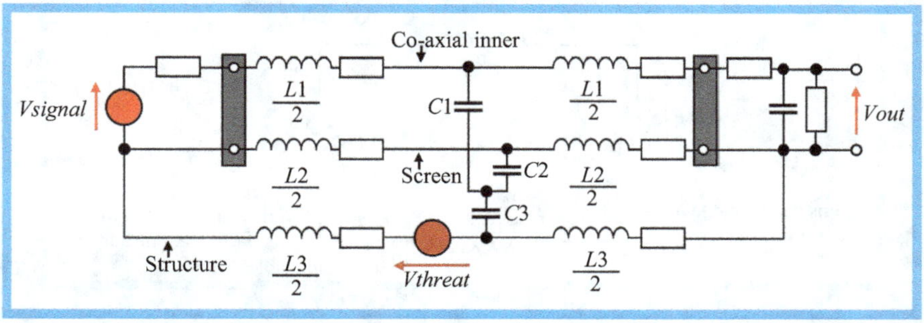

Figure 9.5 Circuit model of the link between pre-amplifier and power amplifier

A circuit model for a signal link can be defined as one which simulates the interference coupling mechanisms. It allows both the susceptibility and emission characteristics to be analysed.

In many cases, the mere act of creating a circuit model allows the characteristics of the signal link to be assessed. It is possible to view such a model in much the same way that an experienced designer can look at the circuit diagram of a functional module and identify its strengths and weaknesses.

In the case of the link-under-review, the only coupling between differential-mode and common-mode loops is via the transfer impedance of the screen. It would be useful to know the frequency response characteristic of this impedance. If this information is available, then the component values for the circuit model can be deduced. Section 2.10 gives details on the method of modelling co-axial cables.

There may be problems when resonance conditions arise and the current in the screen reaches a high amplitude. If, as in this example, the bandwidth of the differential-mode signal is restricted to audio frequencies, then there should be no problems. There should be a low-pass filter at the audio amplifier interface to ensure that any high-frequency content is rejected.

9.2.6 Deriving component values

Initial values for the components of the circuit model can be derived from a review of the physical assembly of the wiring harness.

The common-mode loop is essentially made up of four sections: the co-axial cable between the pre-amplifier and the audio amplifier, the mains cable delivering power to the power amplifier from the mains trunk cable, a short section of trunk cable, and the mains cable delivering power to the pre-amplifier. Figure 9.6 illustrates the magnetic properties of this assembly.

The partial inductance of each section of the loop can be derived by treating the section as a composite conductor made up of several conductors in parallel.

Section 3.1 describes how to derive the inductance of a single composite conductor. Summing the values derived for the four partial inductors gives a value for the loop inductance ($L3$ in the circuit model of Figure 9.5).

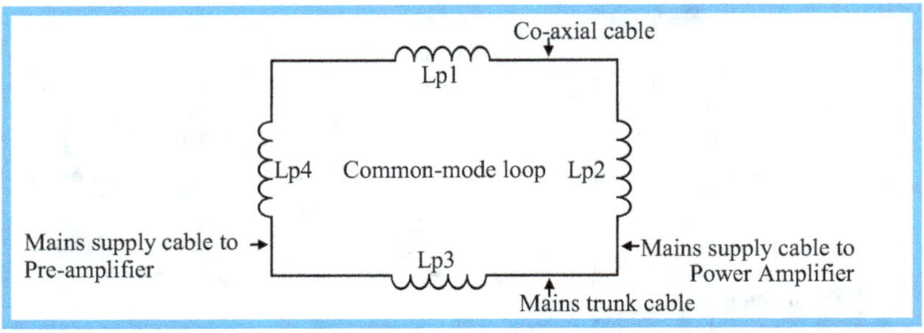

Figure 9.6 Inductance of common-mode loop

Since the value of capacitance *C*3 is closely related to the inductance *L*3, this parameter can also be defined. Section 2.3, on the duality between L and C, provides the relevant formulae.

With a properly terminated co-axial cable, the impedance common to the common-mode and differential-mode loops is the transfer impedance of the screen. As indicated previously, the model can be created if information on the frequency response of this parameter is available. Such information could be provided by the manufacturer's data sheets. Given knowledge of the diameter of the two conductors of the co-axial cable, the inductance per metre of the differential-mode loop, *L*1, can be defined. Since the characteristic impedance of the cable is (to a first approximation) equal to the square root of the ratio between inductance and capacitance, it is possible to derive a value for *C*1 in the circuit model.

If the signal link is more complex (for example, if the conductors form part of a multi-conductor cable with an overall screen), then the technique described in Chapter 3 can be used to determine the coupling parameters.

Alternatively, data from a set of bench tests can be used to construct the model. This could be provided by tests carried out on the prototype equipment, or by data from a library of 'representative circuit models' as described in Section 9.6.6.

Values for the components at each interface are either defined by the designer or specified in the relevant drawings of each equipment unit.

9.2.7 Analysing the signal link

The most appropriate EMC test for this particular configuration is the conducted susceptibility test. A voltage transformer is used to inject a known voltage at a set of known frequencies into the common-mode loop and the amplitude of the differential-mode current is monitored. Figure 4.18 illustrates a setup which could be used.

Given knowledge of the values of the components of the circuit model, it is possible to predict the response of the signal link to such a test during the initial stage of development.

9.2.8 Testing the link

If the results of such a simulation indicate that possible interference problems might occur, then bench testing of this link could be included in the development program for the product. Correlating the test results with the response of the model will allow an accurate circuit model to be created, one which can be used to define the performance of that particular link.

Such an approach will provide a high level of confidence in the ability of the system to pass the formal EMC tests on the first attempt. Significant cost savings can be achieved by avoiding the need to modify production-standard equipment and re-submit it to the qualification test procedure. Delays inherent in this process, which affect the time-to-market of a new product, can also be avoided.

9.3 Printed circuit boards

Detailed design of printed circuit boards is well covered in existing books; so, there is no need for this section to reprise that information. However, there are a few points that are worth making.

In Section 8.7, various ways of maintaining the integrity of an equipment shield were identified. In every case, the purpose was to provide a barrier against the incursion of external fields and to prevent the egress of internal fields into the environment. Figure 8.25 illustrates the concept.

The same concept can be applied to the design of printed circuit boards. If each interface between a board and the external wiring is limited to the transmission or reception of one signal, then that interface will act as a buffer between the board and interference carried by that wiring. Sections 8.4, 8.5 and 8.6 describe a number of designs which can achieve this purpose.

A properly designed buffer will attenuate any interference signal arriving via the interconnecting cable to a level which is within acceptable limits. This could be a requirement that the normal function of the signal processing circuitry should not be upset, or a requirement that no damage should occur and that the system should be able to recover after the event. The buffer should include filter components to limit the signal bandwidth to that which is necessary for the intended signal function.

With multi-layer boards, the ground plane can provide shielding against external fields. If the supply rail is also constructed of a plane surface, then the shielding can be very effective. Even if the board is single-sided, the presence of adjacent boards will provide a measure of protection.

Although it is still possible for interference to be picked up directly by the printed circuit tracks, the fact that buffers exist at every interface means that interference arriving via the cables can be dealt with separately from the assessment of the performance of the board itself. The frequencies of the threat signal would be quite different. During such an assessment, the guidelines of Section 9.1 continue to apply. Concepts that are applicable at the system level are still relevant at the board level.

The concept of 'composite conductors' can be used to analyse the coupling between adjacent tracks on a board, or the coupling between a signal link on a board and the local environment. Chapter 3 describes the technique and Figures 8.1 and 8.3 provide practical illustrations.

The concept of segregation of 'noisy' and 'quiet' circuits is often implemented on printed circuit boards. If one section of the board is used to mount logic circuitry and another section holds analogue components, then it is not a good idea to mount each set of components on its own separate ground plane. The ground plane should cover the entire board, since it provides a return path for each signal trace at precisely the right location. Any separation of ground planes on a board will disrupt the paths for return currents, with an increased probability of interference.

Since the wiring on printed circuit boards is much shorter than that used externally, the frequencies at which problems occur will be much higher. Since the threat level reduces with frequency, interference from distant sources is unlikely (unless the board is completely exposed to the environment). This means that interference problems at very high frequencies are more likely to be due to emanations from nearby boards. Sniffers magneticsciences.com are available to detect sources of high-level emissions.

Sections 8.4, 8.5 and 8.6 are devoted to describing ways of designing the interface circuitry. The most important consideration is that the design of the circuitry at both ends of the cable assembly should be considered at the same time. Books on circuit design [24–26] can provide guidance on the design of the functional circuitry, whilst the book 'EMC Design Techniques' [10] has an excellent chapter on printed circuit boards.

9.4 Susceptibility requirements

Susceptibility requirements can be defined in terms of the threat environment and the acceptable functional behaviour of the circuit interfered with. For example, a $\pm 1\%$ maximum error is harder to design for than a $\pm 10\%$ error.

Although there is a wide range of test specifications defined in a host of EMC regulations, it should usually be possible to define the threat environment in terms of a frequency response curve. Table 9.1 defines the frequency characteristics of the environment to which some UK military equipment can be subjected [27].

It is clear from the table that the most severe threat to military equipment lies in the range between 100 kHz and 100 MHz. So, this is the range depicted in Figure 9.3.1.

In Section 5.5, a method of relating the power density of the external interference to the differential-mode current induced in a signal link was derived. A Mathcad worksheet was used to calculate the frequency response of induced current in that link to a field of constant power density. Electric field strength is related to power density by:

$$E = \sqrt{Zo \cdot S} \qquad (9.1)$$

Frequency	Average field strength (V/m)
10 kHz–100kHz	10
100 kHz–500 kHz	10
500 kHz–1.6 MHz	10
1.6 MHz–5 MHz	560
5 MHz–10 MHz	380
10 MHz–30 MHz	200
30 MHz–100 MHz	200
100 MHz–200 MHz	60
200 MHz–700 MHz	70
700 MHz–1GHz	60

Table 9.1 Frontline and operational support equipment field strength

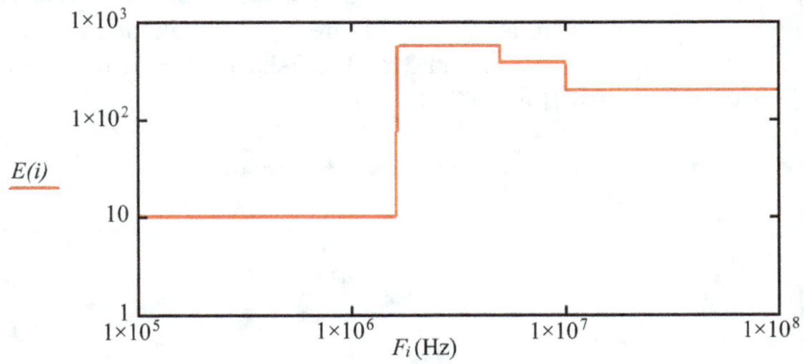

Figure 9.7 Example of military threat environment, 100 kHz to 100 MHz

This being so, the response of Figure 5.36 could also be described as the response of the assembly to a field of constant electric field strength. Since the contractual requirements usually define the threat environment in terms of a field of varying intensity, it is necessary to include this parameter as a variable.

Given knowledge of the length of the assembly-under-review, it is a simple matter to define the threat in terms of the voltage that will be induced in the common-mode loop. Figure 9.8 illustrates the variation of that voltage with frequency for a signal link of 15 m. This is the same length as the link described in Section 5.5; a twin conductor cable routed over a conducting structure. The computations are carried out by Page 1 of Worksheet 9.4 and illustrated by Figure 9.10. It is worth noting that the threat voltage peaks at over 10 kV.

A representative circuit model was created in Section 5.5 to simulate the coupling between the common-mode loop and the differential-mode loop, and a worksheet was derived to calculate the amplitude of the interference current in the signal circuitry.

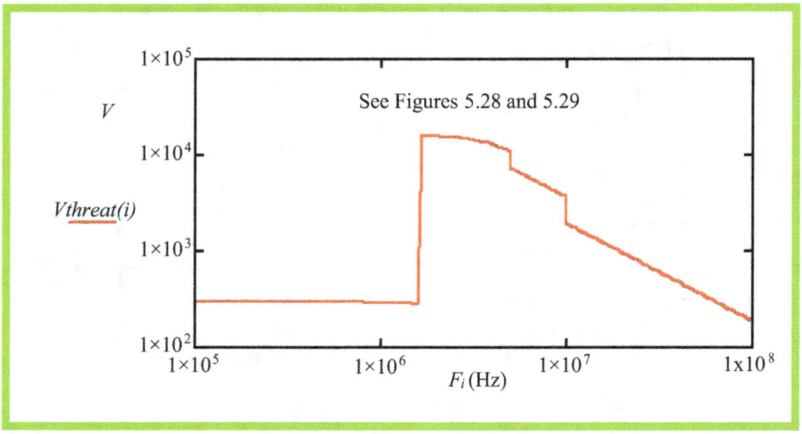

Figure 9.8 Threat voltage induced in the common-mode loop for a 15 m link

By applying the threat voltage *Vthreat* to the culprit loop, the current *Iout* can be calculated. The response is illustrated by Figure 9.9. The computations were carried out by incorporating Pages 1, 2, and 3 of Worksheet 5.5 (Figures 5.30, 5.33 and 5.34), into Worksheet 9.4 (Figure 9.10).

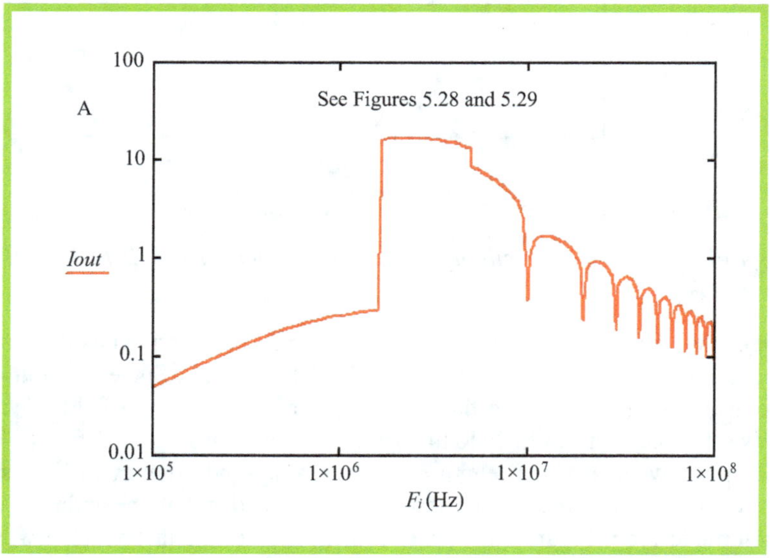

Figure 9.9 Frequency response of current in the differential-mode loop

It is clear from this graph that the critical range of frequencies lies between 1.5 MHz and 10 MHz. So, the design of the interface circuits, the shielding and the signal processing will need to focus on providing protection to the system function over this range of frequencies.

Worksheet 9.4 Page 1 Data from Table 9.1

$$\text{data} := \begin{pmatrix} 0.01 & 10 \\ 0.1 & 10 \\ 0.1 & 10 \\ 0.5 & 10 \\ 0.5 & 10 \\ 1.6 & 10 \\ 1.6 & 560 \\ 5 & 560 \\ 5 & 380 \\ 10 & 380 \\ 10 & 200 \\ 30 & 200 \\ 30 & 200 \\ 100 & 200 \\ 100 & 60 \\ 200 & 60 \\ 200 & 70 \\ 700 & 70 \\ 700 & 60 \\ 1000 & 60 \end{pmatrix}$$

$c := 299.8 \cdot 10^6$ m/s $l := 15$ m Length of cable

$fq := \frac{c}{4 \cdot l} = 4.997 \times 10^6$ Hz

$n := 100$ $N := 20 \cdot n$ $i := 1..N$

$F_i := i \cdot \frac{fq}{n}$

$$\text{fnd (thres)} := \left| \begin{array}{l} j \leftarrow 1 \\ \text{while } \left| \text{data}_{j,1} \right| \le \text{thres} \\ j \leftarrow j + 1 \\ k \leftarrow j - 1 \\ \text{data}_{k,2} \end{array} \right.$$

$$E(i) := \left| \begin{array}{l} f \leftarrow F_i \\ \text{thres} \leftarrow f \cdot 10^{-6} \\ \text{fnd (thres)} \end{array} \right.$$

See Figure 9.7

$$Vthreat(i) := \left| \begin{array}{l} f \leftarrow F_i \\ \lambda \leftarrow \frac{c}{f} \\ Va \leftarrow E(i) \cdot \frac{\lambda}{\pi} \\ Vb \leftarrow Va \sin\left(2 \cdot \pi \cdot \frac{l}{\lambda}\right) \\ Vb \leftarrow Va \text{ if } l > \frac{\lambda}{4} \end{array} \right.$$

Function based on Figure 5.19. Displayed on Figure 9.8

Figure 9.10 Calculating the threat voltage for a 15 m cable

The above example is intended for illustrative purposes only, to show how system design can be related to the susceptibility requirements.

It is also worth noting that this is a worst-case analysis. In practice, there will be a significant reduction in threat level due to shielding afforded by the structure as well as re-radiation of the interference back into the environment. Although this is a limitation of the circuit modelling approach, any errors will be on the side of caution.

9.5 Emission requirements

As with susceptibility requirements, there are a host of formal requirements for controlling the emission characteristics of electrical equipment, and again, it is possible to use a military requirement for illustrative purposes. Table 9.2 is a copy of the emission requirements defined in the UK Defence Standard for signal and secondary power lines [28].

Frequency	Current (dBμA)
20 Hz	130
2 kHz	130
50 kHz	50
500 kHz	20
150 MHz	20

Table 9.2 Emission requirements of DCE02.B

The test involves the use of a current transformer clamped around the cable-under-test and connected to a current measuring device. The values in the second column of the table define points on the characteristic. Figure 9.11 illustrates that characteristic. It defines the upper limit for measured emissions.

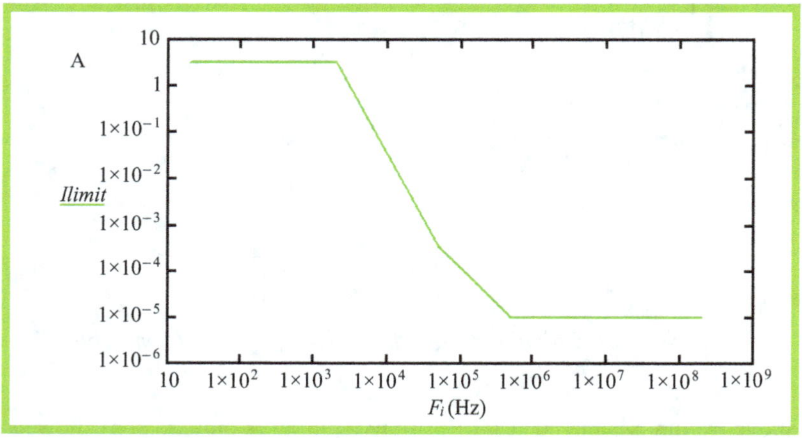

Figure 9.11 Maximum acceptable emission

Creating this graph was a simple matter of converting the values of dBμA in the second column of the table to values in Amperes and plotting these values at the appropriate frequency. Figure 9.12 reproduces the first section of Worksheet 9.5 because some readers might be unfamiliar with the relationship between dBμA and A.

The emission requirements illustrated by Figure 9.11 can be compared to the predicted response of any given signal link; for example, the link depicted by Figure 9.13.

Worksheet 9.5 Page 1

$$\text{Amps}(\text{dBmicroA}) := \left| \begin{array}{l} \text{dBA} \leftarrow \text{dBmicroA} - 120 \\ y \leftarrow \dfrac{\text{dBA}}{20} \\ A \leftarrow 10^{y} \end{array} \right.$$

$$\text{data} := \begin{pmatrix} 20 & 130 \\ 2000 & 130 \\ 50 \cdot 10^{3} & 50 \\ 500 \cdot 10^{3} & 20 \\ 200 \cdot 10^{6} & 20 \end{pmatrix} \qquad \begin{array}{l} Fl := \text{data}^{\langle 1 \rangle} \quad \text{dB} := \text{data}^{\langle 2 \rangle} \\ i := 1 .. \text{rows}(\text{data}) \quad \text{Ilimit}_i := \text{Amps}(\text{dB}_i) \end{array}$$

Figure 9.12 First page of worksheet 9.5

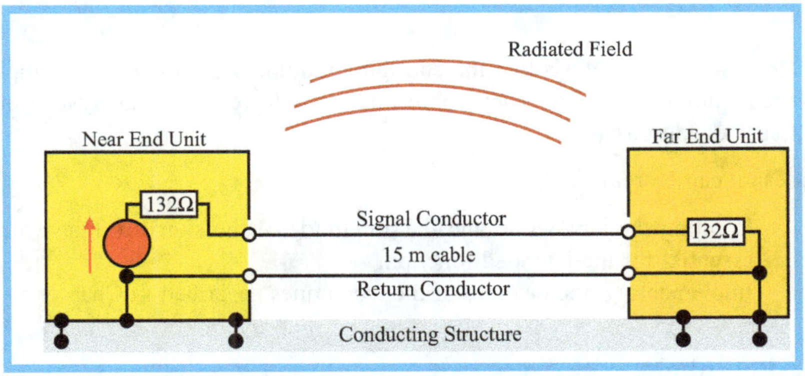

Figure 9.13 Estimating the level of the radiated field

This particular link is used as an example because it is the one which was subjected to a susceptibility analysis in the previous section.

Analysing the emission characteristics of this link can be carried out by copying Pages 1, 2, and 3 of Worksheet 5.5 (Figures 5.30, 5.33 and 5.34), adding them to Worksheet 9.4 (Figure 9.10), and making a few minor changes to the program.

Basically, these changes are:

- Setting the value of the radiation resistance *Rrad* to zero. (Inserting a 50 Ω resistor in series with the structure is only necessary when susceptibility is being analysed).
- Setting the voltage source in loop 2 of the model to zero. (When simulating emission, it is assumed that there is no voltage source in the common-mode loop.)
- Setting the voltage source in loop 1, *Vout*, to unity. (The voltage source is assumed to be in the differential-mode loop).

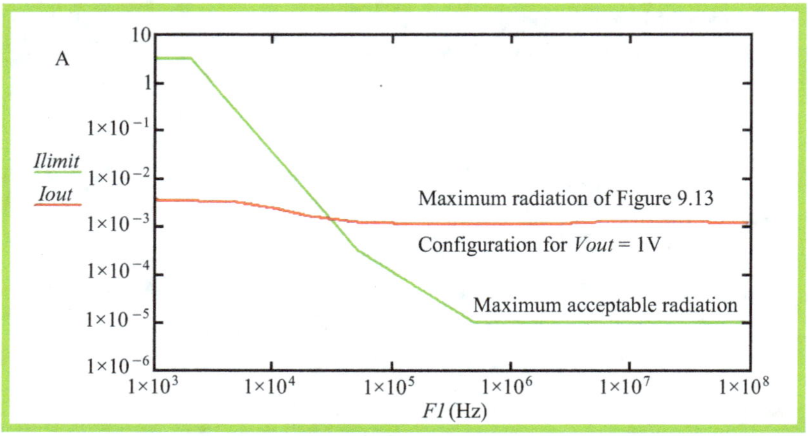

Figure 9.14 Comparing the response of the proposed design with the emission requirements

It is clear from this analysis that the configuration-under-review is non-compliant with the requirements at frequencies above 50 kHz. However, this analysis is of the worst-case conditions.

The design can be remedied by

- Limiting the bandwidth and the amplitude of the signal to that which is essential for the intended function.
- Implementing one or more of the techniques described in Chapter 8.

9.6 Planning

9.6.1 Performance requirements

These usually take the form of a set of EMC test requirements for the particular system being developed. Several groups have developed independent sets of such requirements: military, aircraft, spacecraft, automotive, ships, commercial, etc. Different countries have different regulations. It is usually the case that the manufacturer involved in designing the product already has a full set of regulatory documents in place. If not, then there are several consultancies which specialise in offering advice. The International Journal of Electromagnetic Compatibility interferencetechnology.com identifies a number of such organisations.

It is sometimes the case that external EMC requirements are inadequate to meet the needs of safety or reliability. In which case, in-house EMC requirements may be called for.

Once the EMC requirements have been defined, the task is to design the equipment to meet those requirements. The normal procedure is to carry out formal EMC testing after all other tests have been completed. If the result of the formal test process is a number of non-compliances, then the equipment will need to be modified and re-submitted. This can be an expensive process, since at this stage all the manufacturing drawings will have been completed, all the manufacturing processes will have been written, and all the components will have been purchased. Nor is there any guarantee that the modified equipment will be compliant.

The previous chapters of this book have shown that a clear correlation can be established between test requirements and the detailed design of the equipment. This allows EMC to be treated in the same way as requirements such as frequency response, functional performance, mass, size, reliability and cost. This being so, analysis and test of the EMC requirements can be integrated into the normal design process.

9.6.2 Bench test equipment

Two essential items of equipment are needed to supplement the general-purpose test equipment: the current transformer and the voltage transformer. Although these transducers are available commercially, they are tailored to meet the needs of the formal regulatory requirements rather than the needs of the equipment-under-development. They are also expensive.

Chapter 7 provides details of the design, manufacture, and calibration of simple, low-cost devices which can be constructed in-house. These particular devices were designed to cover a frequency bandwidth of about 20 kHz to 20 MHz. The coverage can be extended downwards by using larger devices or upwards by using smaller devices. The devices described in Sections 7.1 and 7.2 use ten-turn windings on the toroids to ensure that the impedance reflected by the test equipment into the loop-under-test is as small as possible.

Some commercial transformers extend the frequency range by employing just one turn on the primary winding.

It would be a worthwhile exercise to manufacture a set of tri-axial cables, as described in Section 7.3. It is important that the test equipment does not create spurious coupling with the equipment-under-test.

The use of such devices need not be restricted to bench testing during the development process; they can also be used for in-situ testing during trials and troubleshooting. The availability of bench test equipment to check EMC enables newly purchased equipment to be assessed before it is integrated into the system.

9.6.3 Software

Any organisation involved in the design of electronic equipment will already have a number of personal computers devoted to the design and manufacturing process. SPICE software will be installed on many of these to analyse the functional performance of the circuitry on the printed circuit boards.

Such software is based on the use of nodal analysis and the existence of a reference node at zero volts.

If it is assumed that all connections to the reference plane are at zero volts, then extremely complex boards can be analysed. However, this assumption carries with it the hidden assumption that there is no such thing as interference.

This does not mean that SPICE software is incapable of analysing interference coupling. It is entirely possible for this software to analyse any of the lumped parameter models of Chapter 2. Moreover, such models can be combined with models of complex printed circuit boards. Provided the user is aware of the limitations of such software, valuable information can be gleaned from the exercise.

However, there will always be occasions where such an approach is inadequate, making it necessary to use mathematical software such as Mathcad. This type of software has the advantage that the computations are not restricted to circuit models, as many of the worksheets in this book can confirm. If both types of software are installed on the same computer, it becomes possible for them to exchange data.

9.6.4 Critical signal links

In any complex system, it is necessary to select those signal links which may be critical in terms of high susceptibility or high emission. So, it is useful to identify the criteria used during the selection process.

The most obvious criterion is the length. Equation (5.15) indicates that the power received by a dipole receiver is proportional to the square of the wavelength. If the link-under-review is treated as one monopole, with the structure acting as the other monopole, as illustrated in Figure 5.14, then maximum power will be received when the actual length is equal to one-quarter wavelength of the interfering signal.

If the cable is routed along the structure, then the response of Figure 5.20 should provide guidance on the level of interference to expect at different frequencies.

As far as susceptibility is concerned, the most critical links are often, but not always, those which carry very low-level signals, such as those from remote transducers.

As far as emission is concerned, the most critical links are those which carry the highest transient currents. Power supplies fall into this category, since they almost always contain switch-mode converters. The transmission of fast digital signals between separate items of equipment can also create high levels of emission.

It is normal practice to create a block diagram at the early stage of a project to identify the main items of equipment and to define their function. Since the links between blocks at this level of definition are likely to be the longest, as well as carrying the most important signals, such a diagram can be used to identify the most critical links.

9.6.5 Critical frequencies

Experience has shown that both emission and susceptibility are characterised by a series of peaks and troughs. Since EMC requirements are usually defined in terms of limits which should not be exceeded, it is invariably at peaks in the response that these levels are exceeded. Any analysis of the interference coupling mechanisms needs to focus on these peaks; on the frequencies at which they occur as well as on their amplitude.

Since the peaks occur at frequencies corresponding to multiples of the quarter-wave frequency of the line and since the quarter-wave frequency is a function of the length of the signal link, it is not difficult to make a first estimate as to what the critical frequencies are. Figure 5.35 illustrates this for the configuration of Figure 5.28.

The relationship between power and frequency ensures that there is less power available at frequencies higher than that corresponding to a quarter-wavelength. If it is assumed that the threat level is constant with frequency, the frequency at which the first resonance occurs is likely to be the most critical. Even so, some EM environments have much higher power levels at very high frequencies. Hence, it is important to consider the entire range of frequencies when assessing the requirements.

9.6.6 Characterisation

Although the methodology described in this book can be used to create circuit models of any particular type of cable coupling, using the manufacturer's data, there are always uncertainties. For example, the relative permittivity and the radiation resistance are usually unknown variables. Unless one has to hand a field solver which can accurately predict permittivity and radiation resistance from dimensions and material characteristics.

If the link-under-review is constructed from different types of cable, or if the effect of the terminations is uncertain, tests on a representative rig can be used to create a representative circuit model which can be used with confidence.

However, it is not necessary to characterise every signal link in the system. The representative circuit model of a particular type of cable can be applied to every cable of that type. Since the *L*, *C*, *R*, and *G* parameters of a cable are proportional to length, tests on a long cable can be used to define the characteristics of a short cable.

Tests of long cables over a low-frequency range can be used to predict the response at high frequencies. When carrying out such a test, care should be taken that the cross-section is uniform along the entire length and that the effect of the terminations is minimal. The test described in Section 7.5 is an example.

One observation made during these tests was that it did not seem to matter how the cable was laid out, or whether or not there were tight bends. These features might cause local hot-spots in the field distribution, but they did not seem to have any effect on the overall response.

Another observation was that when two conductors are twisted together to create a twin-core cable and subsequently separated, the lengths of the two conductors are different. The more turns per metre, the greater the difference. If the lengths are different, then so are the impedances. If the impedances are different, then the coupling between common-mode and differential-mode signals must increase. Specifying a high number of turns per metre is not a good idea. There are other ways of achieving close spacing between conductors, for example, ribbon cable.

Capacitors are really open-circuit transmission lines coiled into a small volume. Just like transmission lines, they have inductive properties. Unlike transmission lines, there is only one resonant frequency, as the response of Figure 7.56 illustrates. This could be due to capacitive coupling between adjacent turns.

Interference within the component seems to cancel out the higher frequency resonances which would occur with a normal transmission line.

Tests can be devised to measure the value of each 'component-within-a-component' as illustrated by Figure 7.57.

By applying an input signal via a 50 Ω cable and monitoring the output via another 50 Ω cable, the component can be measured over a range which includes a peak or a null in the frequency response. A circuit model can be postulated to simulate the component. Analysis of this model allows the theoretical curve to provide a faithful simulation of the measured response. Adjusting component values of the model allows the actual and simulated responses to coincide. The end result is a representative circuit model of the component, valid over the frequency range of the tests carried out.

By creating a library of representative circuit models to simulate signal links and components, it becomes possible to simulate any signal link in the system. This is essentially the approach used with SPICE software but adapted to meet the needs of EMC.

Several examples of representative circuit models are provided in the previous pages, and these are listed in Table 9.3.

Description	Figures
Screened pair example	3.17
Cross-coupling example	4.9
Conducted emission test setup	4.17
Conducted susceptibility test setup	4.19
Exposed cable – simplified	5.23
Radiation coupling via the structure	5.31
Transient emission from cable	6.38
Voltage transformer	7.6
Current transformer	7.16
Cable under test	7.49
Murata 100 nF capacitor	7.57

Table 9.3 Representative circuit models

These have been derived from the small set of general circuit models listed in Table 9.4.

Description	Figures
Three-conductor signal link	4.6
Isolated conductor	5.7 and 7.23
Isolated cable	5.13 and 7.37
Exposed cable and structure	5.15
Exposed cable and structure – simplified	5.21
Transient emission	6.34

Table 9.4 General circuit models

9.6.7 General approach

Since there is a multitude of regulatory test specifications, there is no point in attempting to compile a list of such regulations. In any case, the relevant set of regulations is dependent on the type of equipment being assessed. However, these requirements have one common factor; they all specify the use of electronic equipment to carry out the tests.

Previous chapters of this book have shown that it is always possible to create a model of the coupling mechanism between two independent circuits. So, it is always possible to relate the performance of the equipment-under-review to that of the formal test setup.

Formal requirements are not the only factor to consider when assessing the EMC of the product-under-review. Cross-coupling within the system will inevitably cause problems. Nor is there any guarantee that a product which meets the formal requirements will experience no interference problems during its lifetime.

Whether it is or not a standard design procedure, it is useful to carry out a Failure Mode Effects and Criticality Analysis (FMECA) on the system-under-review, it is a worthwhile exercise to spend some time assessing the possible problems which could arise due to EMI. The section 'Banana Skins' in every issue of the *EMC Journal* [29] provides some hair-raising examples of what can go wrong.

When dealing with any problem relating to electromagnetic interference, the approach of the circuit designer should always be to create an interface diagram of the link-under-review such as the one illustrated by Figure 9.4, and assess its characteristics using the guidelines of Section 9.1. This task should take less than a day for even the most convoluted link.

If this does not immediately identify the cause of the problem, the diagram can be used to create and assess a circuit model of the link. The results of this initial assessment can determine whether there is a simple solution or whether a more detailed analysis is necessary. Tests on a prototype assembly can be devised to check that the solution is viable before implementing the modification.

The objective should be to acquire a high degree of confidence that the system-under-review will meet the defined requirements without over-designing the protection measures. It is much easier to do this when the design is based on the results of circuit modelling than when it is based on the consensus of opinion at a design review meeting.

If further development is needed to improve the design of a critical assembly, then the information acquired by the circuit modelling approach can be used as a baseline for more sophisticated analyses using three-dimensional field solvers.

Appendix A
Mathcad worksheets

As far as circuit modelling is concerned, Mathcad software eliminates most of the tedium involved in developing programs. It can handle tasks that include the calculation of values of circuit components, analysis of the response of the circuit model, analysis of test data to define the response of an assembly-under-test, and display a single graph showing the responses of the model and the associated test results. This can all be done on the same worksheet; that is, without the need to invoke subroutines that are recorded on separate files. Moreover, the equations used in the computations are displayed in exactly the same way as they would appear in a textbook on circuit theory.

In a programming language, equations look something like:

$$x = (-B + \mathrm{SQRT}(B ** 2 - 4 * A * C))/2 * A$$

With Mathcad software, the same equation looks the same way as it does in a reference book:

$$x = \frac{-b + \sqrt{b^2 - 4 \cdot a \cdot c}}{2 \cdot a}$$

This makes the programs much easier to read and understand. Mathcad software also avoids the need to use a rigorous procedure to write a program. If the programmer gets the syntax wrong, the software refuses to accept the entry and provides a message indicating what is wrong.

On a worksheet, the equations can be set out from left to right, top to bottom. Text can be included anywhere on the page.

Doubtless there are other software packages which provide the same sort of facility. However, for consistency in presentation, Mathcad is the one used here.

Since this software has been available for over a decade, there is a fair probability that the reader is already using it. If not, it should be a fairly easy task to modify the programs in the worksheets of this document to convert them into programs which work with other software packages.

A few symbols in Mathcad have a special meaning:

a := b + c '*a*' is defined as the sum of *b* and *c*
a = 2 the value of '*a*' is 2

If the 'equals' sign is in bold typeface, the software interprets this as the 'Boolean equal' and returns a zero or a one.

The programming operator looks like that shown in Figure A.1.

$$\text{root}(a, b, c) := \left| \begin{array}{l} \text{discr} \leftarrow b^2 - 4\cdot a\cdot c \\ \text{num} \leftarrow -b + \sqrt{\text{discr}} \\ \text{denom} \leftarrow 2\cdot a \\ \dfrac{\text{num}}{\text{denom}} \end{array} \right.$$

Figure A.1 Simple Mathcad program

The programming operator behaves like a function, taking input variables and returning an output. This output is the last variable to be declared. In the example above, it is the value of the ratio of 'num' divided by '*denom*'. The program can return a single variable, a vector, or an array.

Local variables defined in the function are not visible outside. However, variables declared in the worksheet above the program function are visible within the function.

Included in the software are a number of built-in functions. As far as circuit modelling is concerned, the most important is 'lsolve(*M*,**v**)'. The argument '*M*' is a square matrix, whilst '**v**' is a vector. '*M*' contains as many rows as there are elements in '**v**'. This function returns a solution vector '**x**' such that $M\mathbf{x} = \mathbf{v}$.

Another useful characteristic of the software is that it can distinguish between the '*j*' operator as used with complex numbers and the variable '*j*'. It can be defined as an imaginary number by typing the characters '1*j*'. It appears on the worksheet as '*j*'.

Armed with this information, there should be no real problems in understanding the programs contained in the Mathcad worksheets of this book. Since the figures in the book which depict Mathcad worksheets have been hand-copied from copies of the actual worksheets, it is possible that errors have been introduced in the transcription process. Further errors could be introduced by the reader hand-copying the text onto his or her own computer. To prevent the chance of such errors creeping in, copies of the original Mathcad files are available in a zip folder which can be downloaded from designemc.info. These files can be run by any computer which has Mathcad version 15.0 (or higher) software installed.

The techniques described in the previous pages can be used to simulate the interference-coupling characteristics of any signal link. Every electrical or electronic system will have its own particular set of interference problems.

It should be possible for individual designers to carry out tests and create circuit models of their own critical links. Such information can be shared. Mathcad users have access to PlanetPTC, a mix of dynamic channels that enables PTC customers and product development professionals to actively participate in exchanging ideas.

However, it does not really matter which software is used to carry out the calculations. The key feature of any circuit model is the fact that it is a shorthand method of defining a set of equations. It becomes possible to describe the hardware, the tests, the model, and to illustrate the results in a single report. There are many forums and communities which can be used to publish such reports.

Appendix B
MATLAB®

Translating Mathcad worksheets into MATLAB® m-files was a fairly simple task. That being so, it should not be difficult for MATLAB users to understand the function of the Mathcad worksheets replicated in this book. There are some differences which could be puzzling at first.

With MATLAB, there are five windows: Command Window, Command History, Current Folder, the Workspace and the Editor. The program is really a text file in the Editor Window, and needs to be saved to a file before it is run, a task performed by typing and entering a command in the Command Window. All variables and their computed values are accessible in the separate Workspace. The Current Folder holds the files which are accessible to the Command Window, and the Command History keeps a record of all the actions taken to date. If a computation is carried out to produce a graphical picture, the picture is displayed in a separate Figure Window.

With Mathcad, there is only one window, the Worksheet. Variables and expressions can be defined anywhere on the page and intermingled with text, in the same way that calculations are recorded in a laboratory notebook. There is no list of numbers down the left-hand side of the page. Of course, there is a basic rule; the sequence of commands must be left to right; top to bottom. Graphs are created and modified by invoking menu commands. The 'run' command is invoked by clicking anywhere on an empty portion of the worksheet. This means that everything, such as definitions, expressions, functions, numerical results, graphs and explanatory text, is displayed on the worksheet. Any part of the worksheet can be copied and pasted onto a Microsoft word document, which is how the worksheet figures in the book have been created.

Both Mathcad and MATLAB work with arrays and vectors, and the related operations are called up by similar expressions. For example, $A_{i,j}$ in a Mathcad worksheet can be translated into $\boldsymbol{A}(i, j)$ in MATLAB, $V(k) := 0$ into $V = \text{zeros}(k,1)$, lsolve(Z,V) into linsolve(Z,V), etc.

Similarly, ln(f) can be translated into log(f), $|x|$ into abs(x), $\sqrt{x}$ into sqrt(x), etc.

In Mathcad, all vectors are column vectors; in MATLAB, vectors can be defined as row vectors or column vectors.

In Mathcad function statements, the parameters on the worksheet are 'visible' to the computations in that function. In MATLAB, they are not; every variable used by a function must be included in the set of input variables.

In Mathcad, the function statements form part of the worksheet. In MATLAB, these are stored as separate Function Files. This means that those MATLAB m-files which invoke the use of special functions must have the relevant Function Files available in the Current Folder.

If in doubt about the method of computation employed by any worksheet, the reader who is familiar with MATLAB can easily clarify its purpose by examining the text of the equivalent m-file. Each Mathcad worksheet has been translated into one or more MATLAB files, and the set of files is stored in a zip folder. This folder can be downloaded from designemc.info.

Appendix C
The hybrid equations

It is entirely possible that the hybrid equations of Section 4.1 are unfamiliar to engineers, even those who are experienced in the application of electromagnetic theory. If the starting-point equations are viewed with scepticism, then designers are unlikely to have any confidence in the succeeding mathematics. That being so, it is necessary to provide proof of the validity of (4.4) and (4.5).

It is even more likely that the average circuit designer has never seen them before. This section provides a succinct derivation of these equations. It is also made clear that both conductors of the line possess the properties of inductance, resistance, capacitance and conductance. By defining the variables in terms of loop parameters, it avoids the use of the concept of the equipotential ground. It is basically a set of lecture notes copied from a blackboard at Glasgow University in 1959.

Figure C.1 illustrates how voltage and current can vary along the length of a transmission line.

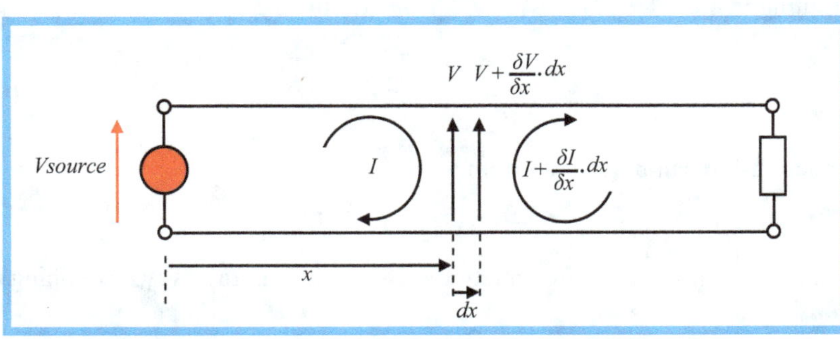

Figure C.1 Transmission line voltages and currents

For the purpose of deriving the transmission line equations, the following definitions apply:

R = resistance per unit length: Ω/m
L = inductance per unit length: H/m
C = capacitance per unit length: F/m
G = conductance per unit length: S/m

Voltage and current are functions of both x and t. For the section dx:

$$(R + j \cdot \omega \cdot L) \cdot dx \cdot I = V - \left(V + \frac{\delta V}{\delta x} \cdot dx\right) = -\frac{\delta V}{\delta x} \cdot dx \tag{C.1}$$

$$(G + j \cdot \omega \cdot C) \cdot dx \cdot V = I - \left(I + \frac{\delta I}{\delta x} \cdot dx\right) = -\frac{\delta I}{\delta x} \cdot dx \tag{C.2}$$

Hence:

$$\frac{\delta V}{\delta x} = -(R + j \cdot \omega \cdot L) \cdot I \tag{C.3}$$

$$\frac{\delta I}{\delta x} = -(G + j \cdot \omega \cdot C) \cdot V \tag{C.4}$$

Differentiating (C.3) gives:

$$\frac{\delta^2 V}{\delta x^2} = -(R + j \cdot \omega \cdot L) \cdot \frac{\delta I}{\delta x} = (R + j \cdot \omega \cdot L) \cdot (G + j \cdot \omega \cdot C) \cdot V \tag{C.5}$$

Differentiating (C.4) gives:

$$\frac{\delta^2 I}{\delta x^2} = -(G + j \cdot \omega \cdot C) \cdot \frac{\delta V}{\delta x} = (R + j \cdot \omega \cdot L) \cdot (G + j \cdot \omega \cdot C) \cdot I \tag{C.6}$$

Substituting γ^2 for $(R + j \cdot \omega \cdot L) \cdot (G + j \cdot \omega \cdot C)$ in (C.5):

$$\frac{\delta^2 V}{\delta x^2} = -\gamma^2 \cdot V \tag{C.7}$$

This relationship can be derived from:

$$V = C \cdot e^{\gamma \cdot x} + D \cdot e^{-\gamma \cdot x} \tag{C.8}$$

This can be confirmed by differentiating (C.8) twice. Another way of defining this relationship is:

$$V = A \cdot \cosh(\gamma \cdot x) + B \cdot \sinh(\gamma \cdot x) \tag{C.9}$$

This is because:

$$\begin{aligned} V &= A \cdot \left(\frac{e^{\gamma \cdot x} + e^{-\gamma \cdot x}}{2}\right) + B \cdot \left(\frac{e^{\gamma \cdot x} - e^{-\gamma \cdot x}}{2}\right) \\ &= \frac{A + B}{2} \cdot e^{\gamma \cdot x} + \frac{A - B}{2} \cdot e^{-\gamma \cdot x} \\ &= C \cdot e^{\gamma \cdot x} + D \cdot e^{-\gamma \cdot x} \end{aligned}$$

From (C.3) and (C.9):

$$I = -\frac{1}{R + j \cdot \omega \cdot L} \cdot \frac{\delta V}{\delta x} = -\frac{\gamma}{R + j \cdot \omega \cdot L} \cdot [A \cdot \sinh(\gamma \cdot x) + B \cdot \cosh(\gamma \cdot x)]$$

Now as:

$$\frac{(R + j \cdot \omega \cdot L)}{\gamma} = \frac{R + j \cdot \omega \cdot L}{\sqrt{(R + j \cdot \omega \cdot L) \cdot (G + j \cdot \omega \cdot C)}} = \sqrt{\frac{R + j \cdot \omega \cdot L}{G + j \cdot \omega \cdot C}} = Zo$$

Then:

$$I = -\frac{1}{Zo} \cdot [A \cdot \sinh(\gamma \cdot x) + B \cdot \cosh(\gamma \cdot x)] \quad \text{(C.10)}$$

Figure C.2 illustrates the boundary conditions of the line.

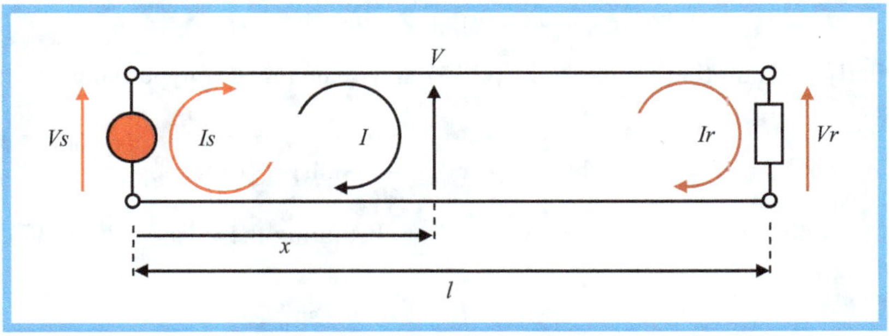

Figure C.2 Transmission line – boundary conditions

From Figure C.2, the boundary conditions are:

$$V = Vr \text{ at } x = l$$
$$I = Ir \text{ at } x = l$$

At the receiving end, x can be replaced by l in (C.9) and (C.10) giving:

$$Vr = A \cdot \cosh(\gamma \cdot l) + B \cdot \sinh(\gamma \cdot l) \quad \text{(C.11)}$$

$$-Zo \cdot Ir = A \cdot \sinh(\gamma \cdot l) + B \cdot \cosh(\gamma \cdot l) \quad \text{(C.12)}$$

Multiplying (C.11) by $\sinh(\gamma \cdot l)$ and (C.12) by $\cosh(\gamma \cdot l)$ gives:

$$Vr \cdot \sinh(\gamma \cdot l) = A \cdot \cosh(\gamma \cdot l) \cdot \sinh(\gamma \cdot l) + B \cdot \sinh(\gamma \cdot l) \cdot \sinh(\gamma \cdot l) \quad \text{(C.13)}$$

$$-Zo \cdot Ir \cdot \cosh(\gamma \cdot l) = A \cdot \sinh(\gamma \cdot l) \cdot \cosh(\gamma \cdot l) + B \cdot \cosh(\gamma \cdot l) \cdot \cosh(\gamma \cdot l) \quad \text{(C.14)}$$

Subtracting (C.14) from (C.13) gives:

$$Vr \cdot \sinh(\gamma \cdot l) + Zo \cdot Ir \cdot \cosh(\gamma \cdot l) = B \cdot \left(\sinh^2(\gamma \cdot l) - \cosh^2(\gamma \cdot l)\right) = -B$$

Hence:

$$B = -Vr \cdot \sinh(\gamma \cdot l) - Zo \cdot Ir \cdot \cosh(\gamma \cdot l) \tag{C.15}$$

Similarly:

$$A = Vr \cdot \cosh(\gamma \cdot l) + Zo \cdot Ir \cdot \sinh(\gamma \cdot l) \tag{C.16}$$

Substituting A and B in (C.9) gives:

$$\begin{aligned} V &= [Vr \cdot \cosh(\gamma \cdot l) + Zo \cdot Ir \cdot \sinh(\gamma \cdot l)] \cdot \cosh(\gamma \cdot x) \\ &\quad -[Vr \cdot \sinh(\gamma \cdot l) + Zo \cdot Ir \cdot \cosh(\gamma \cdot l)] \cdot \sinh(\gamma \cdot x) \\ &= Vr \cdot [\cosh(\gamma \cdot l) \cdot \cosh(\gamma \cdot x) - \sinh(\gamma \cdot l) \cdot \sinh(\gamma \cdot x)] \\ &\quad + Zo \cdot Ir \cdot [\sinh(\gamma \cdot l) \cdot \cosh(\gamma \cdot x) - \cosh(\gamma \cdot l) \cdot \sinh(\gamma \cdot x)] \end{aligned}$$

Hence:

$$V = Vr \cdot \cosh[\gamma(l - x)] + Zo \cdot Ir \cdot \sinh[\gamma(l - x)] \tag{C.17}$$

Similarly, substituting A and B in (C.12) and reducing the hyperbolic products gives:

$$I = Ir \cdot \cosh[\gamma \cdot (l - x)] + \frac{Vr}{Zo} \cdot \sinh[\gamma \cdot (l - x)] \tag{C.18}$$

At the sending end, $x = 0,\ V = Vs,\ I = Is$. Substituting these values in (C.17) and (C.18) gives:

$$\begin{aligned} Vs &= Vr \cdot \cosh(\gamma \cdot l) + Zo \cdot Ir \cdot \sinh(\gamma \cdot l) \\ Is &= \frac{Vr}{Zo} \cdot \sinh(\gamma \cdot l) + Ir \cdot \cosh(\gamma \cdot l) \end{aligned} \tag{C.19}$$

Where the propagation constant is:

$$\gamma = \sqrt{(R + j \cdot \omega \cdot L) \cdot (G + j \cdot \omega \cdot C)} \tag{C.20}$$

And the characteristic impedance is:

$$Zo = \sqrt{\frac{R + j \cdot \omega \cdot L}{G + j \cdot \omega \cdot C}} \tag{C.21}$$

It is worthwhile emphasising the fact that, in this derivation, the parameters *R, L, G,* and *C* are defined in terms of Ω/m, H/m, S/m, and F/m. The concept of 'per-unit-length' parameters is inherent in all the transmission line equations to be found in textbooks. Even so, this appendix is the only place in this book where such parameters are invoked. Section 4.1 shows that the analysis of transmission line behaviour can be related to the actual resistance, inductance, capacitance and conductance of the conductors.

Appendix D
Definitions

These definitions are those used in this book. They are not necessarily the same as those appearing in other documents.

Antenna-mode current	Unidirectional current in the conductors of a cable. That which flows between the cable and the environment.
Buffer circuit	A circuit which forms an interface between the conductors of a signal link and the processing circuitry of the equipment unit.
Chassis	The conducting elements of electrical and electronic equipment sub-frames and casing, upon which PCBs and components are mounted, which is usually connected to earth or ground.
Circuit equations	A set of equations which relate voltages to currents in a circuit model.
Circuit model	A model which obeys the rules of circuit theory and which simulates the behaviour of the assembly-under-review.
Circuit parameter	A parameter used in a circuit model.
Common-mode current	A current which flows in the loop formed by the cable and the structure.
Common-mode gain	Is the ratio of the output voltage of a buffer circuit to the common-mode input voltage, when the differential-mode input is zero.
Common-mode rejection	Is the ratio of the differential-mode gain to the common-mode gain. Usually quoted in terms of decibels (dB).
Composite conductor	A set of elemental conductors, aligned in parallel, which enables the distribution of currents or voltages in the actual conductor to be simulated.
Conducted emission	The current induced in the common-mode loop by a voltage source in the differential-mode loop. See Figure 4.17.
Conducted susceptibility	The current induced in the differential-mode loop by a voltage source in the common-mode loop. See Figure 4.19.
Culprit loop	The loop in which the interfering voltage source is located.
Current transformer	A transducer which monitors the magnetic field surrounding a group of conductors and generates a voltage proportional to the sum of the currents in those conductors.
Delay-line model	A circuit model which simulates the differential-mode behaviour of a transmission line.

(Continues)

(*Continued*)

Differential-mode current	A current which flows in the loop formed by the signal and return conductors.
Differential-mode gain	Is the ratio of the output voltage of a buffer circuit to the differential-mode input voltage, when the common-mode input voltage is zero.
Distributed parameters	Are the parameters which are derived from the use of per-unit-length parameters.
Earth	The conductor(s) designated to carry fault current in an AC distribution network. Always connected to the conducting structure.
Elemental conductor	A conductor which represents a small segment of the surface of a composite conductor.
EMC	Electromagnetic compatibility. The ability of a device, unit of equipment, or system to function satisfactorily in its electromagnetic environment without introducing intolerable electromagnetic disturbances to anything in that environment.
EMI	Electromagnetic interference.
Floating configuration	A wiring configuration where the signal link is isolated from the structure at one (or both) ends.
General circuit model	A model which includes interface circuit components as well as cable coupling components, but which does not specify component values.
Ground	Another name for the conducting structure. The terms 'Ground', 'Earth', 'Chassis' and 'Structure' are interchangeable in the analyses described in this book.
Grounded configuration	A wiring configuration where the return conductor of the signal link is bonded to the local structure at each end.
Loop equations	A set of equations, derived from the primitive equations, which relate loop voltages to loop currents.
Loop parameter	A parameter which has been derived from groups of primitive parameters or partial parameters. Loop parameters can be measured by electrical test equipment.
Lumped parameter	A resistor, capacitor or inductor which represents the relevant properties of a defined length of conductor. The term also applies to circuit components.
Partial capacitance	Is the capacitance of a composite conductor, or the capacitance of a loop segment.
Partial current	A portion of the current flowing in a conductor.
Partial inductance	Is the inductance of a composite conductor, or the inductance of a loop segment.
Partial inductor	Is an inductor associated with partial current or partial voltage.
Partial parameters	Those parameters associated with the behaviour of composite conductors as antennae. The term is also used to distinguish between incident and reflected currents in transmission lines.

(*Continued*)

Partial voltage	A portion of the voltage in a circuit loop.
Per-unit-length parameter	Resistance, inductance, capacitance or conductance which is defined in terms of Ω/m, H/m, F/m or S/m.
Primitive capacitance	A component which relates the voltage on an isolated conductor to the energy contained by the electric field in which it is immersed.
Primitive current	Unidirectional current flow in a single conductor of a multi-conductor assembly.
Primitive equations	A set of equations relating primitive voltages to primitive currents.
Primitive inductance	A component which relates the current in the conductor to the energy contained by the magnetic field in which it is immersed.
Primitive parameters	Are the parameters associated with conductors which are treated as antennae.
Primitive voltage	Is the voltage along a conductor due to electromagnetic coupling.
Radiated emission	The current induced in the antenna-mode loop by a voltage source in the differential-mode loop.
Radiation resistance	A mathematical construct which equates to the apparent resistance of an antenna when it is carrying maximum current during resonance conditions.
Radiation susceptibility	Is the current induced in the differential-mode loop by a voltage source in the antenna-mode loop.
Relative permeability	A pure number which depends on the magnetic properties of the environment.
Relative permittivity	A pure number which depends on the dielectric properties of the environment.
Representative circuit model	A circuit model which simulates the interference coupling mechanisms of a specific signal link.
Return conductor	The conductor designated to complete the loop carrying the differential-mode current, which also carries the common-mode current in the return/structure loop.
Signal conductor	Is the conductor which is exclusively allocated to carry the differential-mode current.
Signal link	Is the conductors and interface circuitry involved in transmitting a signal from one location to another in a particular system.
SPICE	Simulation Program with Integrated Circuit Emphasis.
Structure	The conducting elements of the structure can be referred to as either 'Ground', 'Earth' or 'Chassis'.
Threat environment	Is the frequency response of the power density of the worst-case external radiation to which the equipment can be subjected. It can also be defined in terms of the waveform of a transient current or voltage.
Threat voltage	Is the amplitude of the voltage in the culprit loop.

(Continues)

(*Continued*)

Time-step analysis	A method used to predict currents and voltages in the circuit model a discrete time after a step change has occurred in any current or voltage in that model.
Transfer admittance	Is the ratio of the current in the victim loop to the source voltage in the culprit loop, when there are no other voltage sources.
Transfer impedance	Is the ratio of the amplitude of the voltage appearing in the output loop to the amplitude of the current in the input loop. With screened cable, this is the impedance of the screen.
Victim loop	The loop carrying the signal which is regarded as being susceptible to interference.
Virtual conductor	Is an imaginary conductor which enables the coupling between cable and environment to be simulated.
Voltage transformer	A transducer which uses magnetic field coupling to induce a voltage in the loop-under-test and which allows the induced voltage to be measured.

Appendix E

Links to Internet resources and references

Obituary
https://www.theiet.org/membership/member-news/obituaries/ian-darney-bsc-miet-november-1937-december-2021

Radiation resistance
https://en.wikipedia.org/wiki/Radiation_resistance

designemc.info
http://cmemc.one/

ediss-electric.com
http://www.ediss-electric.com/

ediss-electric.com/bypass_caps_pdf/1_100n0201X5R_murata.pdf
http://www.ediss-electric.com/bypass_caps_pdf/1_100n0201X5R_murata.pdf

Gaussian frequency-shift keying modulated
https://en.wikipedia.org/wiki/Frequency-shift_keying#Gaussian_frequency-shift_keying

DMX512 Standard
https://en.wikipedia.org/wiki/DMX512

data from the James Web Space Telescope (JWST)
https://www.aanda.org/articles/aa/pdf/2022/10/aa44517-22.pdf

interferencetechnology.com
http://www.interferencetechnology.com/

EMC Academy
https://www.emcstandards.co.uk/emcacademy

EMC Journal
https://www.emcstandards.co.uk/

References

[1] Tim Williams. *EMC for Product Designers*. Second Edition. Section 1.1.1 (Newnes, Oxford, 1996), page 4.

[2] Europe EMC Guide. *The International Journal of Electromagnetic Compatibility*. (ITEM Publications) interferencetechnology.com.

[3] Keith Armstrong. EMC design of SMP and PWM power converters (*EMC Journal*, 2011), page 28.

[4] Tim Williams. *EMC for Product Designers*. Fourth Edition (Newnes, Oxford, 2006).

[5] Tesche Frederick M, Ianoz Michel V, and Karlsson Torbjörn. *EMC Analysis Methods and Computational Models* (Wiley, New York, 1997).

[6] Defence Standard 59 411 Part 5 Issue 1 Amendment 1. *Electromagnetic Compatibility. Part 5. Code of Practice for Tri-Service Design and Installation. Section 8.9. Single Point Reference Connection* (Ministry of Defence, January 2007), page 29.

[7] Shepherd J, Morton A H, and Spence L F. *Higher Electrical Engineering. Section 7.28. Equivalent Phase Inductance of a Three-Phase Line* (Pitman Publishing Ltd., London, 1985), pages 234–235.

[8] Shepherd J, Morton A H, and Spence L F. *Higher Electrical Engineering. Section 7.16. Equivalent Phase Capacitance of an Isolated Three-Phase Line* (Pitman Publishing Ltd., London, 1985). pages 216–219.

[9] Burrows B J C. *The Computation and Prediction of Induced Voltages in Aircraft Wings. CLSU Memo 18*. 1974 (Culham Lightning, Units 13/15, Nuffield Way, Abingdon, OX14 1RL).

[10] Keith Armstrong. *EMC Design Techniques for Electronic Engineers* (Armstrong/Nutwood UK 2010), from emcacademy.

[11] Paul Clayton R. *Introduction to Electromagnetic Compatibility*. Second Edition (Wiley-Interscience, New York, 2006).

[12] Skitek G G and Marshall S V. *Electromagnetic Concepts and Applications*. Section 2.5 Electric Field Intensity of a Line of Charge (Prentice-Hall, Englewood Cliffs, NJ, 1982).

[13] Page L and Adams N I. *Principles of Electricity*. Inductance of Straight Conductors. page 325 (D Van Nostrand, Princeton, NJ, 1958).

[14] Skitek G G and Marshall S V. *Electromagnetic Concepts and Applications*. Section 8.3 Magnetostatic Field Intensity from the Biot–Savart Law: Magnetic Field Due to a Filamentary Current Distribution of Finite Length (Prentice-Hall, Englewood Cliffs, NJ, 1982).

[15] Skitek G G and Marshall S V. *Electromagnetic Concepts and Applications*. Section 12.12 Skin Effect, and High and Low Loss Approximations (Prentice-Hall, Englewood Cliffs, NJ, 1982).

[16] Skitek. G G and Marshall S V. *Electromagnetic Concepts and Applications*. Section 7.4 Image Solution Method: Capacitance between Two Cylindrical Conductors (Prentice-Hall, Englewood Cliffs, NJ, 1982).

[17] Skitek. G G and Marshall S V. *Electromagnetic Concepts and Applications*. Section 12.2 General Equations for Line Voltage and Current (Prentice-Hall, Englewood Cliffs, NJ, 1982).

[18] Skitek. G G and Marshall S V. *Electromagnetic Concepts and Applications*. Section 14.4. The Half-Wave Dipole (Prentice-Hall, Englewood Cliffs, NJ, 1982).

[19] Ordnance Board Pillar Proceeding P101 (Issue2) Principles for the design and assessment of electrical circuits incorporating explosive components. Annex E. Appendix 1. The Radio Frequency Environment, pages E1–E3 (The Secretary, Ordnance Board, Walnut 2c, MOD Abbey Wood #67, PO Box 702, Bristol BS12 7DU, UK).

[20] Savant, Roden, Carpenter. *Electronic Design – Circuits and Systems*. Second Edition. Appendix A. SPICE. Section A.2.4.3 Transient Analysis (Benjamin Cummings Publishing Company Inc., New York, 1991).

[21] Gnecco Louis T. *The Design of Shielded Enclosures* (Newnes, Butterworth-Heinemann, Reed Elsevier, Oxford, 2000).

[22] Ordnance Board Pillar Proceeding P101 (Issue2) Principles for the design and assessment of electrical circuits incorporating explosive components. Annex E. Appendix 1. Section 28 : Shielding Assessment, pages E1–14 (The Secretary, Ordnance Board, Walnut 2c, MOD Abbey Wood #67, PO Box 702, Bristol BS12 7DU, UK).

[23] A Guide to BS EN 62305:2006. Protection against Lightning. 2008 (Thomas & Betts Limited, Furse, Wilford Road, Nottingham, NG2 1EB, UK).

[24] Horowitz P and Hill W. *The Art of Electronics*. Second Edition (Cambridge University Press, 1989).

[25] Savant, Roden, Carpenter. *Electronic Design – Circuits and Systems*. Second Edition (Benjamin Cummings Publishing Company Inc., New York, 1991).

[26] Ludwig R and Bretchko P. *RF Circuit Design: Theory and Applications* (Prentice-Hall, Englewood Cliffs, NJ, 2000).

[27] Defence Standard 59-411 Part 2 Issue 1 Amendment 1. *Electromagnetic Compatibility. Part 2 The Electric, Magnetic & Electromagnetic Environment. Table 18 Front Line and Operational Support Equipment Field Strength* (Ministry of Defence, January 2008), page 28.

[28] Defence Standard 59-411 Part 3 Issue 1 Amendment 1. *Electromagnetic Compatibility. Part 3. Test Methods and Limits for Equipment and Sub Systems. Appendix B.2. DCE02.B Conducted Emissions Control, Signal and Secondary Power Lines. 20 Hz–150 MHz. Figure 51. DCE02 – Limit for Air Service Use* (Ministry of Defence, January 2008), page 84.

[29] *The EMC Journal* (Emstrey House North, Shrewsbury Business Park, Shrewsbury, United Kingdom, SY2 6LG). emcjournal. Use '*The EMC Journal*' in the Search Box.

Index

admittance 194–223
  of current transformer 196, 221
  of isolated conductor 199–204
  of open circuit line 205–6
antenna gain 108–10, 129, 136
antenna-mode
  current 163–70, 194, 223
  damping 255
  definition 303
  monitored on scope 210–12
  propagation velocity 205
  simulation 176–83, 222, 294
antenna models 105
  coupling via structure 128–36
  half-wave dipole 106
    field strength 109–10
    power density 109
    power received 110–11
    radiated power 106–9
  radiated emission 138–40
  radiation susceptibility 136–8
  threat current 122–7
  threat voltage 118–22
  virtual conductor 111–18

bench test equipment 287
bench testing 2, 185, 271
  analysis 230–1
  cable characterisation 208–20
  cable transients 221–6
  capacitor characterisation 227–9
  current transformer 193–9
  isolated conductor 201–8
  tri-axial cable 199–200
  voltage transformer 186–93
bench test models 99–104
Biot–Savart equation 19
black art 3
block diagrams 275
bridge network 46
broadband over power lines 233–4, 254–61
bubble plot 72–9

cable characterisation 208–20
cable losses 175–6
cable transients 221–6
capacitance 41–2
capacitive coupling
  combined effect of 6
  between three parallel conductors 6
capacitor characterisation 227–9
capacitors 45, 83
carbon 271
centre-tapped transformer 248
characterisation
  of cable 208
  of capacitor 227
  of components 186
  of voltage transformer 195
characteristic impedance
  definition 85

of delay line mode 149–50, 155
derivation 302
of free space 109
inner and outer braids 200
measurement 225–6
of virtual conductor 176–7
circuit capacitances
of composite pair 72
of screened pair 81
of virtual conductor 112
circuit capacitors
for conductor pair 42
for three conductors 50
circuit diagrams 272–3
circuit equations
capacitive coupling 42
definition 303
inductive coupling 40
and loop equations 29
for three conductors 47
circuit inductance
composite pair 72
screened pair 81
three-conductor assembly 48
virtual conductor 116
circuit modelling 3, 233
circuit models 185, 276–7
calculating parameter values 130
of capacitive coupling 50
of conductor pair 45
of coupling parameters 115
for grounded configuration 103
of half-wave dipole 112
for resistive coupling 44
of right-hand section of conductor pair 114
of transformer coupling 61
circuit parameters 7, 28, 40
basic assumption 44
capacitance 41–2
cross coupling 93
definition 7, 28, 307
exposed cable 122
inductance 40–1
and loop parameters 35
maintaining duality 42
resistance 43–4
circuit theory 28
co-axial coupling 57–9
commercial filters 252–3
common-mode choke 246–7
common-mode current
definition 172, 303
common-mode damping 253, 271
common-mode filter 257–8
common-mode resistor 253–4
minimising pickup 254–5
transformer-coupled resistors 258–9
transformer screen-winding 256–7
tri-axial cable 255–6
common-mode filter 257–8
common-mode rejection 243
centre-tapped transformer 248
common-mode choke 246–7
differential amplifier 243–4
differential analogue driver 245–6
differential logic driver 244–5
opto-isolator 248
transformer coupling 247
common-mode resistor 253–4
composite conductors 8, 63–4
composite pair 68–76
computational electromagnetics 3
conducted emission
circuit model 14, 15, 101
damping 248–9, 270

- definition 303
- test setup 100–1

conducted susceptibility
- circuit model 14, 101
- definition 303
- test setup 100–3

conducted susceptibility test 14–15
- representative circuit model of 102
- setup for 101

conducting structure 233
conductivity 43
conductor resistance 43
critical frequencies 288–9
critical signal links 288
cross-coupling 91–8
cross-over frequency 43, 123, 204
cross-section 63
- composite pair 68–76
- screened pair 76–81
- single composite conductor 65–8

current linking transformer 170–2
current transformer 15–16, 185, 193–9
- assembly 195
- calibration 189
- circuit model 190
- as current source 194
- definition 277
- setup for calibration 196
- use of 194
- as voltage source 195

damping 224
delay-line model 149–56, 224
derivation process 7–8
differential amplifier 243–4
differential analogue driver 245–6
differential driver chip (DDC) 264
differential logic driver 244–5
differential logic receiver 244
differential-mode current 304
- definition 51, 304

differential-mode damping 248, 270
- commercial filters 252–3
- mains filtering 250–1
- solenoid switching 251–2
- transient damping 249–50
- use of carbon 253

differential-mode gain 304
dipole model 17–18
distributed parameter
- circuit model 213
- concept 88
- definition 6
- max frequency 88
- model 10–11, 83, 304
- networks 91
- single-T model 10, 89
- transformation equations 88–9
- triple-T model 11, 90

earth
- definition 304
- conductor 128, 193–4, 250, 259, 262
- loops 3, 24, 128, 269

electrical energy 233
electrical length 9–10
electrical tests 22, 73
electric charge 142, 172
electric field 119
- and threat voltage 18, 101, 114, 258
- at a point 30
- power density 109, 119, 280

electro-explosive device (EED) 257
electromagnetic compatibility (EMC) 1, 208, 233, 267
- academic approach 2

circuit modelling 3
computations 4
definition 1, 304
essence of approach 4–5
importance of testing 22–3
inter-system interference 17–20
intra-system interference 12–17
managerial approach 2–3
misleading concepts 3
model developing 5
basic model 5–6
composite conductors 8
derivation process 7–8
distributed parameters 10–11
electrical length 9–10
parameter types 6–7
proximity effect 9
need for 1
practical design techniques 23–4
pragmatic approach 1–2
system design 24
formal EMC requirements 25
guidelines 24–5
top-down approach 25
testing 4
transients 20–1
electromagnetic coupling 209
electromagnetic field theory 28
electromagnetic interference (EMI) 1, 291
electromagnetic wave 172
electrostatic discharges 141
elemental conductors 63, 65, 304
calculating currents in 74
concept 8
definition of 77
deriving co-ordinates of 73
primitive equations for 66
radii 65
spacing 65
envelope curve 105
Environmental Frequency Losses (EFL) 214–15, 219
Environmental Line Losses (ELL) 215, 219
equipment shielding 260–1
Equipment-Under-Test (EUT) 199
equipotential ground plane 8, 233–4
equivalent circuits 46, 191

Failure Mode Effects and Criticality Analysis (FMECA) 291
field strength 109–10
floating configuration
definition 304
floating transformer 247
Fourier transforms 20
frequency domain 4, 143, 185
frequency-response models 179

general circuit model 177, 290–1
general-purpose test equipment 177, 290–1
grounding 234–8
ground loops 241–3, 269–70
ground plane 59–62, 129
concept 59
on printed circuit board 239

half-wave dipole
radiated power 106
half-wave frequency 20
Heaviside transforms 143
hybrid equations 299–302

impedance 66
induced voltage 142
inductance 33, 40–1, 66

inductive coupling
  combined effect of 6
  between three conductors 5
inductors 28, 45, 83
injected voltage 209
interface diagrams 275–6
internet resources and references 307
inter-system interference 17
  dipole model 17–18
  threat voltage 19
  virtual conductor 18–19
  worst-case analysis 20
intra-system interference 12–17
  conducted emission 14
  conducted susceptibility 14–15
  current transformer 15–16
  equivalent circuits 13
  representative circuit model 16–17
  signal link 12
  simulating structure 12–13
  voltage transformer 15
Iout function 132
isolated conductor 201–8

Laplace transforms 20, 143
line characteristics 156–63
Line Input Simulation Network 251
line parameter measurements 176
line voltage 172–3
local variables 294
loop capacitance
  composite pair 73
  conductor pair 39
loop currents 69
loop equations 8, 304
  definition 304
  derivation process 7, 12
  general formula 68
  for triple-T model 90
loop impedances
  and circuit impedances 48–9, 81
  and primitive impedances 49, 68
loop inductance
  co-axial cable 58
  conductor pair 38
  three conductor assembly 77
loop parameters 7, 27, 37–40, 304
  definition 304
  in derivation process 7
  and test equipment 7, 27
lumped parameter
  in single T model 84
  in triple T model 88
lumped parameter models 29
  circuit parameters 40
    basic assumption 44
    capacitance 41–2
    inductance 40–1
    maintaining duality 42
    resistance 43–4
  co-axial coupling 57–9
  duality of *L* and *C* 36–7
  ground plane 59–62
  loop parameters 37–40
  optimum coupling 51–4
  primitive capacitance 29–32
  primitive inductance 32–5
  three-conductor model 47–51
  transfer admittance 54–7
  twin-conductor model 44–7
lumped parameters 29
  definition 304
  and distributed parameters 10

magnetic field strength
  and electric field strength 109
  at monitor antenna 139–40

at a point 32
and power density 109, 140
and radiated emission 140
mains filtering 250–1
maintaining duality 42
managerial approach 2–3
Mathcad program 84, 130, 185
Mathcad software 63, 196, 293
Mathcad worksheets 4, 293–4, 297
MATLAB® 297–8
Maxwell's equations 4, 107
mesh analysis 143
mesh equations 8
minimising pickup 254–5
mutual inductance 190

nickel 271
nodal analysis 51, 143, 287

Ohm's law 206
open-circuit transfer admittance 96
operational amplifiers 217
optimum coupling 51–4
optimum transmission 233
opto-isolator 248

parallel LCR circuit 146–9
partial capacitance 68, 304
in composite conductor 68–9
definition 304
partial current
antenna-mode current 172
in composite conductor 73, 79
definition 304
due to reflections 6, 9, 142, 150
and mesh analysis 8
partial impedance 67, 71
partial inductance
definition 304
composite conductor 67–8
screened pair 81
section of loop 277
partial inductors 64
partial parameter 7
partial voltage
in composite conductors 64, 67, 79
definition 305
in transient analysis 19, 136, 144
permeability 35
permittivity 31
PlanetPTC 295
power density 109
and field strength 109
and frequency 120
and power received 106, 110, 288
and radiated emission 138–40
and radiation susceptibility 136
and threat voltage 123, 134
and transmitted power 108
power density vector 107, 109
power received 106, 110, 288
power supply
loops 242
returns 246
transient damping 231, 239
practical design 233
common-mode damping 253
common-mode filter 257–8
common-mode resistor 253–4
minimising pickup 254–5
transformer-coupled resistors 258–9
transformer screen-winding 256–7
tri-axial cable 255–6
common-mode rejection 243
centre-tapped transformer 248
common-mode choke 246–7
differential amplifier 243–4
differential analogue driver 245–6

differential logic driver 244–5
opto-isolator 248
transformer coupling 247
conductor pairing 238–40
differential-mode damping 248
commercial filters 252–3
mains filtering 250–1
solenoid switching 251–2
transient damping 249–50
use of carbon 253
grounding 234–8
ground loops 241–3
shielding 259–63
using PCB example 263–5
pragmatic approach 1–2
primary inductance 190
primitive capacitance 29–32
definition 305
general formula 31
of virtual conductor 111, 117
primitive currents 64, 69–70
primitive equations
definition 305
derivation process 7
for elemental conductors 66
for three conductors 47
for two conductors 36, 113
primitive inductance 29, 32–5
definition 30
general formula 35
of virtual conductor 111
primitive parameters 7, 27, 38
printed circuit boards (PCBs) 24, 279–80
propagation constant 85
proximity effect 9, 236

quarter-wave frequency 20, 95, 208, 255

radiated emission 217
of cable and model 217
definition 304
test setup 137
radiated power 106–9
radiated transients
test setup 163, 221
radiation resistance 305
of dipole 107
measured value 215, 220
radiation susceptibility 136–8, 305
assessment 106
definition 305
test setup 136
reflection coefficient 150
relative permeability 35, 73, 305
relative permittivity 305
definition 37, 305
measured value 210
representative circuit model 16–17, 22, 305
concept 16–17
table 290
resistance 43–4
resistive effects 44
resistivity 43
resistors 28, 83
resonant frequency 37, 121, 201, 289
return conductors 269, 305

Schmitt triggers 217
screened pair 76–81
bubble plot 80
with differential analogue driver 245
with differential logic driver 244
representative circuit model 80
series LCR circuit 144–6, 227
shielding 259

of buildings 261–3
equipment shielding 260–1
use of carbon 263
signal conductor 305
signal generator 189
signal link
basic building block 25
circuit model 14, 91
wiring diagram 12
simple Mathcad program 294
Simulation Programs with Integrated Circuit Emphasis (SPICE) software 4, 16, 84, 143, 197
single composite conductor 65–8
single-point ground 3, 217, 224, 225, 247, 248
single-T model 84–8
sinusoidal voltage 17
skin effect 20, 96
formulae 43
and proximity effect 9
software 287–8
solenoid switching 251–2
SPICE
definition 305
steady-state resistance 92
structure as a shield 268
susceptibility requirements 280–3
symmetry 96
system assessment 271
system design 267
design guidelines 268
bench tests 271
common-mode damping 271
current balance 270
differential-mode damping 270
ground loops 269–70
return conductors 269
shield structure 268–9
system assessment 271
diagrams 272
analysing signal link 278
block diagrams 275
circuit diagrams 272–3
circuit models 276–7
deriving component values 277–8
interface diagrams 275–6
testing link 279
wiring diagrams 273–4
emission requirements 284–6
planning 286
bench test equipment 287
characterisation 289–90
critical frequencies 288–9
critical signal links 288
general approach 291
performance requirements 286–7
software 287–8
printed circuit boards 279–80
susceptibility requirements 280–3

threat current 122–7
threat environment 305
threat voltage 19, 118–22, 305
definition 305
and electric field 105, 120
and frequency 122
and threat environment 281
three-conductor assembly 47, 64
circuit impedances for 48
triple-T circuit model of 51
voltage balance in 54
three-conductor model 47–51
time constant 36, 149
three conductor signal link
general circuit model 91
time-step analysis 143, 306
basic concept 143

basic equations 143–4
parallel LCR circuit 146–9
series LCR circuit 144–6
T-network circuit model 28–9, 83
top-down approach 25, 268
transceiver PCB layouts 264
transfer admittance 29, 54–7, 97, 306
of critically damped cable 97
definition 306
duality 56
of floating configuration 102
of grounded configuration 103
for open-circuit, short-circuit, and critically damped lines 98
with open-circuit terminations 97
transfer emission 177, 183, 290
transfer function 189–93
transfer function analyser 227
transfer impedance 306
of capacitor 227, 228
of co-axial cable 29, 58, 277, 278
of current transformer 197–9
definition 306
transformation equations 88–9
transformer centre-tapped 248
transformer-coupled resistors 258–9
transformer coupling 60, 247
transformer driver 248
transformer screen-winding 256–7
transient analysis 141
antenna-mode current 163–70
delay line model 149–56
line characteristics 156–63
radiated emission 170
cable losses 175–6
current linking transformer 170–2
line parameter measurements 176
line voltage 172–3
radiated current 174–5
source current and voltage 173–4
time-step analysis 143–9
transient emission model 176–83
transient damping 249–50
transient emission model 176–83
transients 20–1
transmission line models 83
bench test models 99–104
cross-coupling 91–8
single-T model 84–8
triple-T model 88–91
transmission line theory 10, 149
tri-axial cable 199–200, 255–6
triple-T model 13, 88–91
try-it-and-see approach 22
twin-conductor model 44–7
two-conductor assembly 34

Vcross coupling 256
vector algebra 91
Vector Network Analyser (VNA) 265
vertical branch 28
victim circuit 268
victim loop 306
virtual conductor 18–19, 111–18, 306
capacitance 116
concept 18
definition 306
inductance 116
voltage source 205, 210
voltage transformer 15, 185–93, 306
assembly 188–9
characterisation of 189
definition 306
representative circuit model of 190
transfer function of 190
use of 187

Vthreat curve 121
Vthreat(*f*) function 132

wire-to-wire coupling test 212
wiring diagrams 273–4
worst-case analysis 20

Ycrit curve 97
Ymodel function 205

Zbranch(*f*) function 94, 131, 205
Zener diodes 262
Zloop(*f*) function 94, 96, 131

www.ingramcontent.com/pod-product-compliance
Lightning Source LLC
LaVergne TN
LVHW011843200926
848484LV00002B/397
* 9 7 8 1 8 3 7 2 4 2 0 6 1 *